AF828928

Annals of Bryozoology

Annals of Bryozoology: aspects of the history of research on bryozoans

Edited by
Patrick N. Wyse Jackson
&
Mary E. Spencer Jones

International Bryozoology Association
2002

This volume is dedicated to the memory of our friend Alan Stanley Horowitz 1930-1999 - a fine bryozoologist and palaeontologist and supporter of the International Bryozoology Association.

© Individual authors 2002

ISBN 0-9543644-0-6

First published 2002 by the International Bryozoology Association,
c/o Department of Geology, Trinity College, Dublin 2, Ireland.

Printed by Colour Books, Baldoyle, Dublin 13.

All rights reserved. No part of this book may be reproduced or stored in any form by any means, electronic or mechanical, including photography, photocopying, recording or by any other means, without the prior written permission of the publisher.

Cover illustrations

Front: Plate from George Allman's 1856 monograph on freshwater bryozoans.
Back: Portion of a plate from Alicide d'Orbigny's *Paléontologie française* (1850-1852) showing *Retepora royana*, from the Cretaceous.
Background: Structure of *Flustra* from Robert Hooke's *Micrographia* (1665).

CONTENTS

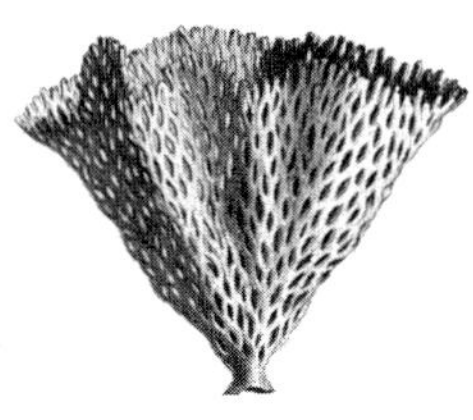

PREFACE

Bryozoans are complex colonial invertebrates, that have a long geological range of nearly 500 million years, and which are found in most marine and freshwater ecosystems.

For a number of years we have developed an interest in the historical hertiage of persons and their collections connected with research on bryozoans. We were aware of the publications of George Johnston, Sidney Harmer and Leslie Pitt on this subject, but felt that a modern treatment of the subject would be worthwhile undertaking. As the International Bryozoological Association [IBA] was due to meet for its 12th International Conference, this gathering was seen as an ideal opportunity to present some papers with a historical content. To this end we invited the Society for the History of Natural History to co-sponsor a meeting with the IBA and this was convened on Saturday 21st July 2001 in the historic surroundings of Trinity College, Dublin.

Fifteen papers and three posters were presented at the Dublin meeting, and these, together with a number of subsequently commissioned papers are gathered together in this volume. A number of papers document in great detail the lives and work of some bryozoologists, including the Austrian August Emanuel Reuss, and the hitherto neglected George Vine and Eliza Jelly, two amateur Victorian enthusiasts, who against all the odds, between them made major contributions to bryozoology. The now somewhat eccentric ideas of W.D. Lang are recalled, as is the friendship and inter-continental research of the Frenchman Ferdinand Canu and the American Ray Smith Bassler, and the similar friendship and illuminating correspondence of Mary Rogick and Ernst and Eveline Marcus. An assessment of the collective contributions of scientists who worked in France prior to 1816 and in Vienna during much same period is presented here. So too is the work of bryozoologists who worked on Ordovician faunas in America, Recent bryozoans in both the north and southern Pacific Oceans, in the Bering Sea, and in the Netherlands, Luxembourg and Belgium. Bryozoans have a complex physiology that has fascinated scientists for over four centuries. An overview of both colonial and individual behaviour and feeding mechanisms in bryozoans is presented in two papers. The International Bryozoology Association and its triennial conferences acts as the forum for the discussion and presentation of research on bryozoans and its dissemination in the form of the IBA conference proceedings. The story of the foundation and early history of the IBA and an analysis of its conferences is given here.

The feedback and response to this meeting has been most positive and encouraging, and it is hoped that this volume will be followed soon by a second compilation of papers on the same theme.

The International Bryozoology Association and the organisers of the Dublin 2001 meeting thank the Society for the History of Natural History for its support, and gratefully acknowledge the generous support and sponsorship of the Department of Geology, Trinity College, Dublin, the Geological Survey of Ireland, and Foras na Mara (the Marine Institute of Ireland) which allowed this volume to be published.

In this volume we have used English as against American spelling except where it seemed more appropriate to use the latter, for example "Cincinnati paleobryozoologists".

Patrick Wyse Jackson (Trinity College, Dublin)
& Mary Spencer Jones (The Natural History Museum, London)

6th September 2002

George Robert Vine (1825-1893): stay maker, bryozoologist and fossil dealer

Caroline J. Buttler,[*] Patrick N. Wyse Jackson[†] and Tom Sharpe[*]
[*]*National Museums and Galleries of Wales, Cathays Park, Cardiff CF10 3NP, Wales, UK.*
[†]*Department of Geology, Trinity College, Dublin 2, Ireland*

1. Introduction

George Robert Vine (1825-1893) (Figure 1), was a Sheffield-based stay or corset maker and amateur naturalist who specialised in the study fossil bryozoans and Coal Measures plants. He published about 75 papers on Bryozoa between 1877 and 1893. In particular an ambitious series of British Association reports on British fossil and Recent faunas are his most valuable contribution to bryozoological research, and he is principally remembered today as the author of the Order Cryptostomata.

This paper presents a brief biography of this illusive amateur palaeontologist, gives an assessment of his bryozoological work, and describes his activities as a fossil dealer who specialised in micropalaeontological preparations. Examination of 59 letters sent to Dr Philip B. Mason (a doctor from Burton-on-Trent) by Vine, and now housed in the National Museums and Galleries of Wales gives a valuable insight into Vine's fossil dealership and some of the problems that he experienced during the late 1880s and 1890s. Much of our present information comes from these letters, which are herein referred to as the "Vine-Mason letters". The date cited in footnotes and elsewhere refers to the date the letter was written by Vine.

Figure 1. George Robert Vine, senior (1825-1893)
(Photograph courtesy Sheffield Libraries, Archives, and Information; Local Studies Section)

Additionally we discuss the present whereabouts of his collections and manuscripts associated with him and his family, and give an extensive bibliography.

2. Biographical sketch

Very little is known of the background and early life of George Robert Vine. His son in two obituaries and a subsequent book about Attercliffe, Sheffield where they lived, provides some information[1] while more is available in a short notice of his death published in *The Sheffield and Rotherham Independent*.[2] These are the main sources of our information given below. Further gleanings have been obtained from an autobiographical piece which Vine recounted his early life[3] and from the Census return for 1881.

George Robert Vine was born in Hassops, Portsea, just north of Portsmouth on the south coast of England on 20th October 1825, into a poor family. His father was a mariner. He received no formal education beyond national school; however George Robert developed an interest in poetry, and many years later he could still recite long portions of his favourite passages. He was not keen on mathematics - a failing that was to haunt him later in life.

Vine left school before the age of thirteen and began a series of jobs, that included a butcher's boy and then a grocer's boy. However, these occupations did not give him satisfaction, and he would indulge his passion for learning by purchasing as many books as his meagre wages could allow. He taught himself Greek and Latin, and became interested in poetry, history and science. He then turned his hand to journalism. Vine produced papers on Martin Luther, published poetry: *Poems by a Factory Lad*, and wrote semi-political articles for the paper *The Free Inquirer*. By all accounts he did not find this financially rewarding and after a period of some instability when he took up a trade - that of stay making, he accepted a position as the manager of a corset factory in Athlone, Ireland in 1855.

In Athlone Vine developed an interest in geology and he spent some time studying the Carboniferous Limestone fossils found in the local quarries,[4] as well as researching the nature of the peat bogs that were plentiful in the Irish midlands. Athlone is situated on the River Shannon which flows north-south through the midlands of Ireland. During the late 1850s when Vine was in residence it was an important centre for woollen manufacturing.[5] This was because the town had the benefits of a close indigenous source of wool, plenty of water necessary for manufacturing and dying processes, links to foreign markets via a navigable river and the port of Limerick, as well as links to the domestic markets of Dublin via Shannon Harbour and the Grand Canal.

While in Ireland he married Elizabeth McKenzie,[6] and their eldest child Elizabeth was born there in 1859, probably at their home on Castlemaine Street. In 1861 Vine and his wife and daughter left Ireland and settled in Middlesex, where their son George Robert (Figure 2) was born in 1862, and their younger daughter Alice on 13th February 1864.[7] Vine still remained somewhat unsettled and became involved in the struggle for workers' rights.

Figure 2. George Robert Vine, junior (1862-1940)
(Photograph courtesy Sheffield Libraries, Archives, and Information; Local Studies Section)

During the late 1840s and early 1850s Britain experienced a rapid industrial development and associated urban expansion. Large numbers of men and women were brought into the industrial workforce as rural migrants, and sought work in the cities. Various elements in this workforce sought social and educational improvements, and Vine, being young and idealistic was one of them. He came into contact with other radicals and reformers such as William James Linton (1812-1898), Giuseppe Mazzini (1805-1872) and Lajos Kossuth (1802-1894). Linton was an engraver who did much fine work for the London Illustrated News and subsequently became a noted illustrator of children's books in the United States. Mazzini, an Italian patriot who sought a united republic of Italy while in exile in London, shared republican views with Kossuth, a Hungarian activist against Austrian rule, who lived in London. Vine was also involved in the Chartist movement which was active in industrialised Britain from the 1830s. This movement sought votes for all men over 21 years of age and fairer parliamentary representation. By the late 1840s the movement which had been taken over by radicals had simply fizzled out. In the 1850s, as his son mentions in his obituary, Vine himself recognised his need to settle down and provide for his family.

He moved briefly to his hometown of Portsmouth, before taking a position as manager

Figure 3. Grave and headstone of George Robert Vine, senior and his wife Elizabeth. (Photograph: P.N. Wyse Jackson)

of a stay and crinoline warehouse in Granville Street, Park in Sheffield in 1865. He was to remain in Sheffield for the remainder of his life. After only two years the warehouse business folded and he moved to the adjacent suburb of Carbrook where he established his own stay manufactory at Alexandria Terrace Dun Street (now Dunlop Street), while still continuing to write articles for *The Sheffield Post* on topics such as "Wages and Prices". Two years later he moved to another house at Luther Place in Attercliffe where he initiated his scientific researches, which ultimately became a consuming passion and a source of some valuable income. By 1875 he was in residence and business at 112 Hill Top, Attercliffe which remained his home and his workshop for the last eighteen years of his life.[8] He was later joined in the business by his two daughters who were both described as "Stay Makers Assistant" in the returns of the 1881 census. We are uncertain as to how he sold his products, but suspect that he would have supplied local shops such as John Banner Ltd, a General Drapers and Ladies' Outfitters, situated close by at 620-626 Attercliffe Road.[9]

He was a hard and dedicated worker; his evenings were spent studying bryozoans and he rarely took a holiday. In 1875 he spent a few days at the holiday resort at Cleethorpes near the Humber Estuary, and returned twice, in 1879 and 1882, for day visits. Cleethorpes was a popular holiday destination for the Yorkshire hinterland and its industrial towns.

From Sheffield it was easily reached via the railway which passed eastwards through the towns and villages of Worksop, Blyton and Kirkton in Lindsey, north above the Lincolnshire Wolds, and southwards to the port of Grimsby, and beyond where it terminated at Cleethorpes. One can imagine Vine disembarking the train with his collecting paraphenalia and heading directly to the sea shore and rocky pools, where he must have cut an unusual sight. Even on his infrequently-taken holidays Vine engaged in bryozoan research – he wrote up the bryozoan fauna of this part of the east coast of England in one of his final papers.

From an early age his parents instilled in him a strong sense of the importance of religion and this never left him. He was an active member of the Methodist congregation at Clifton Street in Sheffield for a number of years between 1867 and 1875 when he would occasionally preach at Sunday worship.[10] He always acknowledged the importance of Sunday school training in the furtherance of his education and in later life he served his local community as a teacher and superintendent at the Leigh Street Baptist Church Sunday School for seventeen years from 8th October 1876 until 1893.[11]

According to his son he was still working on his beloved bryozoans up to two days before his death. On Tuesday 8th August 1893 at 1.20 am George Robert Vine died at home in Sheffield after an illness that had lasted several months, and which had resulted in his being confined to his bed from the preceeding 14th July. Surrounded by his family, he whispered "All right; good night" just before his death. He was buried two days later on Thursday 10th August at Tinsley Park Cemetery, close to his home (Figure 3).[12] This cemetery was one of many in Sheffield at the time, having opened *c.* 1880.

3. Work on bryozoans

During the mid-1800s there developed a proliferation of provincial literary and scientific societies,[13] to which men such as Vine became attracted. He was to become closely associated with a number of Sheffield-based organisations later in life. He was co-founder and first President of Sheffield Microscopical Society and a member of the Sheffield Naturalists' Club (1882-1891),[14] the Yorkshire Geological and Polytechnical Society (1883-1893), and the Sheffield Literary and Philosophical Society. He would no doubt engage in the geological debates of the day, and was a popular lecturer himself who gave talks on diverse subjects such as Coal Measure plants, Carboniferous geology, Hugh Miller, origins of coal, zoophytes,[15] collecting postage stamps (a pastime he took up in the latter years of his life),[16] and wells and well-boring.[17] Towards the end of his life his contribution to local science was appropriately acknowledged by the Yorkshire Naturalists' Union which elected him to Honorary Membership at their meeting in Scarborough on 14th November 1891.[18]

As we have seen Vine's interest in palaeontology was first fostered when he worked in Ireland – during which time he investigated the local geology and assembled a collections of Lower Carboniferous fossils. We know from his letters that he collected a great deal of material over the next 25 years.

However his interest in micropalaeontology and in particular in bryozoans was ignited around the year 1875 when he received some fossil material from Scotland and Northumberland from the Rev. Walter Howchin (*c.* 1845-1937) of Morpeth.[19] These were given in exchange for some theological magazines. With the assistance of Howchin, Vine familiarised himself with the nature of foraminiferans, but he was unable to solicit much help with the bryozoans, a group which initially puzzled him.[20]

What is somewhat surprising is that a humble stay or corset maker of Sheffield should have rapidly become an influential figure in the study of bryozoans, at a time when such leisurely activities were generally preserved for the wealthier classes. He published his first scientific paper in 1876 and by the time of his death some seventeen years later had published over seventy papers (see Appendix 1). This publication record is quite remarkable considering that he continued to run his own business at the same time.

On 18th February 1881 Vine was in London to receive one of two awards from the Lyell Fund for that year from the Geological Society of London. Its President, Robert Etheridge seems to have been a strong supporter of Vine's initial forays into the world of bryozoology, as was Professor Philip Duncan of University College, London the recipient of the Wollaston Medal at the same ceremony. In making the presentation Etheridge said of Vine in his citation:

> In making this award the Council were actuated in part by the wish to express their sense of the value of your researches upon the fossil Bryozoa of the Palaeozoic rocks, as evinced especially by your published writings on the Diastoporidae, an exceedingly difficult group, and in part by their desire to assist you in the further prosection of your investigations.[21]

It must have been a very proud moment for George Vine. In the company of eminent geologists such as Archibald Geikie and H.G. Seelcy, he stood up and responded:

> I receive through you this token of recognition, on the part of the Council, of my humble labours, with mingled feelings of pleasure and pride. I cannot, at present understand the reason why I should be selected as one of the recipients of the Lyell Fund. ...I shall do my utmost to fulfil the higher promise which my hitherto crude efforts have awakened. When, some years ago, I began to study the Carboniferous Polyzoa, I wrote to Prof. Duncan, F.R.S., asking for certain information respecting these Polyzoa; he wrote me back word saying that he did not know much about the species himself, and he did not think that there were many men in England who did; but he counselled me to go and work and find out for myself, and then he and others would be glad to know.

I may differ from many of you in the future, but I shall hope to do so in such a manner as to show that I do not forget that we are fellow-students and at the same time gentlemen.[22]

Two years earlier Duncan had persuaded the British Association to establish a committee to report on fossil Polyzoa. Its members were himself and Vine as Secretary. It would not have been difficult for Duncan to persuade his eminent colleagues to fund the work of this committee – he was after all President of Section C (Geology) for that year. What is surprising is that Vine should have found patronage from this body. In 1879 the British Association awarded the Fossil Polyzoa committee the sum of £10-0-0, and renewed this figure for the next four years up until 1884. Duncan acted as Chairman of the committee in 1879 and 1880 and was replaced by the Sheffield naturalist Dr Henry Clifton Sorby between 1881 and 1884.[23] In that year a similar sum was granted for the study of Recent Polyzoa and in the three years between 1889 and 1891 £10-0-0 was awarded for the study of Cretaceous Polyzoa, under the chairmanship of H. Woodward. In total the committee received £90 from the British Association, most of which was probably given to Vine to support his research. He was later to acknowledge:

"most of my scientific work has been carried on by means of small grants from the Royal Society, and from the British Association – you will understand I am not in an independent position."[24]

There can be no doubt that the Secretaryship of the British Association committees that reported on bryozoans gave George Vine a position of some authority in bryozoological circles. He was soon corresponding widely and exchanged material with other bryozoologists including George Busk,[25] Eliza Jelly[26] and John Young in Britain and E.O. Ulrich and J.M. Nickles in the United States. He also received material for identification from University academics who wished to improve the holdings of their institutions. These included Professors Judd of London and Hughes of Oxford.[27] He also corresponded with various amateur enthusiasts including George Shrubsole a chemist from Chester, with whom he collaborated on a paper concerning Silurian bryozoans,[28] with Mr Thomas Jesson who sent him specimens from the Greensand of Cambridge,[29] Mr Thomas John George (*fl.*1880-1904) who supplied material from the Jurassic of Northampton[29] and Mr H.J. Euson, F.G.S. who had obtained specimens from the boring at Gayton also in Northamptonshire.[31] From these fellow scientists he received much bryozoan material.

It is perhaps easy to dismiss the work of Vine as being somewhat poor in quality. Indeed, when one compares his work with contemporary works published by George Busk and Arthur William Waters it appears poorly executed and badly illustrated. Most of Vine's papers were illustrated with drawings executed by his son, who had studied Art, as well as science. While none of the published versions are actually signed, a number of characteristic mauve coloured copies interleaved in offprints now in Cardiff and London bear the younger Vine's signature. Vine junior had distinctive handwriting which was somewhat blocky in style and contrasted sharply with that of his father, which was an untidy copper-plate style. Other unpublished illustrations, of inferior draftmanship, also found interleaved in some of these offprints, were probably drawn by Vine himself. Good quality illustrations, drawn by professional draftsmen (including Arthur Humphries Foord who later worked on Irish Carboniferous cephalopods), were used to illustrate his papers published by the Geological Society. One must imagine that these were paid for by the Society and not by Vine.

Recent examination and assessment of his work on Carboniferous faunas has shown that many of his taxonomic determinations are doubtful; his descriptions were short, and his illustrations are very poor. He erected a large number new species and varieties (see Appendix 3), many of which were based on limited populations and sample sizes – some of his taxa are now regarded as tenuous.[32] There is also some evidence to suggest that in his eagerness to describe British faunas he created nomenclatural problems elsewhere.[33] A detailed reassessment of his specimens which may be held in various museum collections (see below) is required to fully assess the validity of many of his taxa.

However, before dismissing his work as worthless, one must remember that he worked purely in his own free time – which must have been limited – and given his output his contribution is really rather remarkable. He wrote extensively on Carboniferous bryozoans particularly from Lancashire and Yorkshire. He is best remembered as being the author of the Order Cryptostomata which he proposed in 1884, and the series of comprehensive reports for the British Association that described all the fossil bryozoans of the Great Britain known at that time. Such an undertaking was ambitious. The reports totalled some 570 printed pages, and remain as a testimony to Vine's energy and infectious enthusiasm for bryozoans. He is also remembered in the specific name *Pinnatopora vinei*, erected by Ulrich in 1888, and in the name of the genus *Vinella* Ulrich, 1890, a Silurian ctenostome.[34] This would suggest that Ulrich forgave Vine his publishing his manuscript name *Streblotrypa* a few years earlier.[35]

4. Fossil dealership

George Vine ran a business dealing in fossils from his home at 112 Hill Top, Attercliffe, Sheffield from about 1880. The business was jointly run with his son George Robert, junior[36] who was a mathematics teacher and later headmaster of Hunters Bar Primary School, and subsequently of Huntsman's Gardens School, both of which were in Attercliffe. The latter school was of an unusual design, in that all 1,250 pupils could be seen simultaneously from the headmaster's desk![37] Vine and his son specialised in the preparation and supply of micro-palaeontological specimens. Whereas Vine senior was primarily concerned with bryozoans, his son was interested in forams and ostracodes (see Appendix 2), they also supplied echinoderms, gastropods, sponges and annelids.

Figure 4. Selection of Vine's bryozoan slide mounts. The middle cavity slide was prepared by John Young of the Hunterian Museum, Glasgow. (National Museums and Galleries of Wales)

It is difficult to see where Vine got the time to collect his own material, given his dependence on his stay-making business to provide a living for him and his family. We know that he had collected fine Carboniferous Limestone fossils from Ireland in the 1850s, but these he never intended selling. From the 1860s he had been assembling material, so that by the middle of the 1880s he was able to state that:

> "When you consider that my stock is the accumulation of hard upon 30 years and in some cases more than that one need not be surprised at the accumulation."[38]

Additionally Vine as we have mentioned earlier, was in correspondence with many naturalists across the world, with whom he exchanged material. In late 1887 he traded some Lower Palaeozoic fossils for a suite of fine fossils from the Cenomanian of France.[39] Examination of the museum holdings of his collections indicate the wide geographical provenance of his material. He had specimens from the Ordovician of North America, the Miocene of Australia, as well as material from Oceania, northern Italy, and closer to home from western Ireland. Not only did he offer bryozoans and other fossils for sale, but also sold natural history specimens such as sea weeds, diatoms, lower plants and sections of various timbers, on the rare occasions when he had these in stock.[40]

The bulk of the Vine's business comprised selling fossil bryozoans and other micropalaeontological organisms. These were offered for sale mounted either as strews or more often mounted in cavity slides (Figure 4). In an attractive printed catalogue

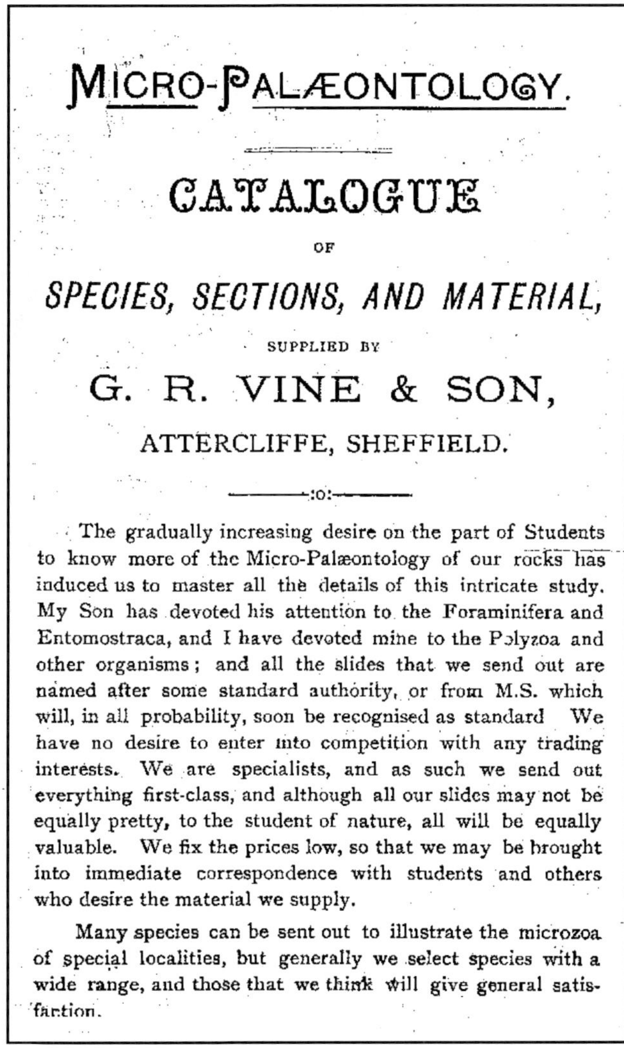

MICRO-PALÆONTOLOGY.

CATALOGUE

OF

SPECIES, SECTIONS, AND MATERIAL,

SUPPLIED BY

G. R. VINE & SON,

ATTERCLIFFE, SHEFFIELD.

:o:

The gradually increasing desire on the part of Students to know more of the Micro-Palæontology of our rocks has induced us to master all the details of this intricate study. My Son has devoted his attention to the Foraminifera and Entomostraca, and I have devoted mine to the Polyzoa and other organisms; and all the slides that we send out are named after some standard authority, or from M.S. which will, in all probability, soon be recognised as standard We have no desire to enter into competition with any trading interests. We are specialists, and as such we send out everything first-class, and although all our slides may not be equally pretty, to the student of nature, all will be equally valuable. We fix the prices low, so that we may be brought into immediate correspondence with students and others who desire the material we supply.

Many species can be sent out to illustrate the microzoa of special localities, but generally we select species with a wide range, and those that we think will give general satisfaction.

Figure 5. First page of catalogue of micro-palaeontological specimens available from G.R. Vine and Son. (Department of Geology, National Museums and Galleries of Wales)

(Figure 5) some 154 different samples are listed for sale.[41] These include 81 Carboniferous taxa comprising Foraminifera, Ostracoda and Bryozoa. Additionally various Mesozoic and Tertiary foraminifera are also listed, together with dredge material from the voyages of H.M.S. *Challenger* and H.M.S. *Alert*.

In this catalogue Vine explained that he and his son were specialists and although all their slides may not be equally pretty, all were equally valuable. They described their prices as low so that they could be brought into contact with students and others who desired their material. Material was supplied identified with determinations either printed

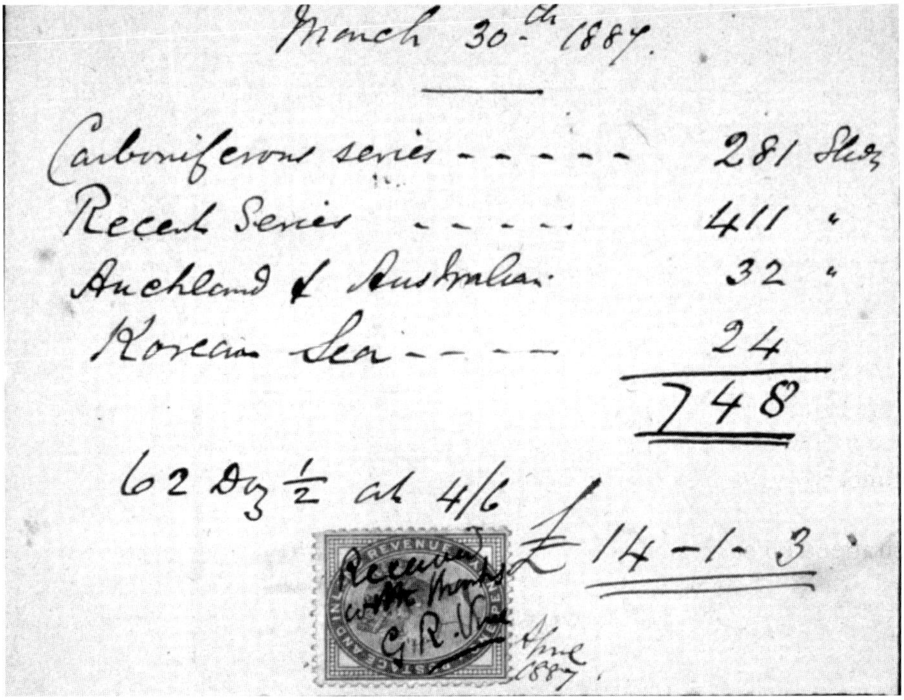
March 30th 1887.

Carboniferous series ----- 281 Slides
Recent Series ----- 411 "
Auckland & Australian. 32 "
Korean Sea ---- 24
748

62 Doz 1/2 at 4/6 £14-1-3

Received with thanks G.R. Vine
June 1887

Figure 6. Receipt from G.R. Vine to P.B. Mason 1887
(Department of Geology, National Museums and Galleries of Wales)

on a label marked "G.R. Vine" or some similar wordage, and sometimes dated. The cost of the slides ranged from 5s 6d (27.5 pence) for a dozen bryozoan slides to 4s 6d (22.5 pence) a dozen in the 1880s (Figure 6). Vine regarded his material as being "first class".[42]

Among his customers were Professor John W. Judd of Imperial College, London who in 1886 purchased about 1000 bryozoan and foraminiferan slides.[43] Vine claimed to have sent material to George Busk which was used by him for his monograph on the bryozoans of the Crag,[44] but given the date of publication of the monograph this is unlikely.

Not only did he sell material but he and his son also prepared, picked, mounted and identified material for others. For this service he charged a rate of between 1 and 2 shillings (5-10 pence) per dozen slides. The higher rate was for forams and ostracodes. We suspect that he was sent a good deal of such material by gentlemen collectors who enjoyed collecting the material and enjoyed examining the curated specimens in their study cabinets, but who did either not enjoy or did not have the time to spend many tedious hours preparing it.

Given the size and nature of the material Vine was dealing in, all of it had to be collected, prepared, picked, mounted and identified in natural light. This often caused him problems particularly during the short winter days when it was often dark when he returned home from work. He complained that picking Foraminifera in such conditions was virtually impossible[45] – particularly given his failing eyesight.[46] Thin sections were prepared in his wife's kitchen or scullery sink.

Between 1886 and 1892 Vine supplied Philip Brookes Mason (2nd January 1842-6th November 1903) a doctor in Burton-on-Trent, with a large quantity of palaeontological material. Mason was an enthusiastic collector of natural history, interested particularly in ornithology and entomology.[47] Rapidly Mason purchased from Vine a huge collection of fossil and Recent bryozoans. An indication of his enthusiasm for developing his collection can be gauged by the number of specimens that he purchased. In 1886 Vine supplied 16 dozen slide preparations at a cost to Mason of £4/10/9. The following year Mason purchased a staggering 270 dozen slides at a total cost of £60/17/3. This was a large sum of money at that time.

Soon after their business relationship started Vine must have realised that Mason would purchase almost anything that he offered him. Whenever Vine obtained new material a collection was dispatched to Burton-on-Trent. When Vine had decided to no longer study Upper Carboniferous plants he offered Mason his collection together with a series of lantern slides of some specimens – these were naturally purchased.

In 1886 Mason sent Vine a box of Crag and Chalk bryozoans, which we assume he collected himself. These were prepared by Vine and the choicest specimens returned. Certainly Vine would have kept residues of such material sent to him for preparation, and certainly these were offered for sale to others. Later Mason sent material that he had dredged from his yacht while sailing at Roundstone in Connemara in Western Ireland. However, Vine later admitted that the material may have become mixed up, but was relieved when Mason told him that he would be returning to Roundstone the following year and that new material could be obtained.[48]

He was at several times in his life in serious financial difficulty and at one dark period in 1889 he was close to selling all his prized fossil collections. At that time he was supporting a family on an income of 16 to 17/- a week after rent and taxes.[49] He wrote to Mason in a state of deep despair stating that:

> "I find now I shall be obliged from circumstances over which I have no control – failing circumstances &c – to give up almost entirely the study of all but a very small group of organisms. My age (65 next birthday) is one of the causes – and limited means without any pecuniary help from societies is another – and now as age creeps on – unless I give more of my time to my small business – failure will be my last reason for my resolution to stop in time - I have a few papers in hand that I shall try to complete, then I must say goodbye to Palaeontology."[50]

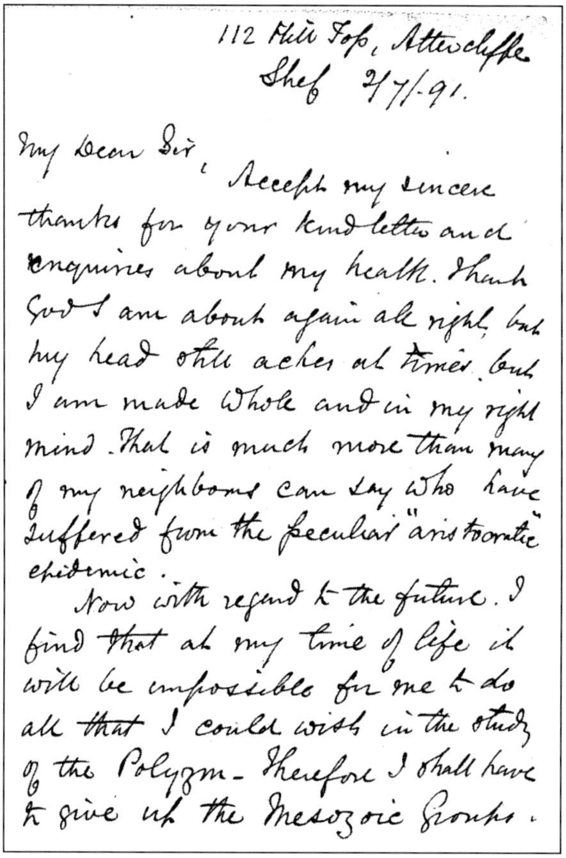

112 Hill Top, Attercliffe
Shef 2/7/91.

My Dear Sir,

Accept my sincere thanks for your kind letter and enquiries about my health. Thank God I am about again all right, but my head still aches at times. but I am made Whole and in my right mind. That is much more than many of my neighbours can say who have suffered from the peculiar "aristocratic" epidemic.

Now with regard to the future. I find ~~that~~ at my time of life it will be impossible for me to do all ~~that~~ I could wish in the study of the Polyzoa. Therefore I shall have to give up the Mesozoic Groups.

Figure 7. Portion of letter from G.R. Vine to P.B. Mason 2/7/1891 (Department of Geology, National Museums and Galleries of Wales)

He offered Mason his collection of macrofossils, which included his prized Irish Carboniferous suite, at a price of 1d or 2d each. Happily Vine's financial plight was discovered by some of his scientific friends who encouraged him to prevail. He was told that he would be a "loss to science", and it is probable that these events forced the reactivation of the Committee on Fossil bryozoans to which the British Association awarded £10 in 1889 and in each of the following two years.

During 1891 when both himself and his wife were suffering from the "peculiar 'aristocratic' epidemic" Vine realised that he would never complete all the projects that he had set out to undertake (see Figure 7). Soon he was trying to sell Mason his personal collection of offprints and other fine bryozoological reference volumes including his copy of Busk's monograph (which he had purchased for 23/-) and a copy of d'Orbigny. Although keen to off load this material he was slightly reluctant to do so, as he hoped that Mason might become a serious student of bryozoans, and was unsure if this would happen.[51]

During a period of seven years George Vine supplied Philip Mason with a large volume of fossil material, and over time their relationship went from being purely that of a business relationship to that of friendship. In a letter of mid-1891 Vine opened up and showed his true feelings and appreciation of Mason's patronage:

> "The fact is you have been so kind to me in your dealings – that I look upon you as a good personal friend. Had I been able to hourd (sic) up or put aside the money paid to me by you I should be rich to day in a moderate way. But that had to go to wipe off all my debts contracted in the interval between 1879- & 1885 – all by your kindly help swept away leaving me now a free man in every sense. I have longed to tell you the above so as to prevent you imagining my importunity in offering you things had other meanings – than those which I have now disclosed to you."[52]

In the last letter of their correspondence Vine wrote:

> "I may now say that your collection of Fossil Polyzoa is now a valuable one and a student ought to be able to make a good head in the study of that most peculiar group."[53]

It is a measure of how much the family valued the custom of Mason that after the death of Vine, his son sold his father's personal collection of type specimens to him. These specimens are now in Cardiff, while an inferior set was sold to the British Museum in London (these remain at The Natural History Museum).

George Robert Vine was unusual for his time in that he sold micropalaeontological fossil specimens. Judging by the diversity and volume of material supplied to one customer, he must have conducted a reasonably successful business albeit a sideline to his main source of income as a stay-maker. It would be interesting to source other customers and discover if they too, like Mason, were active purchasers of material from the fossil dealer of Attercliffe.

5. Collections and archives

Vine material is held at a number of museums in Britain,[54] and was acquired either by purchase from Vine himself, or from his son, or from collectors who had bought material from Vine for their own collections. Major collections are held in Bolton Museum; The Natural History Museum, London; The National Museums and Galleries of Wales, Cardiff; Glasgow Museum; and at Sheffield City Museum. Smaller collections of Lower Greensand specimens were presented to the Geological Society by Vine in 1885. These passed to the Geological Museum in 1911 and are now probably at the British Geological Survey at Keyworth in Nottingham if not at The Natural History Museum, London. A collection of French Bathonian bryozoans described by Vine (1888) was presented to Northampton Museum, where we assume they still may be found. In addition the Hunterian Museum in Glasgow contains a collection of approximately 100 Silurian fossils from Shropshire, including bryozoans, brachiopods, gastropods, crinoids, and arthropods, which was acquired in the 1880s, most likely as a result of Vine's professional friendship with Mr John Young.

Bolton Museum: The Vine specimens in Bolton Museum were acquired as part of the P.B. Mason collection in 1909.[55] On his death in 1903 Mason's widow dispersed his collections, partly by auction and by direct sales to museums. A letter now at Bolton, from Mrs Mason to Jansen, natural history dealers of London, lists material that she wished to dispose of. Subsequently she bypassed this dealer and liaised with various museums directly herself. The letter refers to 'four microscope cabinets [that] contain the type collection of Bryozoa formed by G.R. Vine Sheffield referred to in his papers of the reports of the British Association for the Advancement of Science. Afterwards he sold to the British Museum a second collection but one greatly inferior to this'. From this, Hancock *et al.* concluded 'that it seems likely that some of Vine's best material may be present at Bolton'.[56] This is not actually correct as most of the best material, including one cabinet, is now at Cardiff. The collection of 250 Vine bryozoans from Britain and North America purchased by Bolton Museum from Mrs Mason in 1909 is listed in a manuscript catalogue in the museum. Not all this material can now be recognised with certainty in its collections.

The Natural History Museum, London: This institution purchased a collection of 1695 bryozoan specimens from Vine in 1893, which according to Woodward[57] included specimens referred to in Vine's BAAS reports. It also obtained bryozoans described by Vine (1890) from the Red Chalk in the collection of Thomas Jesson (1850-1928).[58] This latter collection originally formed part of the Institute of Geological Science collection. Vine's specimens principally comprise thin sections and small zoaria housed in cavity slides – many are labelled on purposely-printed labels in Vine's handwriting. The Bryozoa Section of the Department of Zoology holds a number of Vine's letters to George Busk that are pasted into Busk's 1883-1886 bound volume of Vine's offprints. It also holds a bound volume of his BAAS reports, and some annotated offprints in the Harmer Library (Volumes 20 and 46). These were purchased by S.F. Harmer from the book dealer Dulau and Co. of Soho Square, London. It is possible that these were originally sold by Vine or his son to Dulau. The Palaeontological Library contains a letter written by Vine concerning his collection. In the offprint holdings of the Bryozoa Section of the Department of Palaeontology are found many of Vine's annotated offprints.

National Museums and Galleries of Wales: The Department of Geology at the National Museums and Galleries of Wales (NMW) possesses a microscope slide cabinet (Figure 8) with about 2000 specimens mostly fossil Bryozoa, along with letters, manuscript catalogues, and bound volumes of Vine's papers.

On 17th November 1914, the NMW purchased four cabinets of microscope slides from

Figure 8. Cabinet purchased from P.B. Mason's widow, containing micro-palaeontological specimens from the G.R. Vine collection (Department of Geology, National Museums and Galleries of Wales)

Mrs Mason (only one cabinet remains utilised to house its original material). These include 1270 specimens of fossil Bryozoa (NMW accession number 21.490G); fossil Foraminifera (NMW accession number 21.491G); fossil crustacean (NMW accession number 21.492G); and 124 specimens of rock sections and deep-sea deposits (NMW accession number 21.493G).

At presumably the same time, the NMGW acquired various archives. These included

59 letters and receipts from G.R.Vine to P.B. Mason covering the period 1886 – 1892 referring to the material being purchased by Mason for his collection. There are also two bound volumes of Vine's published papers and a bound volume of Vine's catalogues with the note by P.B. Mason stating 'This was Mr Vines Type Collection and was sold to me before the British Museum has bought his duplicates'. The papers also include a bound volume of manuscript 'Notes on Cyclostomatous Polyzoa Faeroe limestone, Denmark by G R Vine'; a bound volume of British Association Polyzoa Reports 1880-1890, and an index of Polyzoa manuscripts.

Hunterian Museum, Glasgow: This includes Silurian bryozoans, brachiopods, gastropods, crinoids, worms, and arthropods from Shropshire; *c*. 1880s; 100 specimens.

Sheffield City Museum: Cleevely[59] notes that Vine material is present at Sheffield City Museum which was presented by G.R. Vine, junior in 1936. FENSCORE lists a considerable volume of material at Sheffield, including *c*. 600 slides of bryozoans, 250 formainifera slides and some coelenterates and hydroids. These are predominantly British in origin, but material from Ireland, continental Europe, Africa, Australia, America and Tahiti are also present. 55 slide mounts of foraminiferans are derived from the Challenger expedition, and other material is associated with Miss Jelly and Mr Waters.

Further work is needed to compare Vine's published material with the manuscript catalogues at Cardiff and Bolton, and with the collections housed at these various museums.

Unfortunately there is no trace of Vine's molluscan specimens collected from the Carboniferous of the midlands of Ireland which his son remarked were most important and highly prized by his father.

6. Conclusions

George Robert Vine was unusual in that he became an acknowledged expert in the study of fossil bryozoans, and a proponent of microscopical techniques in their study during the late Victorian period. He was unusual because he hailed from a working class background, yet was a recipient of awards and financial aid from the premier geological and scientific societies in Britain including the Geological Society of London, the British Association for the Advancement of Science, and the Royal Society. He produced a large body of published work including a fine series published by the British Association (see Appendix 1).

However, he was unable to rise above his working class roots and join the scientific establishment through membership of these august societies. This may have been because he, or they felt that it was not proper for such men to gain, for example, Fellowship of the Geological Society. This would explain why none of his papers were actually communicated by him in person to the Society.[60] Social exclusion is difficult to prove, and Vine's non-membership of such organisations may have been purely as a result of his financial state. Membership of the British Association cost £2 in the first year and £1 subsequently – a considerable sum for a Sheffield corset-maker.

Surprisingly his death passed with little notice, other than two obituaries penned by his loyal son and a notice published in the Sheffield and Rotherham Standard. Even the Yorkshire Geological Society, to which he submitted many fine papers on bryozoans, published no obituary. The Geological Society of London who had honoured him with the Lyell Fund made no mention in their Proceedings of his passing.

Nevertheless, Vine's fine body of work on bryozoans remains encapsulated in his large number of publications and in his collections which are now housed in various museums in Britain. To bryozoologists his name is forever linked with the stenolaemate Order Cryptostomata which he erected in 1884.

7. Acknowledgements

We are most grateful to Gaynor Boon (Sheffield City Museum), Dr Paul Taylor (The Natural History Museum), Dr Matthew Parkes (Geological Survey of Ireland), Michael Holmes (Sorby Natural History Society) Alan Howell (formerly of Bolton Museum) and Mary Spencer Jones (The Natural History Museum) for help during the preparation of this paper. Ryan O'Connell of London, who is a direct descendant of George Robert Vine, kindly supplied us with considerable information on him and his family. Gaynor Boon brought to our attention the portrait of Vine now in the Sheffield Local Studies Library. Permission to reproduce this portrait and that of George Vine junior was kindly granted by Mike Spick.

Appendix 1. Bibliography of G.R. Vine, senior (only papers on bryozoans and other fossil organisms are listed here)

1860

Mountain Limestone Fossils. *The Geologist* **1860**, 463.

1876

On the discovery of macrospores in Carboniferous sandstone. *Hardwicke's Science Gossip* **12**, 247-248.

1877

Chapters on Carboniferous Polyzoa. *Hardwicke's Science Gossip* **13**, 108-110, 152-156, 220-222, 271-274.

1878

The genus *Fenestella*: its history, development, and range in space and time. *Hardwicke's Science Gossip* **14**, 247-251, 274-276.

1879

Physiological character of *Fenestella. Hardwicke's Science Gossip* **15**, 50-54.

On *Palaeocoryne* and the development of *Fenestella. Hardwicke's Science Gossip* **15**, 225-229, 247-249.

On Carboniferous Polyzoa and *Palaeocoryne. Report of the Forty-ninth meeting of the British Association for the Advancement of Science (Sheffield, 1879).* John Murray, London, pp. 350-351.

1880

A Review of the Family Diastoporidae for the purpose of classification. *Quarterly Journal of the Geological Society of London* **36**, 356-361.

A Review of the Family Diastoporidae for the purpose of classification. *Geological Magazine* (n.s.) Decade 2 **7**, 323. [Abstract of communication to the Geological Society of London, 12th May 1880]

A Review of the Family Vincularidae, Recent and Fossil for the purpose of classification. *Geological Magazine* (n.s.) Decade 2 **7**, 381. [Abstract of communication to the Geological Society of London, 23rd June 1880]

Report of the Committee, consisting of Prof. P.M. Duncan and Mr. G.R. Vine, appointed for the purpose of reporting on the Carboniferous Polyzoa. *Report of the Fiftieth meeting of the British Association for the Advancement of Science (Swansea, 1880).* John Murray, London, pp. 76-87.

On the Carboniferous Polyzoa. *Geological Magazine* (n.s.) Decade 2 **7**, 501-512. [Reprinted version of previous item - British Association report presented at Swansea.]

1881

Second report of the Committee, consisting of Prof. P.M. Duncan and Mr. G.R. Vine, appointed for the purpose of reporting on Fossil Polyzoa. *Geological Magazine* (n.s.) Decade 2 **8**, 471-477; 509-519. [Abridged version of British Association report, published in full in 1882]

Notes on the Carboniferous Polyzoa of North Yorkshire. *Proceedings of the Yorkshire Geological and Polytechnic Society* (n.s.) **7**, 329-341.

Further notes on the family Diastoporidae, Busk, species from the Lias and Oolite. *Quarterly Journal of the Geological Society of London* **37**, 381-390.

Silurian uniserial Stomatoporae and *Ascodictya. Quarterly Journal of the Geological Society of London* **37**, 613-619.

1882

Notes on the Polyzoa of the Wenlock shales, Wenlock limestone, and shales over the Wenlock limestone. *Quarterly Journal of the Geological Society of London* **38**, 44-68.

Notes on the Polyzoa of the Wenlock shales, Wenlock limestone, and shales over the Wenlock limestone. From material supplied by G. Maw, Esq., F.L.S., F.G.S. *Geological Magazine* (n.s.) Decade 2 **9**, 43. [Abstract of communication to the Geological Society of London, 7th December 1881]

The Silurian species of *Glauconome* and a suggested classification of the Palaeozoic Polyzoa. *Proceedings of the Geological Society of London* **38**, 245. [G.W. Shrubsole and G.R. Vine]

Notes on the *Annelida tubicola* of the Wenlock shales, from the Washings of Mr. George Maw, F.G.S. *Quarterly Journal of the Geological Society of London* **38**, 377-392.

Notes on the *Annelida tubicola* of the Wenlock shales, from the Washings of Mr. George Maw, F.G.S. *Geological Magazine* (n.s.) Decade 2 **9**, 377. [Abstract of communication to the Geological Society of London, 7th June 1882]

The Silurian species of *Glauconome* and a suggested classification of the Palaeozoic Polyzoa. *Geological Magazine* (n.s.) Decade 2 **9**, 381. [Abstract of communication to the Geological Society of London, 21st June 1882] [G.W. Shrubsole and G.R. Vine].

The Diastoporidae, or the natural history of a family type. *Hardwicke's Science Gossip* **18**, 81-83, 146-147, 244-247.

Second report of the Committee, consisting of Prof. P.M. Duncan and Mr G.R. Vine, appointed for the purpose of reporting on the Carboniferous Polyzoa. *Report of the Fifty-first meeting of the British Association for the Advancement of Science (York, 1881)*. John Murray, London, pp. 61-176.

1883

Notes on the corals and bryozoans of the Wenlock shales. *Proceedings of the Geological Society of London* **39**, 69-70.

Third report of the Committee, consisting of Dr H.C. Sorby and Mr G.R. Vine, appointed for the purpose of reporting on fossil Polyzoa, Jurassic species—British area only. *Report of the Fifty-second Meeting of the British Association for the Advancement of Science (Southampton, 1882)*. John Murray, London, pp. 249-266.

Notes on the Carboniferous Polyzoa of West Yorkshire and Derbyshire - (an attempt to identify Phillip's species). *Proceedings of the Yorkshire Geological and Polytechnic Society* (n.s.) **8**, 161-174.

Notes on the Carboniferous Entomostracea and the Foraminifera of the North Yorkshire shales. *Proceedings of the Yorkshire Geological and Polytechnic Society* (n.s.) **8**, 226-239.

Notes on the corals and bryozoans of the Wenlock shales. (Mr Maw's Washings). *Geological Magazine* (n.s.) Decade 2 **10**, 190. [Abstract of communication to the Geological Society of London, 21st February 1883]

1884

Micro-palaeontology of the Northern Carboniferous shales. I. Introduction: Foraminifera, &c. *The Naturalist* (n.s.) **10** (110), 37-40.

Fourth report of the Committee, consisting of Dr H.C. Sorby and Mr G.R. Vine, appointed for the purpose of reporting on fossil Bryozoa. *Report of the Fifty-third meeting of the British Association for the Advancement of Science (Southport, 1883)*. John Murray, London, pp. 161-209.

Further notes on new species, and other Yorkshire Carboniferous Polyzoa described by Prof. John Phillips. *Proceedings of the Yorkshire Geological and Polytechnic Society* (n.s.) **8**, 377-393.

Micro-palaeontology of the Northern Carboniferous shales. II. Polyzoa of the Redesdale Shales, Northumberland. *The Naturalist* (n.s.) **10** (111), 61-66.

Micro-palaeontology of the Northern Carboniferous shales. III. The Ostracoda, *Monticulipora*, and Miscellaneous Forms: Redesdale Shales, Northumberland. *The Naturalist* (n.s.) **10** (113),

97-103.
The Silurian species of *Glauconome* and a suggested classification of the Palaeozoic Polyzoa. *Quarterly Journal of the Geological Society of London* **40**, 329-332 [with G.H. Shrubsole].
Polyzoa (Bryozoa) found in the boring at Richmond, Surrey, referred to by Prof. J.W. Judd. *Quarterly Journal of the Geological Society of London* **40**, 784-794.
Notes on some Cretaceous Lichenoporidae. *Quarterly Journal of the Geological Society of London* **40**, 850-864.
Polyzoa (Bryozoa) found in the boring at Richmond, Surrey, referred to by Prof. J.W. Judd. *Geological Magazine* (n.s.) Decade 3 **1**, 377. [Abstract of communication to the Geological Society of London, 25th June 1884]
Notes on some Cretaceous Lichenoporidae. *Geological Magazine* (n.s.) Decade 3 **1**, 384. [Abstract of communication to the Geological Society of London, 25th June 1884]
Notes on species of *Ascodictyon* and *Rhopalonaria* from the Wenlock shales. *Annals and Magazine of Natural History* **80**, 77-89.

1885

Fifth and last report of the Committee, consisting of Mr H.C. Sorby, F.R.S. and Mr G.R. Vine, appointed for the purpose of reporting on fossil Polyzoa. *Report of the Fifty-fourth meeting of the British Association for the Advancement of Science (Montreal, 1884).* John Murray, London, pp. 97-219.
On *Phyllopora* and *Thamniscus* from the Lower Silurian rocks near Welshpool, Wales. *Quarterly Journal of the Geological Society of London* **41**, 108-113.
Notes on species of *Phyllopora* and *Thamniscus* from the Lower Silurian rocks near Welshpool, Wales. *Geological Magazine* (n.s.) Decade 3 **2**, 90. [Abstract of communication to the Geological Society of London, 14th December 1884]
Notes on the Polyzoa and Foraminifera of Cambridge Greensand. *Proceedings of the Geological Society of London* **41**, 101-102.
Notes on the Polyzoa and Foraminifera of Cambridge Greensand. *Proceedings of the Yorkshire Geological and Polytechnic Society* (n.s.) **9**, 10-29.
Notes on the Polyzoa and Foraminifera of Cambridge Greensand. *Geological Magazine* (n.s.) Decade 3 **2**, 280-281. [Abstract of communication to the Geological Society of London, 15th April 1885]
Notes on the Yoredale Polyzoa of North Lancashire. *Proceedings of the Yorkshire Geological and Polytechnic Society* (n.s.) **9**, 70-98.
Micro-palaeontology of the Northern Carboniferous shales. IV Polyzoa, Entomostraca, Gastropoda, and Miscellaneous organisms of the Skelly Gate Shales, Northumberland. *The Naturalist* (n.s.) **10** (117), 207-212.
Micro-palaeontology of the Northern Carboniferous shales. V Upper Carboniferous Shales, Northumberland, Yordale: Foursknes, 'Inghoe', Lowick. Polyzoa and Entomostraca. *The Naturalist* (n.s.) **10** (122), 312-318.
Luminous wood. *The Naturalist* (n.s.) **10** (123), 340-341.

1886

Notes on the Polyzoa of the Wenlock shales, Etc. *Proceedings of the Yorkshire Geological and Polytechnic Society* (n.s.) **9**, 179-201.
Notes on the Palaeontology of the Wenlock shales of Shropshire. (Mr. Maw's washings, 1880).

Proceedings of the Yorkshire Geological and Polytechnic Society (n.s.) **9**, 224-248.

Report of the Committee, consisting of Dr H. Woodward, Mr G.R. Vine, Dr P.M. Duncan, H.C. Sorby and C.E. de Rance, appointed for the purpose of reporting on Recent Polyzoa. *Report of the Fifty-fifth meeting of the British Association for the Advancement of Science (Aberdeen, 1885)*. John Murray, London, pp. 481-680.

1887

Notes on the distribution of the Entomostraca in the Wenlock Shales. Mr. Maw's washings. *Proceedings of the Yorkshire Geological and Polytechnic Society* (n.s.) **9**, 393-405.

Jurassic Polyzoa in the neighbourhood of Northampton. *Journal of the Northamptonshire Natural History Society and Field Club* **4**, 202-211.

Notes on classifications of Cyclostomatous Polyzoa; old and new. *Proceedings of the Yorkshire Geological and Polytechnic Society* (n.s.) **9**, 316-362.

Notes on a species of *Entalophora* from the Neocomian clay of Lincolnshire. *Annals and Magazine of Natural History* Series 5, **19**, 17-19.

Notes on the Polyzoa and other organisms from the Gayton Boring, Northamptonshire. *Journal of the Northamptonshire Natural History Society and Field Club* **4**, 255-266.

Fossil Polyzoa in Lincolnshire. *The Naturalist* (n.s.) **12** (138): 18.

1888

Notes on the classification of the Palaeozoic Polyzoa. *Proceedings of the Yorkshire Geological and Polytechnic Society* (n.s.) **11**, 20-44.

A Monograph of Yorkshire Carboniferous and Permian Polyzoa. Part I. *Proceedings of the Yorkshire Geological and Polytechnic Society* (n.s.) **11**, 68-85.

Notes on the Polyzoa of Caen and Ranville now preserved in the Northampton Museum. *Journal of the Northamptonshire Natural History Society and Field Club* **5**, 1-24.

1889

Notes on British Eocene Polyzoa. *Proceedings of the Yorkshire Geological and Polytechnic Society* (n.s.) **11**, 154-169.

A monograph of Yorkshire Carboniferous and Permian Polyzoa. Part II. *Proceedings of the Yorkshire Geological and Polytechnic Society* (n.s.) **11**, 184-200.

Further notes on the Polyzoa of the Lower Greensand and the Upper Greensand of Cambridge. Part II. *Proceedings of the Yorkshire Geological and Polytechnic Society* (n.s.) **11**, 250-275.

1890

Notes on the Polyzoa and Microzoa of the Red Chalk of Yorkshire and Norfolk. *Proceedings of the Yorkshire Geological and Polytechnic Society* (n.s.) **11**, 363-398.

A monograph of the Polyzoa (Bryozoa) of the Red Chalk of Hunstanton. *Quarterly Journal of the Geological Society of London* **46**, 454-486.

A monograph of the Polyzoa (Bryozoa) of the Hunstanton Red Chalk. *Geological Magazine* (n.s.) Decade 3 **7**, 234. [Abstract of communication to the Geological Society of London, 26th March 1890]

1891

Report of the Committee, consisting of Dr H. Woodward, Mr G.R. Vine, T. Rupert Jones and Dr

H.C. Sorby, appointed to prepare a report on the Cretaceous Polyzoa. *Report of the Sixtieth meeting of the British Association for the Advancement of Science (Leeds, 1890)*. John Murray, London, pp. 378-396.

Notes on some new or but little known Eocene Polyzoa from localities. *Proceedings of the Yorkshire Geological and Polytechnic Society* (n.s) **12**, 52-61.

British Palaeozoic Ctenostomatous Polyzoa. *Proceedings of the Yorkshire Geological and Polytechnic Society* (n.s.) **12**, 74-93.

1892

Notes on Polyzoa of the zones of the Upper Chalk. *Report of the Sixty-first meeting of the British Association for the Advancement of Science, (Cardiff, 1891)*. John Murray, London, pp. 656-659.

Notes on Polyzoa found at Cleethorpes in 1875, 1879, and 1882. *The Naturalist* **17** (198), 5-11.

Fossil Polyzoa; further additions to the Cretaceous lists. *Proceedings of the Yorkshire Geological and Polytechnic Society* (n.s.) **12**, 149-161.

1893

Report of the Committee, consisting of Dr H. Woodward, Mr G.R. Vine, T. Rupert Jones and Dr H.C. Sorby, appointed for the completion of a report on the Cretaceous Polyzoa. *Report of the Sixty-second meeting of the British Association for the Advancement of Science (Edinburgh, 1892)*. John Murray, London, pp. 301-337.

Notes on the Polyzoa, *Stomatopora* and *Proboscina* groups, from the Cornbrash of Thrapston, Northamptonshire. *Proceedings of the Yorkshire Geological and Polytechnic Society* (n.s.) **12**, 247-258.

Appendix 2. Bibliography of G.R. Vine, junior (only scientific papers and papers of local history are listed here)

1880

Natural History of the district (of Sheffield). *Report of the Forty-ninth meeting of the British Association for the Advancement of Science (Sheffield, 1879)*. John Murray, London, p. 101.

1898

On some Miocene Foraminifera from Australia. *Transactions and Annual report of the Sheffield Microscopical Society* for 1897-1898, pp. 6-9.

1913

Tinsley forty years ago. *Ward's Local Annual* for 1913, p. 9.

1914-1916

Attercliffe: past and present. *Ward's Local Annual* for 1914, pp. 5-19; 1915, pp. 5-31; 1916, 5-31.

1932-1936

The story of old Attercliffe. Ward Bros, Sheffield (1932-1936), 333 pp.

Old Attercliffe and Carbrook. *Transactions of the Hunter Archaeological Society* **3**, 80-84.

Appendix 3. Some new bryozoan taxa described by G.R. Vine, senior. [Modern accepted names where known are given in square brackets]

ORDER

Cryptostomata Vine, 1884

GENERA

Arcanopora Shrubsole and Vine, 1882 [= *Sulcoretepora* d'Orbigny, 1849]
Pinnatopora Shrubsole and Vine 1884 [= *Penniretepora* d'Orbigny, 1849]

SPECIES

Diastopora cricopora Vine, 1881 [= *Reptomultisparsa cricopora* (Vine, 1881)]
Diastopora oolitica Vine, 1881 [= *Reptomultisparsa oolitica* (Vine, 1881)]
Diastopora ventricosa Vine, 1881 [= *Reptomultisparsa ventricosa* (Vine, 1881)]
Stomatopora dissimilis Vine, 1881 [= *Corynotrypa dissimilis* (Vine, 1881)]
Stomatopora dissimilis Vine var. *compressa* Vine, 1882 [= *Corynotrypa dissimilis* (Vine, 1881)]
Stomatopora dissimilis Vine var. *elongata* Vine, 1882 [= *Corynotrypa dissimilis* (Vine, 1881)]
Stomatopora elongata Vine, 1882 [= *Corynotrypa elongata* (Vine, 1882)]
Ascodictyon radiciforme Vine, 1882
Ascodictyon stellatum Nicholson and Etheridge, 1877 var. *siluriense* Vine, 1882
Spiropora intermedia Vine, 1882 [= *Mitoclema intermedia* (Vine, 1882)]
Spiropora regularis Vine, 1882 [= *Mitoclema regularis* (Vine, 1882)]
Hornera? *delicatula* Vine, 1882
Polypora? *problematica* Vine, 1882
Ptilodictya lonsdalei Vine, 1882
Ptilodictya interporosa Vine, 1882
Phylopora phillipsia Vine, 1883
Thamniscus dubius (King) var. *carbonaria* Vine, 1883
Heterotrypa delicatula Vine, 1883
Diplotrypa petropoliatana var. *petropolitanaforma* Vine, 1883
Streblotrypa nicklisi Vine, 1884
Pinnatopora ornata Vine, 1885
Pinnatopora simplex Vine, 1885
Thamniscus gracilis Vine, 1885
Rhombopora vinculiformis Vine, 1885 [= *Baculopora megastoma* (M'Coy, 1844)]
Streblotrypa minuta Vine, 1885
Entalophora striatopora Vine, 1885
Entalophora gigantopora Vine, 1885
Diastopora cretacea Vine, 1885
Diastopora cretacea var. *lineata* Vine, 1885
Diastopora fecunda Vine, 1885
Lichenopora? *paucipora* Vine, 1885
Membranipora dumerelli Audouin var. *cantabrigensis* Vine, 1885

Notes

1 G.R. V[ine, jr.], 'In memoriam. George Robert Vine', *The Naturalist* (n.s.) 18(220) (1893), 333-335; Anon. 'Notes and News [Notice of death of G.R. Vine]', *The Naturalist* (n.s.) 18(218) (1893), 282; Anon. 'A Prince among Men', *Hartley's Attercliffe Almanak.* Hartley & Son, The Attercliffe Press, Sheffield (1894), 117-121 [Almost certainly also by G.R. Vine, junior]. This useful source of local information was published annually between 1880 and 1898. See also G.R. Vine, *The story of old Attercliffe.* Ward Bros, Sheffield (1932-1936), 333 pp, which remains a valuable source of local history.

2 *The Sheffield and Rotherham Independent*, Friday 11th August 1893. The notice of Vine's death at the age of 68 contrasts sharply with the preceding notice which reported the sad death of Mabel Stevens, aged only three, who drowned in a maidening pot. This was a piece of equipment unique to Yorkshire used for washing linen.

3 Sheffield Literary and Philosophical Society Lecture, 4th March 1890: *A working man's recollections of passing events during the early part of the last half century,* 3 pp.

4 In *The Geologist* for 1860 (p. 463) Vine requested information about identifying Mountain Limestone mollusca. He had by then assembled a collection of some three hundred species from the Athlone area.

5 See H. Murtagh, *Athlone.* Irish historic towns atlas; no. 6. Royal Irish Academy, Dublin (1994), 16 pp; 9 sheet maps. The Ordnance Survey 6 inch to 1 mile map of 1912 still shows three woollen factories situated on the riverside.

6 She was born in 1825 or 1826 in Ireland; Vine was her second husband, previously she had been married to a Mr Beard. She died 29th May 1901 aged 76 years (see note 12).

7 Little is known of Elizabeth his eldest daughter. His younger daughter Alice was born at 44 Cherry Garden St, Bermondsey, London. She married William Henry Wood (1862-1944) an iron moulder of 129 Attercliffe Common on 5 April 1887 at the Baptist Chapel, Townhead Street, Sheffield. She died in 1924 and he twenty years later in Leeds. They had at least one daughter Hilda Evelyn Wood (1898-1984).

8 Hill Top Road no longer exists. Hill Top Chapel which no doubt was close-by is still in use as an Anglican Chapel. Attercliffe was a village east of Sheffield which became encorporated into the city during the 1870s. It is reputed to have the oldest haunted public house in Britain. See R. Harman and S. Ogden, *The Attercliffe village trail.* Sheffield East End history trail 2. Hallamshire Press, Sheffield, (1997), 64 pp.

9 Harman and Ogden, note 8, p. 30 reproduce an advertisement for Banner's store taken from *The Attercliffe Almanack.* The shop is still standing, and named Banner's, although it is now occupied by smaller retail units, and a museum of wartime Sheffield.

10 This church was subsequently destroyed during World War I when it took a direct hit from a bomb dropped during an air raid in September 1916 (Vine 1936, p. 35).

11 A mounted photographic portrait of Vine now in the Sheffield Local Studies Library bears the following inscription: "Presented to Mr G.R.Vine BSc / By the teachers of the Attercliffe Baptist Sunday School / As a token of their Appreciation of his services & in memory of his Revered Father / Who was Teacher & Superintendent for seventeen years (Oct 8th, 1876 to 1893) / May 1909". This suggests that his son was also a Sunday School teacher or indeed Superintendent.

12 His grave and that of his wife is located in the front row on the right-hand section of the

graveyard. Walk approximately 100 m up the path to the right from the ornate double chapel at the cemetery entrance, until the junction with a cross-path. Vine's grave is ten up on the right from this point. It is marked with a large headstone with the following inscription: In loving memory of / GEORGE ROBERT VINE / (native of Portsmouth) / who was called home August 8th 1893 / Aged 68 years / "Until the day dawn.' / Also of / ELIZABETH. / Wife of the above named / Who entered into rest May 29th 1901 / Aged 76 years / "Peace Perfect Peace.

13 S.J. Knell, *The culture of English geology, 1815-1851: a science revealed through its collecting*. Ashgate, Aldershot (2000), xxi + 377 pp.

14 These two societies amalgamated in 1918 to form the Sorby Scientific Society named in honour of the Sheffield scientist Henry Clifton Sorby

15 Lecture, 1st June 1887: What are Zoophytes?: a natural history study. *65th Annual Report Sheffield Literary and Philosophical Society for the year ending Dec. 1887*, pp. 1-7. In this lecture Vine outlined the contributions made to the study of bryozoans by early naturalists such as John Ellis.

16 Lecture, 6th September 1887: The postage stamp, or the history of a fascination. 65th *Annual Report Sheffield Literary and Philosophical Society for the year ending Dec. 1887,* p. 40. Vine was a passionate stamp collector.

17 Lecture, 2nd October 1889: Well boring, and the lessons taught by the study of the cores. *67th Annual Report Sheffield Literary and Philosophical Society for the year ending Dec. 1889,* pp. 1-7.

18 Vine was nominated for Honorary Membership by the geologist A.H. Green and the microscopist H.C. Sorby. *The Naturalist* 201, 100-101.

19 R.J. Cleevely, *World palaeontological collections.* London: British Museum (Natural History) and Mansell Publishing Ltd, (1983), 365 pp., notes (p. 156) that Walter Howchin was at one time Professor of Geology at the University of Adelaide.

20 See G.R. Vine junior (note 1), *The Naturalist*220, p. 334 for a brief discussion of Walter Howchin's influential role in the development of Vine as a micropalaeontologist.

21 R. Etheridge jr., 'Anniversary address of the President', *Proceedings of the Geological Society*, 398 (1881), 35. [This was reprinted in the *Geological Magazine* n.s. Decade 2, 8 (1881), 187.]

22 Anon, Report of Annual General Meeting, 18th February 1881 (namely Vine's response on receiving the Lyell Geological Fund) *Proceedings of the Geological Society*, 398 (1881), 35-36. [In contrast to that of the President's citation which was reprinted in the *Geological Magazine* (see note above), Vine's response was not.]

23 It is interesting to note that Sorby was born in Woodbourn Hall, Attercliffe close to where Vine lived, although he came from an upper class background, quite different to that of Vine (see M.J. Bishop, 'New biographical data on Henry Clifton Sorby (1926-1908)', *Earth Sciences History*, 3 (1984), 69-81). Maybe this explains why Sorby was the Chairman of the Committee while Vine was its Secretary.

24 Vine-Mason letters 1.12.1887. One such grant from the Royal Society was acknowledged by Vine who thanked George Busk who had sat on the committee that awarded these grants. In this letter he outlines the value of this money in allowing him to purchase books and other research materials. See Vine-Busk letters, June 1st 1884) [Department of Zoology, Natural History Museum, London].

25 In 1884 Busk sent Vine some specimens of *Lichenopora* from the Greensand (Vine-Busk letters, June 1st 1884) [Department of Zoology, Natural History Museum, London].

26 See H.S. Torrens and J.E. Winston, this volume. Eliza Jelly and Vine shared a love of bryozoans

and were by virtue of social standing and gender unable to gain the full accolade of their contemporaries their work truly deserved. They corresponded, exchanged material, and on occasion lent each other obscure volumes on bryozoans that the other had been unable to obtain (see Vine - Busk letters, June 7th 1884) [Department of Zoology, Natural History Museum, London].

27 See note 22.

28 See G.W. Shrubsole and G.R. Vine, 'The Silurian species of *Glauconome* and a suggested classification of the Palaeozoic Polyzoa', *Quarterly Journal of the Geological Society of London*, 40 (1884), 329-332.

29 See G.R. Vine, 'Notes on the Polyzoa and Foraminifera of Cambridge Greensand', *Proceedings of the Yorkshire Geological and Polytechnic Society* (n.s.) 9 (1885), 10-29, and G.R. Vine, 'Further notes on the Polyzoa of the Lower Greensand and the Upper Greensand of Cambridge. Part II', *Proceedings of the Yorkshire Geological and Polytechnic Society* (n.s.), 11 (1889), 250-275.

30 These were described by Vine in 1887. 'Jurassic Polyzoa in the neighbourhood of Northampton', *Journal of the Northamptonshire Natural History Society and Field Club* 4, 202-211.

31 See G.R. Vine, 'Notes on the Polyzoa and other organisms from the Gayton Boring, Northamptonshire', *Journal of the Northamptonshire Natural History Society and Field Club*, **4** (1887), 255-266.

32 His genera *Pinnatopora* Vine 1883 is now regarded to be a junior synonym of *Penniretepora* d'Orbigny, 1849, while *Arcanopora* Vine, 1883 is identical to *Sulcoretepora* d'Orbigny, 1849. The species *Rhombopora vinculariformis* Vine, 1877 described from the Carboniferous of Great Britain is considered to be a junior synonym of *Baculopora megastoma* (M'Coy, 1844) (see P.N. Wyse Jackson, *Irish Journal of Earth Sciences*, 9 (1988), 198).

33 This is highlighted by his short description of a species of *Streblotrypa*. In 1883 J.M. Nickles sent Vine some specimens of *Streblotrypa*, which he said were named *Streblotrypa nicklesi* by Ulrich in manuscript. Unfortunately the following year Vine published the name, but misspelt the trivial name *nicklisi* (Vine 1884), and in doing so also validated Ulrich's manuscript genus *Streblotrypa*. Ulrich's comprehensive description of the genus and species appeared in 1890. (See P.N. Wyse Jackson, P.N. 1991. *Unpublished Ph.D. thesis,* University of Dublin, and Hageman, S.J. *The University of Kansas Paleontological Contributions*, n.s. 4 (1993), 1-13 for alternative arguments regarding the nomenclature of this taxon).

34 See E.O. Ulrich, *Bulletin of Denison University*, 4 (1888), 77 for the original description of *Pinnatopora vinei* and E.O. Ulrich, *Journal of the Cincinnati Society of Natural History*, 12 (1890), 173 for the original description of *Vinella repens*.

35 See note 26.

36 G.R. Vine, Jr. was in 1881 a science and art student; in 1893 a school master at the Central Schools, 50 Leigh St., Sheffield; and in 1896 a board schoolmaster, living at 98 Cowlishaw Road, Sheffield. He was later a headmaster of several schools in Sheffield (see main text). He and his wife Alice had at least one child, a son Robert Saxelby Vine (1891-1916). During World War I he served in the Cambridgeshire Regiment, and reached the rank of Second Lieutenant. He was killed at the age of 25 on Saturday 14th October 1916 during the Battle of the Somme, and left a widow, Elsie. George Robert Vine, Jr. died on 25th August 1940 at 40 Main Avenue, Totley, Sheffield, and left the sum of £791 0s 5d in his will. He was the author of two mathematics textbooks, both of which were published in September 1912: *Experimental mathematics: a first year course for evening continuation classes and upper standards in day*

schools Book I. Pitman's mathematical series. Pitman, London, pp. vi+66, and *Experimental mathematics: a second year course for evening continuation classes and upper standards in day schools, with answers.* Book II. Pitman's mathematical series. Pitman, London, pp. 72. These retailed at 7d and 9d respectively. He also wrote a valuable local history *The story of old Attercliffe*. (See note 1).

37 Harman, R. and Ogden, S, note 5, p. 35.
38 Vine-Mason letters 19.5.1887.
39 Vine-Mason letters 15.11.1887.
40 Vine-Mason letters 10.10.1888.
41 A copy of this catalogue is in the Vine-Mason archive at Cardiff.
42 Vine-Mason letters 30.11.1886.
43 Note 27.
44 Note 27. George Busk's *A Monograph of the Fossil Polyzoa of the Crag*. Palaeontographical Society, London (1859) remains a valuable source for students of fossil bryozoans even today.
45 Vine-Mason letters 15.2.1888.
46 Vine-Mason letters 4.5.1888.
47 See Cleevely 1983, note 11.
48 Vine-Mason letters 10.7.1888 and 13.7.1888.
49 Vine-Mason letters 2.7.1891. It is interesting to note that as he approached old age he was still living in rented accommodation.
50 Vine-Mason letters 20.6.1889.
51 Vine-Mason letters 9.7.1891. This material was probably purchased by Mason after Vine's death. Much of it was subsequently sold by Mason's widow and is now in Cardiff. See Section 5.
52 Vine-Mason letters 2.7.1891.
53 Vine-Mason letters *c.* 12.12.1892.
54 See Cleevely 1983, note 11 and the series of FENSCORE reports: E.G. Hancock and C.W. Pettitt (eds), *Register of natural science collections in north west England*, (Manchester, 1981); P. Davis and C. Brewer, *Register of natural science collections in north-east England,* (North of England Museum Service, 1986); M.M. Hartley, A. Norris, C.W. Pettitt, T.H. Riley and M.A. Stier (eds), *Register of natural science collections in Yorkshire and Humberside,* (Leeds, 1987); H.E. Stace, C.W. Pettitt and C.D. Waterston, *Natural science collections in Scotland,* (Edinburgh, 1988); G.P. Walley (ed.), *Register of natural science collections in the midlands of England,* (Nottingham, 1993); and J. Bateman and G. McKenna (eds), *Register of natural science collections in south east Britain* (Cambridge, 1993).
55 See E.G. Hancock, A. Howell and H.S. Torrens, 'Geological collections and collectors of note. 11. Bolton Museum', *Newsletter of the Geological Curators' Group*, 1 (1976), 323-335.
56 See note 42.
57 A.S. Woodward, 'Geology', in H.B. Woodward (ed.), *History of the collections contained in the natural history departments of the British Museum.* Volume 1. (London, 1904).
58 See note 15.
59 See note 15.
60 The bulk of his papers were communicated by P.M. Duncan. Others were delivered by H.C. Sorby, G.W. Shrubsole and H. Jesson.

Early bryozoan studies in the Netherlands

Gerhard C. Cadée
Netherlands Institute for Sea Research, PO Box 59, 1790 AB Den Burg, The Netherlands

1. Introduction

In papers dealing with the history of bryozoology hardly any mention is made of contributions from the Netherlands.[1] Nevertheless, bryozoans were not completely overlooked here. In this overview I will highlight some aspects of bryozoology in the Netherlands from 1694 to *c.* 1950.

2. Antoni van Leeuwenhoek

During his long life (1632-1723) this famous Dutch naturalist made an enormous variety of mostly microscopical observations,[2] which he wrote down in his *Letters to the Royal Society in London* to which he was elected as Fellow. From these letters we might get the impression of a chaotically working researcher. However, van Leeuwenhoek directed his attention to a number of central problems among which were phenomena of growth and reproduction, trying to find the same organs and mechanisms in a large number of organisms.[3] He studied reproduction in mussels to refute the Aristotelean idea of spontaneous generation of molluscs from mud. He started to look for ovaries and spermatozoa in a number of molluscs. On the outside of the shells of mussels he observed what he identified as the fixed eggs of the mussels.[4] Another Dutch naturalist Job Baster proved in 1762 that these "eggs" were bryozoans.[5] He named them *Eschara crustacea,* but as Baster did in general not use binominal names, his name did not find general use. Moreover, the figures published by van Leeuwenhoek in 1694 (combined here in Figure

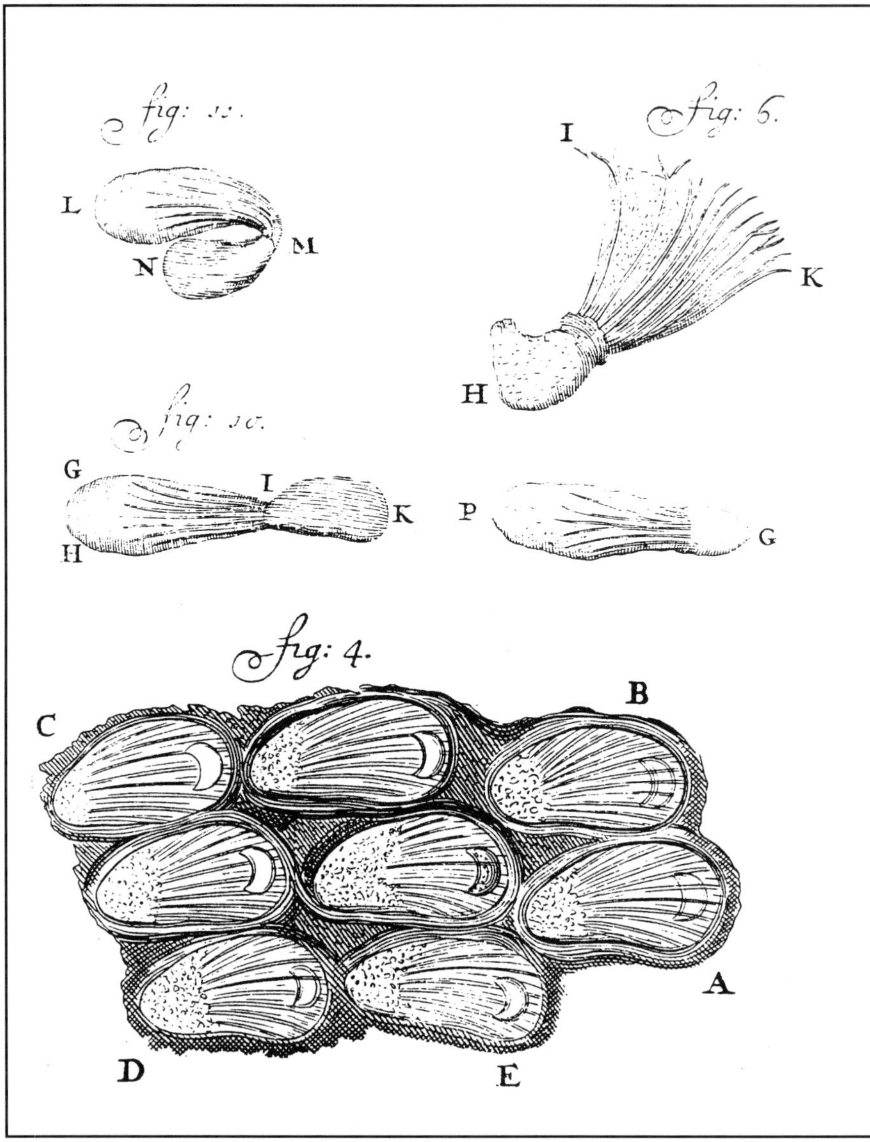

Figure 1. "Eggs of Mytilus edulis *attached to its shell" figured by Antoni van Leeuwenhoek (1694) (these are actually a bryozoan, probably* Conopeum reticulum*)*

1) make identification to the species level doubtful. Cole[6] (1937) wisely only gave a genus name (*Membranipora*) stating this to be a sufficiently close approximation to make any error of minor importance. Palm in 1982 used the name *Membranipora membranacea* (L.),[7] but this is a species normally occurring on soft surfaces such as algae.[8] Most probably it is *Conopeum reticulum* (L.), a species nowadays often found on Dutch living mussels,[9] as well as on many other, particularly hard substrates.[10] Van Leeuwenhoek's

drawing, however, does not show the pair of triangular kenozoids at the distal angles of each autozoid characteristic for this species. His drawing does show distinct opercula, which might indicate that they are opaque as in *Electra crustulenta* (Pallas) which sometimes occurs also on living mussels,[11] but the characteristic median proximal spine is not drawn. To which species van Leeuwenhoek's bryozoan belongs can probably not be solved. Anyway, the drawings and descriptions of van Leeuwenhoek's mussel eggs are in fact an early mentioning of Dutch bryozoans.

3. Job Baster

Job Baster (1711-1775) was a Dutch physician living in Zierikzee, a small town in the province of Zeeland.[12] He studied medicine in Leiden, became there a good friend of Herman Boerhaave and like Boerhaave also showed an interest in plants and animals. During his "grand tour" (1731-1732) together with van Visvliet, he visited Tournefort and Vaillant in the Jardin Royal des Plantes of Paris, and Sloane and the botanists Mead, Miller and Collinson in London. In 1732 he started an ever growing medical practice. He was an arduous collector of books and curiosities such as shells, corals and marine animals in spirits. He also found time to study living marine animals and plants near Zierikzee on which he published. Among these were also Zoophytes, plant-animals. These had been classified as plants, but A. de Peyssonel and later J. Ellis recognised them as animals. Nevertheless, the discussion continued and Baster in 1758 disagreed with Ellis' findings of 1758.[13] According to Baster hydroids consisted of a plant on which (different species of) polyps not belonging to this plant had settled. The polyps mentioned by Baster in fact were bryozoans, which indeed did not belong to the hydroid.[14]

Baster in 1762 was the first to recognize bryozoans as the originators of the "growing stones" in brackish inland lake near Zierikzee.[15] He named them *Eschara lapidescens, conglomerata, foliolis tenuibus, crispis, coadnatis, simplicibus,* but as this was not a binominal name, Baster's name has no priority. The growing stones were first described in 1729.[16] The cores of the growing stones indeed consist of a bryozoan now named *Electra crustulenta*; which is sometimes surrounded by a stromatolitic limestone crust. In the past chemical precipitation and even radioactivity were invented to explain the peculiar lithifications.

4. Abraham Trembley

Abraham Trembley (1710-1784) (Figure 2) is known as the first discoverer of freshwater Bryozoa. He was born in Geneva, Switzerland, but was of French origin. It is hardly possible to exaggerate the influence of Réaumur on Trembley during their 17 years of correspondence until the death of Réaumur.[17] I will not claim him as a Dutch scientist, but his most important work as a naturalist he did in the Netherlands, where he stayed from 1739 - 1747 as a tutor for the two sons of count William Bentinck. In Bentinck's mansion

Figure 2. Abraham Trembley (1710-1784)

Sorgvliet near the Hague he did his famous experiments on asexual reproduction and regeneration of freshwater polyps. These studies brought him great fame already in the 1740s and made him the "founder of experimental biology".[18] Trembley (1744) published his work on *Hydra* in Leiden, the Netherlands[19] and not in France as d'Hondt contended in 1985.[20] The Paris publication by Durand was pirated.[21] Lenhoff and Lenhoff in 1986 published an English translation of Trembley's 1744 work.[22]

In the same pond on Sorgvliet from which he collected his freshwater Hydra, he also collected the first freshwater bryozoan ever seen, *Lophopus crystallinus*. He studied its anatomy and discovered asexual reproduction also to occur in this phylactolaemate bryozoan. It is not always realised that he also observed colonial budding (and subsequent asexual dispersal) in his 'tufted polyps' (= *Lophopus crystallinus)*, a phenomenon recently studied again in cheilostome bryozoans.[23] To cite Trembley himself: 'When the water surrounding tufted polyps abounds with food, the young ones grow in great numbers. I have often seen more than a hundred joined together forming a pretty bouquet. They then separate, but not one by one. The bouquet divides into two or three parts containing varying numbers of polyps'.[24]

Baker in 1952 gave an excellent description of Trembley's observations.[25] The reason

MÉMOIRES
POUR L'HISTOIRE
DES POLYPES.

PREMIER MÉMOIRE.

Où l'on décrit les Polypes, leur Forme, leurs Mouvemens, & une partie de ce qu'on a pu découvrir sur leur Structure.

DES Faits auſſi ſinguliers, auſſi contraires aux idées généralement reçues ſur la nature des Animaux, que le ſont ceux que m'a fait voir l'Inſecte dont je vais donner l'Hiſtoire, demandent, pour être admis, les preuves les plus évidentes. Il eſt arrivé plus d'une fois, que la précipitation, & l'amour

A du

Figure 3. A. Trembley, 'Mémoires, pour servir à l'histoire d'un genre de polypes d'eau douce à bras en forme de cornes', *(Leiden, 1744).*

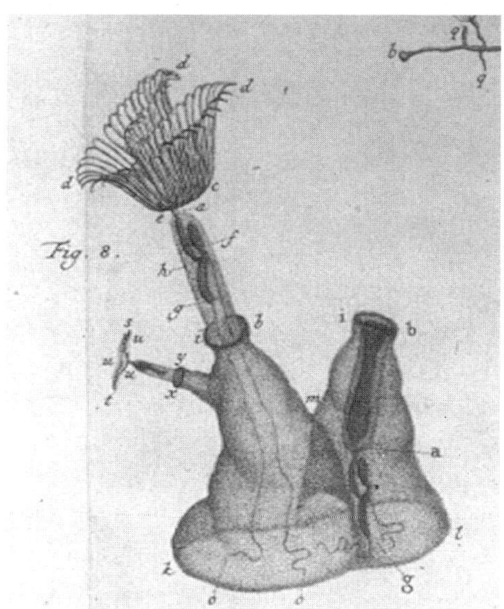

Figure 4. Lophopus crystallinus *from A. Trembley,* 'Mémoires, pour servir à l'histoire d'un genre de polypes d'eau douce à bras en forme de cornes', *(Leiden, 1744).*

Figure 5. Trembley tutoring the two sons of William Bentinck.

I include Trembley's findings in this overview is that apparently only during his years in the Netherlands did he get the opportunity and leisure to do his epoch-making research. His employer William Bentinck gave him the opportunities to do his research at Sorgvliet and Trembley made the two sons of Bentinck partake in collecting his animals and watching his experiments as part of his task as a tutor (as can be seen in the vignettes illustrating his *Mémoires*) (Figure 5). After his stay in the Netherlands he became too much distracted by other tasks to continue his observations. Only much later in 1765 in Geneva he returned to observations on freshwater organisms together with his friend Charles Bonnet. However, he never published again on his polyps. He was aware there was still a lot to discover, but he wanted younger people to carry on the work. Affairs of state absorbed him too much. His friend Bonnet wrote in 1778:

> 'M. Trembley is not less occupied than I am by our political situation, and it would be useless to talk polyps to him'.[26]

He later mainly published on the education of children. For more information on Trembley's life and discoveries I refer readers to Baker's excellent biography and the very informative introduction by Lenhoff and Lenhoff.[27]

5. Leendert Bomme

Leendert Bomme (1727-1788) was a merchant in Middelburg (in the province of

Zeeland), with a great interest in natural sciences. He collected living marine organisms and observed them in glass bottles in which he tried to keep them alive as long as possible for his (microscopical) studies.[28] Some of his observations on marine organisms he published himself.[29] He was the first to observe the cilia on bryozoan tentacles transporting food particles, hereby correcting Trembley (1744) who thought transport to be due to movements of the tentacles themselves. This observation gives Bomme a place in the history of bryozoology.[30]

6. Other Dutch studies on Recent Bryozoa

The exposed sandy Dutch coast, with its strong tidal currents is a poor environment for marine Bryozoa; the shifting sands are an unsuitable substrate. This may partly explain why Bryozoa attracted few students and why there is no monograph on Dutch Bryozoa. Also the (never finished) series 'Fauna van Nederland' (1927-1956) does not have an issue on Bryozoa. However, marine Bryozoa were not completely overlooked in the Netherlands. The studies of the changes in flora and fauna in the Zuiderzee due to its closure in 1932 did include Bryozoa.[31] This proved that even in this sheltered area the number of species was low (only 8). Expectedly, marine and brackish-water species disappeared when the salinity dropped to near zero in the closed part of the Zuiderzee, and fresh-water species gradually immigrated.

In the past lists were published repeatedly of all the species known from the Netherlands. Bennet and Olivier in1828 mentioned 10 species of bryozoans living in the Netherlands.[32] This number increased regularly in the 19th and 20th centuries. Herklots who published in 1870, mentioned 31 species;[33] Maitland, in 1897, already knew 60 species,[34] and Lacourt in 1949 enumerated 71 marine and 8 freshwater species.[35] Of these marine species only 44 were indigenous, the other 27 had reached our coast on drifting material mainly from the Channel area. The list has increased further since, mainly with species on drifting material: in 1948 Bloklander and Leenhouts observed 11 species new to the Netherlands.[36] The residual current along our coast from South to North, added by storms from the Southwest, brings an appreciable amount of drifting algae and other material from the Channel and even from farther away to our beaches. The best examples of trans-oceanic transport are tropical drift-seeds from the Dutch coast.[37] Beached material of southern origin is easy recognisable as such: it contains large amounts of the thong-weed *Himanthalia elongata*, which does not occur along the Dutch coast, but does live on the rocky shores along the Channel coasts.[38] Its disc-shaped structure near the base is a well-known habitat for encrusting organisms such as bryozoans. Also new indigenous species are reported regularly from the Netherlands.[39]

The Dutch *Siboga* Expedition to Indonesian waters (1899-1900) (Figure 6) revealed the high biodiversity of this area, also Bryozoa were collected. The well-known English bryozoologist Sir Sidney F. Harmer was asked to study these collections. The final volume

Figure 6. The Siboga.

was completed by Anna B. Hastings and published 7 years after Harmer's death. Although Harmer's monographs published between 1915 and 1957 are monumental,[40] it becomes clear now that the Indonesian bryozoan fauna is much richer in species.[41] The biologists on board of the *Siboga* were not specialized in Bryozoa, the material collected consisted in general of few specimens per species only (as such already an indication of the high biodiversity), the material was dredged, no bottom samplers were used, and the *Siboga* could not sample all habitats. Harmer in 1915, himself complained that he got no hydroids or algae to search for encrusting bryozoans.[42]

A.W. Lacourt (1910-1987) specialized in freshwater bryozoans. He was administrator at the Laboratory for Pathology of Leiden University and studied bryozoans (as well as marine organisms from west European coasts) in his spare time. After his retirement he became an honorary member of the staff of Natural History Museum in Leiden, now 'Naturalis'.[43] His studies of freshwater bryozoans culminated in a monograph on the Phylactolaemata published in 1968. This monograph has met with criticism since,[44] already foreseen by Lacourt himself, he 'was sure his publication would not be the last word in phylactolaemate taxonomy'.[45]

7. Dutch studies on Fossil Bryozoa

The famous Maastrichtien fossil bryozoan fauna of the type locality was hardly studied by Dutch scientists, however, collecting fossils from these Cretaceaous deposits was quite popular in the 18th century,[46] and these collections were used by others. In 1852 a committee of three geologists (Staring, Miquel and van Breda) was installed to produce a geological description and map of the Netherlands. Some 20 others were added as corresponding members. Among these was J. Bosquet (1814-1880) (Figure 7), an apothecary in Maastricht, because of his large knowledge and collections of fossils in particular of the upper Cretaceous rocks around Maastricht.[47] The work of this committee was completed by Staring between 1855 and 1960[48] and Bosquet contributed long fossil lists including Bryozoa from the Maastrichtien type locality to this publication. In this framework Bosquet himself published separately on brachiopods and crustaceans from these deposits. Bosquet also provided material for the first monograph on the Bryozoa from Maastricht by the German von Hagenow published in 1851.[49] It is almost incredible now that von Hagenow, living in Cassel only 120 miles away from Maastricht, never went to collect there himself, but based his monograph solely on the collections of others!

Another apothecary from Maastricht, also an ardent collector of fossils in this area, Johan Casimir Ubachs (1829-1894) (Figure 8) published two papers on Dutch Upper-Cretaceous Bryozoa, adding new species to those already described by von Hagenow.[50]

Figure 7. J. Bosquet

Figure 8. Johan Casimir Ubachs

Later on his interest changed to the vertebrates of the Upper Cretaceous of Maastricht, in particular *Mosasaurus* and *Chelonia* of which he had nice specimens. In 1866 he opened a private museum in Maastricht and he also traded in fossils.[51] The interest in the Maastrichtien fossils including Bryozoa from the type locality is certainly not diminishing. Since his first publication on Bryozoa from the Maastricht area in 1930[52] Ehrhard Voigt is continuing to enlarge our knowledge of this rich fauna. Contrary to von Hagenow, Voigt was an undefatigable collector of bryozoans around Maastricht himself and he stressed the importance of the burrows present in some hardground layers to collect beautifully preserved bryozoans, which were elsewhere in the bryozoan layers often somewhat eroded by transport.[53]

Another rich bryozoan fauna in the Netherlands is found in the Pliocene deposits. In 1885 Lorié was the first Dutch geologist to study these bryozoans,[54] which he could identify using Busk's 1859 monograph of Bryozoa from deposits of similar age from the British side of the southern North Sea Basin.[55] Robert Lagaaij (1924-1975) (Figure 9), internationally the best known Dutch bryozoologist, published his Ph.D. thesis on the same bryozoan fauna in 1952.[56] Later he became one of the founders of the International Bryozoology Association.[57]

Figure 9. Robert Lagaaij (1924-1975)
(from a copy from A.B. Hastings photograph album)

8. Acknowledgements

I am very grateful to Dr L.C. Palm for his help with the data on van Leeuwenhoek and providing me with the translation of van Leeuwenhoek's letter of June 1712, not yet published in the new edition of 'All the letters'. Dr P. de Wolf made useful comments on earlier drafts of the MS. The NIOZ library's rare book department was very useful for writing this paper. This is NIOZ publication Number 3646.

Notes

1 G. Johnston, *'A History of British Zoophytes',* 2nd ed. (London, 1847), 488 pp.; G.J. Allman, *A monograph of the fresh-water Polyzoa* (London, 1856), 119 pp.; S.F. Harmer. Polyzoa', *Proceedings Linnean Society, London,* 141 (1930), 68-118; J.S. Ryland, '*Bryozoans*' (London, 1970), 175 pp.

2 C. Dobell, *'Antony van Leeuwenhoek and his "little animals"'*, (1932, 1960 reprint Dover publications, New York), 435 pp.; A. Schierbeek, '*Antoni van Leeuwenhoek zijn leven en zijn werken',* (Lochem, 1950), 526 pp.

3 L.C. Palm, 'Antoni van Leeuwenhoek's malacological researches as an example of his biological studies', In *Antoni van Leeuwenhoek 1632-1723,* edited by L.C. Palm & H.A.M. Snelders (Amsterdam, 1982), 153-167.

4 Van Leeuwenhoek letters of 30th April 1694 and 10th June 1712. See the new edition of his letters: A. van Leeuwenhoek, 1979 *'Alle de brieven van Antoni van Leeuwenhoek',* vol X (1694-1695) edited by L.C. Palm, (Lisse, 1979).

5 Job Baster, *'Natuurkundige Uitspanningen behelzende eenige waarnemingen over sommige zee-planten en zee-insecten, benevens derzelven zaadhuisjes en eijernesten',* (Haarlem, 1762), part 1, 169 pp; p. 121-122.

6 F.J. Cole, 'Leeuwenhoek's zoological research', part 1 & 2, *Annals of Science,* 2 (1937), 1-46 and 185-235.

7 See Palm, note 3.

8 J.S. Ryland and P.J. Hayward, 'British Anascan Bryozoans', *Synopses of the British Fauna (New Series),* 10 (1977), 1-188.

9 G.R. Heerebout, 'Verspreidingsoecologie van de Bryozoa uit het Deltagebied, speciaal in relatie tot het brakke water', *Rapporten en Verslagen Delta-Instituut Yerseke,* D1 (1970), 1-53; G.C. Cadée personal observation.

10 See Ryland and Hayward, note 8.

11 See Heerebout, note 9.

12 W.S.S. van Benthem Jutting and C.M. van Hoorn. 'Oude en nieuwe gegevens over leven en arbeid van dr Job Baster (1711-1775)', *Archief Zeeuws Genootschap der Wetenschappen* (1967), 29-70.

13 J. Baster, 'Observationes de corallinis, iisque insidentibus Polypis, aliisque animalculis marinis', *Philosophical Transactions Royal Society,* 50 (1758), 258-280; J. Ellis, 'Remarks on dr Job Baster's Observationes de Corallinis', *Philosophical Transactions Royal Society,* 50 (1758), 280-287. See also Johnston (note 1) for a record of this discussion.

14 See van Benthem Jutting and C.M. van Hoorn, note 12.

15 See Baster, note 5.

16 For a full account of these "stones" and their history see J. Bijma and G.J. Boekschoten, 'Recent bryozoan reefs and stromatolite development in brackish inland lakes, SW Netherlands', *Senckenbergiana Maritima,* 17 (1985), 163-185.
17 J.R. Baker, *'Abraham Trembley, Scientist and philosopher 1710-1784',* (London, 1952), 259 pp.
18 S.G. Lenhoff and H.M. Lenhoff, '*Hydra and the birth of experimental biology - 1744*', (Pacific Grove, 1986) 60 + 188 pp.
19 A. Trembley, '*Mémoires, pour servir à l'histoire d'un genre de polypes d'eau douce à bras en forme de cornes',* (Leiden, 1744), 324 pp.
20 J.-L. d'Hondt, 'Contributions des chercheurs français à la connaissance des Bryozoaires actuels', *Histoire et Nature,* 26-27 (1985), 3-26.
21 See Lenhoff and Lenhoff, note 18.
22 See Lenhoff and Lenhoff, note 18.
23 E. Thomsen and E. Håkansson, 'Sexual versus asexual dispersal in clonal animals: examples from cheilostome bryozoans', *Paleobiology,* 21 (1995), 496-508.
24 See the translation of Lenhoff and Lenhoff , note 18, p. 130.
25 See Baker, note 17.
26 See Baker, note 17.
27 See Baker, note 17, and Lenhoff and Lenhoff, note 18.
28 W.S.S. van Benthem Jutting. 'Leendert Bomme 1727-1788 een 18-de eeuwse bioloog te Middelburg', *Archief Zeeuws Genootschap der Wetenschappen,* (1969), 21-35.
29 L. Bomme, 'Bericht wegens een zonderling zeeinsect, gevonden aan enige zeewieren gevisch op het strand van het eiland Walcheren', *Verhandelingen uitgegeven door het Zeeuws Genootschap der Wetenschappen te Vlissingen,* (1769), 394-402.
30 See Allman, note 1, and Ostrovsky, Shunatova and Antipenko, this volume.
31 T. van Benthem Jutting, 'Bryozoën' In *Flora en Fauna der Zuiderzee,* edited by H.C. Redeke (Den Helder, 1922), 411-417; A.G. Vorstman, 'Bryozoa', In *Flora en Fauna der Zuiderzee, Supplement,* edited by H.C. Redeke, (Den Helder, 1936), 145-149.
32 J.A. Bennet and G. van Olivier, 'Naamlijst van wormen in Nederland aanwezig', *Natuurkundige Verhandelingen Hollandsche Maatschappij der Wetenschappen,* 15(2) (1826), 1-256.
33 J.A. Herklots, '*Natuurlijke Historie van Nederland, de Weekdieren en lagere Dieren*', (Amsterdam, 1870), 466 pp.
34 R.T. Maitland, '*Prodrome de la Faune des Pays Bas et de la Belgique flamande', 1679-1897',* (Leiden, 1897), 62 pp.
35 A.W. Lacourt, 'Bryozoa of the Netherlands', *Archives Néerlandaises de Zoologie,* 8 (1949), 1-33.
36 G. Bloklander and P. Leenhouts, 'Wat de najaarsstormen ons brachten', *Het Zeepaard,* 8 (1948), 13-14, 24-25.
37 See G.C. Cadée, 'Tropical drift seeds from the Dutch coast in a wider perspective, palaeontological implications', *Neues Jahrbuch für Geologie und Paläontologie, Abhandlungen,* 202 (1996), 183-190.
38 See Bloklander and Leenhouts, note 30
39 D. Jebram, 'Zur Bryozoen-fauna der Niederlande', *Netherlands Journal of Sea Research,* 4 (1968), 86-94; J.-L. d'Hondt and G.C. Cadée, '*Bugula stolonifera* nieuw voor Nederland en enkele andere Bryozoën van Texel', *Het Zeepaard,* 54 (1994), 33-37.
40 S.F. Harmer, 'The Polyzoa of the Siboga Expedition', *Uitkomsten Siboga-Expeditie,*

Monographie XXVIII a-d, (E.J. Brill, Leiden, 1915-1957), 1147 pp.

41 G.C. Cadée, 'The shallow soft-bottom fauna of the Java Sea and the Banda Sea', In: *Bryozoa present and past*, edited by J.P. Ross, (Bellingham, 1987), 49-56; K.J. Tilbrook, 'New bryozoan family from the Indo-Pacific shows unexpected diversity', *Journal of the Marine Biological Association, U.K.*, 80 (2000), 1129-1130.

42 See Harmer 1915, note 40.

43 A. Verduin, 'In memoriam A.W. Lacourt', *Correspondentieblad van de Nederlandse Malacologische Vereniging*, 263 (1987), 263-265.

44 F. Wiebach, 'Preliminary notes on a revision of the genus *Hyalinella*', In *Living and Fossil Bryozoa,* edited by G.P. Larwood, (London, 1973), 539-547.

45 A.W. Lacourt, 'A monograph of the freshwater Bryozoa-Phylactolaemata', *Zoologische Verhandelingen, Leiden*, 93 (1968), 1-159.

46 C.O. van Regteren Altena, 'Achttiendeeuwse-verzamelaars van fossielen te Maastricht en het lot hunner collecties'. *Publicaties Natuurhistorisch Genootschap in Limburg'*, 9 (1957), 83-112.

47 W.A.E. van de Geyn, 'Staring's medewerkers uit Limburg', *Verhandelingen Geologisch Mijnbouwkundig Genootschap Nederland en Koloniën*, 14 (1945), 205-214; E.M. Kruytzer, 'J. Bosquet, apotheker en paleontoloog 1814-1880', *Natuurhistorisch Maandblad*, 52 (1963), 95-103.

48 W.C.H. Staring, '*De Bodem van Nederland'*, (Haarlem, 1855, 1860), 441 + 480 pp.

49 F. von Hagenow, '*Die Bryozoen der Maastrichter Kreidebildung',* (Cassel, 1851), 111 pp.

50 C. Ubachs, 'Die Stellocavea-Schichte bei Valkenburg, und neue Bryozoen-Arten derselben', *Palaeontographica*, 9 (1858), 127-131; C. Ubachs, 'Die Bryozoen-Schichten der Maastrichter Kreidebildung, nebst einigen neuen Bryozoen-Arten aus der Maastrichter Tuff-Kreide', *Verhandlungen des naturhistorischen Vereins der preussischen Rheinlande und Westphalens*, 22 (1865), 31-62.

51 W.A.E. van de Geyn, 'Johan Casimir Ubachs (10 Oct. 1829 - 4 Febr. 1894)', *Natuurhistorisch Maandblad*, 33 (1944), 11-14.

52 E. Voigt, 'Morphologische und stratigraphische Untersuchungen über die Bryozoenfauna der oberen Kreide', *Leopoldina*, 6 (1930), 379-579.

53 E. Voigt, 'Die ökologische Bedeutung der Hartgründe ("Hardgrounds") in der oberen Kreide', *Paläontologisches Zeitschrift*, 33 (1959), 129-147.

54 J. Lorié, 'Contributions à la géologie des Pays Bas. I. Résultats géologiques et paléontologiques des forages de puits à Utrecht, Goes et Gorkum', *Archives du Musée Teyler,* (2)3, (1885), 109-240.

55 G. Busk, 'A monograph of the fossil Polyzoa of the Crag', *Monographs of the Palaeontological Society London*, (1859), 1-136.

56 R. Lagaaij, 'The Pliocene Bryozoa of the Low Countries', *Mededelingen van de Geologische Stichting,* Serie C-V-5 (1952), 1-233.

57 See A.H. Cheetham, this volume.

The founding and early history of the International Bryozoology Association, 1965-1974

Alan H. Cheetham

Department of Paleobiology, National Museum of Natural History, Smithsonian Institution, Washington, DC 20650-0121, USA

1. Introduction

When the International Bryozoology Association (IBA) held its first conference in San Donato Milanese on the outskirts of Milan, Italy, in August 1968, the organization was already three years old. Planning for the conference had begun in Stockholm, Sweden, in May 1965, when the IBA acquired its name, adopted its purpose of 'promoting the exchange of information, ideas, and techniques' among bryozoologists (i.e. 'all students engaged in research on ... bryozoans'), and enrolled the first 16 of its members.[1] The founders also elected a chairman and a secretary whom they charged with enlisting the support of a sufficient number of colleagues to achieve the initial goal of the IBA—to hold its first conference approximately three years later.[2] By the conclusion of its third conference, in Villeurbanne, part of greater Lyon, France, in September 1974, the schedule and format of the IBA's conferences, and the responsibilities of the officers and council in carrying them out, had become well enough established to formalise in a constitution. This sequence of events marks the IBA as intermediate among specialist societies in the formality of its organization, with extremes represented by the Society for Vertebrate Paleontology, whose constitution was drafted prior to its organization meeting,[3] and the Pander Society, which has no constitution, but in which 'a meeting could be said to take place when two or more members meet to discuss conodonts'.[4]

What follows is a personal memoir of those formative years from 1965 to 1974, during which we made up procedures as we went along, generally taking a minimalist view with regard to organization. We felt ourselves motivated by internationalism as well as our interest in promoting the subject of our research. We had deliberately chosen 'bryozoology' rather than 'bryozoologists' for our title to put the focus on the subject, not its participants. These motives were very much in evidence in the scholarly community during the two decades following World War II, as the founding of narrowly focused scholarly societies accelerated, especially those with 'International' in their names (Figure 1). When we erred, it was on the side of inclusiveness and perhaps too naively misjudging the effects of international tensions.[5]

In many respects, the IBA's history parallels that of other societies in which the interests of biologists and palaeobiologists were melded, such as the Society for the Study of Evolution.[6] Unlike many of those societies, however, the IBA was never intended to define or redefine a field of study. Nevertheless, after just the first two IBA conferences, 'a general trend toward model building and synthesis'[7] and 'a renewed vigor in the study of bryozoans'[8] were perceived, perhaps evidence of a newly emerging synergism. The long-term effects of the twelve IBA conferences to date on directions and productivity of bryozoan research would make an interesting study. There is little question, for example, that among the papers presented at the first IBA conference, those dealing with models of growth of hard and soft tissues had an immediate impact on bryozoan studies.[9] A proliferation of similar studies, and much spirited discussion, followed at the second conference.[10] It also seems apparent that the IBA has fostered a 'renaissance of interest' in Bryozoa and 'has been instrumental in bringing paleontologists and neontologists to a greater understanding of each other's field of interest'.[11] However, an assessment of the IBA's influence on bryozoology as a science, beyond these general impressions, awaits further analysis.

2. The Stockholm meeting: enthusiasm and expectation

However unsurprising it might be that the IBA was founded in the homeland of the bryozoologists F.A. Smitt, Folke Borg, and Lars Silén, it was fortuitous from an internationalist perspective that this was also the homeland of Dag Hammarskjöld, the second secretary-general of the United Nations, who had died serving the cause of international peace just a few years before. Moreover, more than a third of those attending the Stockholm meeting (Figure 2) were affiliated with Swedish institutions (the non-bryozoologists I. Hessland, F. Adamczak, H. Mutvei, and K. Mori; plus L. Silén, S. Schager, and A.H. Cheetham). The meeting was not intended to be the founding event of a new society, but rather a gathering of a few researchers, mostly palaeobiologists, with interests in the post-Palaeozoic fauna of the areas bordering the North Sea.

Ivar Hessland, head of the Geological Institute of Stockholm University where I was guest professor that year, had given me what was then the far from inconsequential sum of 10,000 Swedish kronor (about 2,000 US dollars) to assemble a 'think tank' of scientists

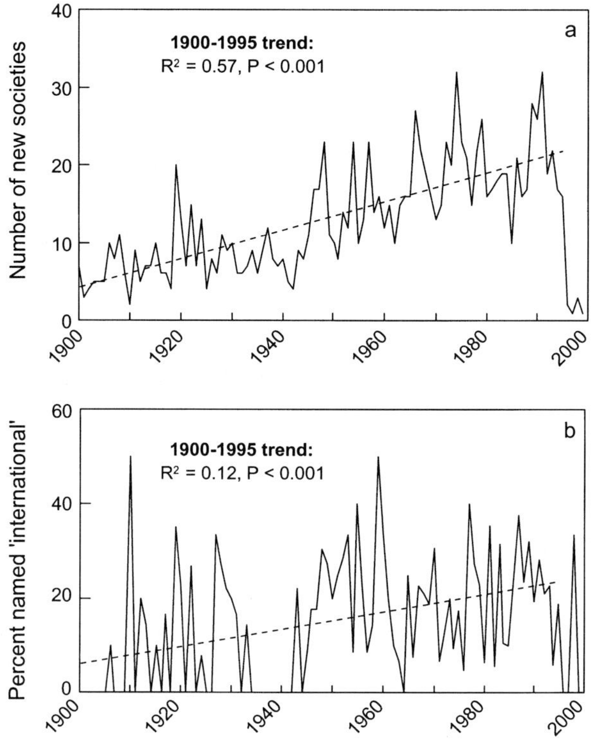

Figure 1a. The number of new scholarly societies founded during each year of the 20th century, based on data from the Scholarly Societies Project (J. Parrott, editor, http://www.lib.uwaterloo.ca/society/overview.html, last accessed 7 December 1999). The increasing trend from 1900 to 1995 is highly significant and explains more than half of the variation. Data after 1995 may be incomplete. 1b, The percent of the new societies in graph a that include 'International' in their names. The 1900-1995 trend is highly significant, but explains just 12 per cent of the variation. Decreases in internationalism are evident during times of international political tension, such as the World Wars and periods during the Cold War.

interested in a subject of my choosing. At the time, I was finishing a study of Eocene bryozoans from England and beginning one on Danian bryozoans from Sweden, so the scope of the meeting seemed logical to me.

The budget allowed me to invite eight 'foreign' bryozoologists, some of whom (e.g. J. Malecki of Poland) declined. The final list included O. Berthelsen (Denmark), P.L. Cook, G.P. Larwood, A.W. Medd, and J.S. Ryland (England), Y.-V. Gautier (Algeria), R. Lagaaij (Netherlands), and E. Voigt (Germany). How did these seven attendees, plus Silén, Schager, and Cheetham, swell into the 16 IBA founders?[12] At least initially, news of the meeting must have spread by word of mouth, and soon E. Annoscia and G. Braga (Italy), R.S. Boardman (USA), M. Labracherie and J. Prud'homme (France), and N. Spjeldnæs (Norway) all planned to come using their own resources. Additional inquiries, some seeking funding, came from abroad. Needless to say, the scope of the discussions

Figure 2. Participants (bryozoologists' surnames in capital letters) in the meeting on post-Palaeozoic Bryozoa of the North Sea region, Stockholm, Sweden, 24th-27th May 1965. Left to right: Sten SCHAGER, Kei Mori, Alan MEDD, Giampietro BRAGA, Enrico ANNOSCIA, Robert LAGAAIJ, Patricia COOK, Yves GAUTIER, Ole BERTHELSEN, Richard BOARDMAN, Franciczek Adamczak, Monique LABRACHERIE, Harry Mutvei, Nils SPJELDNÆS, Janine PRUD'HOMME, Lars SILÉN, Gilbert LARWOOD, Alan CHEETHAM, Ehrhard VOIGT, Ivar Hessland.

that we had originally planned was significantly altered by this infusion of additional expertise.

According to my notes, the topics we discussed for three and a half days (24th-27th May 1965) included questions that resonate even today: Are the problems of bryozoan research different from those facing specialists in other taxa? Are the results of our research being communicated quickly and effectively? With what success can we determine morphological and functional relationships among different groups of Bryozoa? What should be the role of biometrical, numerical, and non-numerical methods in characterising and comparing bryozoan species? Is it possible and worthwhile to make our taxonomic methods standard and 'objective'? What principles should guide us in taxonomy at the various categorical levels? What is our knowledge of the ecological tolerances of living Bryozoa, and how can it be applied to fossil forms? With what success can Bryozoa be used in biostratigraphy? Do their ranges show major discontinuities?

What major phylogenetic trends are indicated? My recollection is that the discussions were quite informal, as befits the 'think tank' model, except for Rob Lagaaij's polished presentation based on the Rhône delta paper that he and Yves Gautier had just published.[13]

It was the scope of these discussions, and the desire to involve workers from more bryozoological and geographic areas, that spontaneously generated the idea of forming the IBA on the last day in Stockholm (27th May 1965). Pat Cook and I were elected secretary and chairman, respectively. The year 1968 was targeted for the first IBA conference, the venue of which was to be determined by the new officers, based on maximising the participation of fellow bryozoologists (both biologists and palaeobiologists) from as many countries as possible. All 16 of us were eager to begin assembling the list of names and addresses to be used in promoting the 1968 conference. I am sure that the nearly 100 letters that I sent from Stockholm that summer represent just a fraction of the effort of the group of 16.[14]

3. Organizing the first conference: inventing a procedure

By the end of 1965, the list of names, which we dubbed the IBA membership list, had grown to more than 150.[15, 16] However, mailings to some of those on the IBA list did not elicit replies for the very fundamental reason that they were no longer alive—more about that later. We received suggestions, but only one attempting to 'correct' the name of the organization, with a replacement 'such as International Association for the Study of Ectoprocts and Entoprocts'.[17] Nevertheless, the IBA survived a possible identity crisis, keeping its simple, easily remembered initials intact, whatever the merits of the Bryozoa versus Ectoprocta argument.[18]

Tom Schopf, ever brimming with ideas, also proposed holding the IBA's first conference in Woods Hole, Massachusetts, under the joint sponsorship of the Marine Biological Laboratory and the Smithsonian Institution (where I had moved in 1966).[19] Moreover, he would have had us apply for US government funding to finance the conference and to 'bring about 15 persons from abroad'.[20] These suggestions, intended to be constructive, might very well have turned the IBA toward a more national orientation at a time when international political tensions were about to escalate in ways that we could not anticipate. At the very least, we hoped to have a much broader international representation than this proposal envisioned.

At the Stockholm meeting, Enrico Annoscia had suggested holding the IBA's first conference in Italy, and in the fall of 1967 the management of the petroleum company AGIP formally offered its facilities in San Donato Milanese for that purpose.[21] This venue promised to be excellent in every way. By April 1968, 55 people planned to attend the conference, representing Algeria, Austria, Brazil, Canada, Chile, Denmark, France, the Federal Republic of Germany, India, Italy, Japan, the Netherlands, Rumania, Spain,

Sweden, the UK, the USA, and the USSR (with five prospective attendees).[22] By the time of the conference, one more country, Norway, was added, but attendance of the participants from Brazil, India, and all those from the USSR was cancelled.[23] The military action ending the 'Prague Spring' (the political reforms undertaken by the Dubcek government in Czechoslovakia in the spring of 1968) began just four days after the conference in Milan ended on 16th August, and we could not help but feel that the situation leading up to this event had prevented our colleagues from Russia and the Ukraine from attending the conference. Although discouraged, we at least had the satisfaction of seeing that our internationalist ambitions for the IBA were partly fulfilled, and we felt convinced that some further work could make prospects even more promising for the future.

Prague was already a factor in the founding of the IBA, as we began to consider in 1965 where and when to hold the first conference. The next International Geological Congress (IGC) was scheduled to meet in Prague in August 1968. Our original plan for the three-year IBA cycle was based in part on the idea that some conferences would fall in years when the IGC meets and others in years that the International Congress of Zoology would be convened. A number of the palaeontologists attending the Milan conference went on to Prague only to find that IGC sessions were aborted, in many cases with the sound of gunfire in the background. (I missed the 'action' because Janine Prud'homme and Monique Labracherie had invited me to go on a most rewarding post-Milan collecting trip in the Aquitaine.) A number of others had cancelled their plans to attend the IBA conference because of involvement in some of the pre-IGC events. Interestingly, the only resignation from the IBA of which I am aware cited the IBA-IGC scheduling 'conflict', suggesting that it should have been resolved by consulting the IBA's 'only living octogenarian'. This correspondent, M.K. Elias, also said that he wished to 'dissociate' himself from an organization that had 'enrolled dead members.' Unfortunately, his letter seems to have been lost, but the IBA was not the first society from which this individual resigned in protest, although apparently for a different reason. In 1947, Elias was the first member to resign from the Society for the Study of Evolution, after his manuscript had been rejected for publication in *Evolution*. 'The actual reason for his manuscript's rejection is unclear, though it appears that [George Gaylord] Simpson's dislike of the manuscript and its author was part of the reason. Simpson referred to the paper as "all nonsense, and part of it malicious nonsense". Later on he wrote, "I am already somewhat at outs with this gentleman, who has made and continues to make large demands on my time and patience." … It appears that Elias was also in disfavour with [Wendell] Camp, who described Elias as a "sorehead"'.[24] Thus, the scorn heaped on us in Elias's missing letter may say more about its writer than the recipients.

Scheduling the Milan IBA conference in the Prague IGC year seems to have produced another interesting phenomenon. Although the number of papers presented at IBA conferences has varied relatively little (Figure 3a), the Milan conference was one of four in which palaeobiological papers outnumbered biological ones (Figure 3b), as pointed out

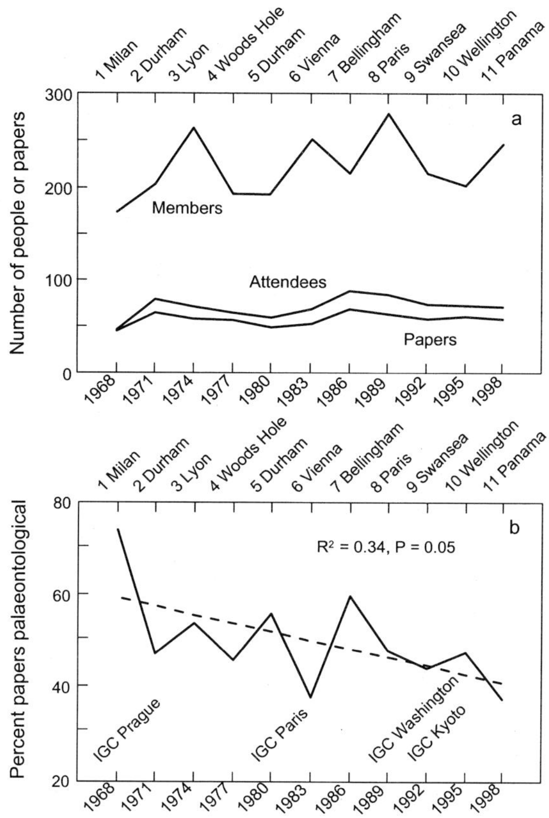

Figure 3a. Variation in numbers of IBA members and numbers of attendees and papers presented at conferences (data from conference programs and IBA newsletters). 3b, Percentages of papers at IBA conferences based on palaeontological material. Data are from Reguant (note 25). Two conferences, Milan 1968 and Durham 1980, were held in the same years on the same continent as International Geological Congresses (IGC). The decreasing trend in palaeontological papers accounts for more than a third of the variation, but is only marginally significant statistically.

by Reguant.[25] Only one other IBA conference (1980 in Durham) was held in proximity to an IGC meeting, and again there seems to have been a correlated, although weaker, increase in the percentage of palaeobiological offerings. However, an overall trend of decreasing contributions from palaeontology seems to have overshadowed whatever effects IGC meetings might have had (Figure 3b), despite the fact that the majority of IBA's presidents have been palaeobiologists. Including the Milan conference, the overall 1968-1998 proportion of palaeobiological offerings has been 50 per cent, but this proportion fell to 48 per cent for the five conferences following Milan and to 47 per cent for the last five conferences (Figure 3b). The role of palaeontology in biology has been an ongoing issue in other organizations, including the Society for the Study of Evolution.[26] It would be interesting to explore whether the decline of palaeobiology at IBA conferences

Figure 4. The 'spotlight' (actually the summer sun reflected from an adjacent building) is on Tom Schopf (right) as he 'lobbies' (right to left) Philip Sandberg, Alan Cheetham, and Bob Pohowsky regarding the IBA draft constitution, and Marge Cheetham looks on. The occasion was the buffet lunch given by the Université Claude Bernard on 6th September 1974.

is reflected in the bryozoan literature in general.

4. Milan's aftermath: providing continuity

It had become apparent to me in the spring preceding the first conference that a procedure was needed for choosing the IBA's next officers. A nominating committee (with representatives from England, France, Germany, Rumania, Sweden, and the USA) was empanelled to present nominees for chairman and secretary at the business meeting at the Milan conference (Table 1).[27] Because no dues were collected for membership in the IBA, its new officers, and the electorate, would be the bryozoologists who had invested the most effort in the organization—i.e. those who attended the conference.[28] When the new officers were elected (16th August 1968), the chairman's title was changed to president 'to avoid confusion with the conference chairman'.[29] In addition, the nominating committee was reconstituted as an advisory council, with the addition of members from Japan, the Netherlands, and the USSR, plus the outgoing IBA chairman and the chairman of the Milan conference (Table 1). Thus I became the only 'past president' without having

Table 1—Officers and council members of the International Bryozoology Association, 1965-1974. 'Official' titles are italicised.

Triennium	Officers	Council
1965-1968	A.H. Cheetham, *Chairman*	(*Nominating Committee*)
	P.L. Cook, *Secretary*	R.S. Boardman
	E. Annoscia, *Conference Chairman*	V. Ghiurca
		G. Illies
		L.J. Pitt
		J. Prud'homme
		S. Schager
1968-1971	N. Spjeldnæs, *President*	E. Annoscia (*past Conference Chairman*)
	P.L. Cook, *Secretary*	G.G. Astrova
	G.P. Larwood, *Conference Chairman*	R.S. Boardman
		A.H. Cheetham (*past President*)
		V.Ghiurca
		R. Lagaaij
		J. Prud'homme
		S. Sakagami
		S. Schager
		E. Voigt
1971-1974	F.J.S. Maturo, *President*	R. J. Cuffey
	P.L. Cook, *Secretary*	G.P. Larwood (*past Conference Chairman*)
	L. David, *Conference Host*	J. Harmelin
		I.P. Morozova
		S. Sakagami
		J.D. Soule
		N. Spjeldnæs (*past President*)
		E. Voigt

been 'president'.[30]

While I was occupied with these procedural matters, Enrico Annoscia was diligently inventing the responsibilities of conference chairman. In addition to organizing the conference's logistics, he established the tradition of obtaining backing for publishing the conference papers.[31] For the most part, the conference chairmen (or 'conference hosts' as they became known by the time of the Lyon conference) found their 'reward' in being the editors of the conference volumes. Despite the magnitude of their accomplishments in ensuring a written record of the conferences, only three of the 16 conference organizers or co-organizers have been elected to the IBA presidency. Worries to the contrary notwithstanding, successive editors of the IBA proceedings volumes were able to maintain an 'open' policy with regard to inclusion of papers presented at the conferences, rather than concentrating on just those with 'marketable value' for publishers.[32]

A different approach to publication was suggested by Schopf.[33] In a page and cost analysis of a number of 'secondary taxon' journals (e.g. *Crustaceana* and *Herpetologica*), he suggested that an IBA journal of about 350 pages per year could be viable. It seems instructive to consider here the experience of organizations, such as the Society for the Study of Evolution, founded largely to publish a journal. Originally visualised as being published simultaneously on both sides of the Atlantic, *Evolution* (subtitled 'International Journal of Organic Evolution') ended up in the USA at least in part because of the post-World War II paper shortage in Britain.[34] Today, approximately three-fourths of the society's membership and most of its journal's authors (65 and 66 per cent, respectively, in the October and December 2000 issues) have US or Canadian addresses. In 1974, Schopf became co-editor of a new journal, *Paleobiology*, published by the Paleontological Society, more than 70 per cent of whose membership is in the USA.[35]

5. Durham 1971 and Lyon 1974: the IBA as we know it

Nils Spjeldnæs also took international participation as a keynote of his 1968-1971 presidency, negotiating IBA membership in both the International Union of Biological Sciences and the International Union of Geological Sciences (the latter as a 'working group' of the International Palaeontological Association). As many as 12 bryozoologists from the USSR were scheduled to attend the conference in Durham, UK, in September 1971, but in the end, perhaps because of renewed tensions during the last two years of US involvement in the war in Vietnam, only three were able to do so.[36] Nevertheless, the 1971 Durham conference is still remembered by its participants as one of the high points in the history of the IBA, at least in part because of the participation of our colleagues from the USSR.[37] In addition, the size of the conferences, in terms of both numbers of attendees and numbers of papers, seems to have been firmly established, despite significant fluctuations in the total numbers of IBA members (Figure 3a).

Although major international tensions seemed to have lessened by 1974, all of the potential participants from the USSR cancelled their plans to attend the September conference in Lyon, France.[38] Frank Maturo focused his attention on formalising the IBA's plan of organization as the hallmark of his 1971-1974 presidency, presenting a draft constitution that was revised and adopted at the business meeting on 11th September 1974.[39] Frank felt that this level of formality was a necessity for 'dealing with funding committees and affiliated associations, etc.'. Some aspects of the draft constitution elicited much thought and discussion (Figure 4), and for the first (and perhaps only) time, an IBA business meeting continued into a second half-day session. In an important sense, this event marked the end of the formative phase of the IBA's history, although the shifting balance away from palaeobiology (Figure 3b) may also prove to be an important factor as IBA's history runs its course.

6. In retrospect: "Our dear child"

Whatever the preservation of an internationalist orientation may or may not have done for the advancement of bryozoology as a science, it did much to 'put a human face' on its practise. Perhaps the following excerpt from a letter from Geneviève Lutaud, cancelling her participation in the Milan conference for medical reasons, says it best: 'I am very much disappointed in losing this opportunity of talking with bryozoologists I already know by their published papers and I very much wished to meet'.[40] The opportunities the IBA gave us to develop friendships, I cannot help thinking, fostered our interest in bryozoology (see, for example, Giampietro Braga's account of his interaction with Salvador Reguant at IBA conferences).[41] Indeed, the IBA even began to take on 'human' qualities of its own, as so poignantly put by Yves Gautier, 'our dear I.B.A. (our child, born 1965 in Stockholm)'.[42]

Still another 'face' of IBA's internationalism has been the experience of a diversity of social aspects afforded to us by meeting in different places. Not the least of these has been the variety of culinary customs we were privileged to experience. Only once, however, did social events seem to overshadow the scientific side of our meetings. At the meeting in Stockholm, we were visited by representatives of the press, not to note what we had to say about Bryozoa, but rather to interview Gunilla Schager who, with the assistance of Marge Cheetham, had prepared a Swedish smörgåsbord in the old traditional way. The recipe for 'Jonsson's temptation', made with herrings, potatoes, and dill, elicited particular comment in the press, although smoked reindeer and other Scandinavian delicacies proved just as memorable to many of the diners. When it came time to wash up the four sets of antique china, silver, and crystal in the elegant apartment of Sten Schager's parents overlooking one of Stockholm's principal squares, it was Rob Lagaaij, ever the consummate gentleman, who was the first to roll up his sleeves and plunge his hands into the dishwater. He, as all of our departed friends, continues to live in our thoughts.

7. Acknowledgements

This account is intended as homage to all of those who contributed to the founding and evolution of the IBA, especially fellow members of the 'group of 16.' I thank Patrick Wyse Jackson and Mary Spencer Jones for inviting me to participate in the session on the history of bryozoan studies, and Enrico Annoscia, Giampietro Braga, Richard Boardman, Patricia Cook, Sten Schager, and Nils Spjaeldnæs for comments on the manuscript. Mary Spencer Jones also kindly provided copies of the IBA constitution and other documents.

I plan to place my correspondence, notes, and other documents on which this brief history has been based in a suitable archive. Perhaps others who possess pertinent documents will join me in determining which archive might be most appropriate.

Notes

1 Anonymous, 'Announcement: International Bryozoology Association', *Journal of Paleontology*, 39 (1965), 1221.
2 Somewhat misleadingly referred to as 'a second conference' (note 1).
3 G.G. Simpson in J.A. Wilson, 'The Society of Vertebrate Paleontology, 1940-1990, a fifty-year retrospective', *Journal of Vertebrate Paleontology,* 10 (1990), 1-39, 5.
4 M. Purnell, *The Pander Society*, http://www.le.ac.uk/map2/pander, last accessed October 15, 1999.
5 Here I include such factors as currency restrictions, as well as the more obvious confrontations of an overtly political nature.
6 V.B. Smocovitis, 'Organizing evolution: founding the Society for the Study of Evolution (1939-1950)', *Journal of the History of Biology*, 27 (1994), 241-309.
7 T.J.M. Schopf and J.S. Ryland, 'Bryozoology', *Science*, 162 (1968), 1034-1035.
8 P. Bretsky, 'Review of *Living and Fossil Bryozoa, Recent Advances in Research*, edited by G. P. Larwood', *Quarterly Review of Biology*, 49 (1974), 157-158.
9 R.S. Boardman and A.H. Cheetham, 'Skeletal growth, intracolony variation, and evolution in Bryozoa: a review', *Journal of Paleontology*, 43 (1969), 205-233.
10 G.P. Larwood (editor), *Living and Fossil Bryozoa, Recent Advances in Research* (Academic Press, London and New York, 1973), ix-xi.
11 R.M. Woollacott and R.L. Zimmer (editors), *Biology of Bryozoans* (Academic Press, New York, San Francisco, London, 1977), xv-xvii.
12 John Ryland had to cancel because of illness.
13 R. Lagaaij and Y. Gautier, 'Bryozoan assemblages from marine sediments of the Rhône delta, France', *Micropaleontology*, 11 (1965), 39-58.
14 P.L. Cook and A.H. Cheetham, 'Report of the officers of the International Bryozoology Association for the triennium 1965-1968' (excerpts published in note 23).
15 Cook and Cheetham, note 14.
16 Thus the ratio of 'members' to 'founders' was almost exactly the same as in the first year of the Society for the Study of Evolution, an organization almost four times our size (see note 6).
17 T.J.M. Schopf, copy of a letter to P. L. Cook, dated 14th March 1966.
18 T.J.M. Schopf, 'Names of phyla: Ectoprocta and Entoprocta, and Bryozoa', *Systematic Zoology*, 16 (1967), 276-278.
19 T.J.M. Schopf, letter to A.H. Cheetham, dated 19th July 1967.
20 Schopf, note 19.
21 T. Rocco, letter to A.H. Cheetham, dated 18th October 1967.
22 A.H. Cheetham, letter to E. Annoscia (with two attachments), dated 4th April 1968.
23 P. Ascoli and P.L. Cook, 'Conference report', *Atti della Società Italiana di Scienze Naturali e del Museo Civico di Storia Naturale di Milano*, 108 (1968), 13-24.
24 Smocovitis, note 6, p. 289.
25 S. Reguant, 'Thirty-one years (1968-1998) of bryozoology: The I.B.A. international conferences', *Memorias de la Real Academia de Ciencias y Artes de Barcelona*, 58 (1999), 141-153; S. Reguant, this volume.
26 Smocovitis, note 6, pp. 288-289.
27 Cook and Cheetham, note 14; until the 1970s, there was relatively little disagreement with the traditional usage of 'chairman' to include individuals of either sex.

28 Membership of the nominating committee had also been determined with this criterion in mind (A.H. Cheetham, letter to S. Schager, dated 1st April 1968).
29 Ascoli and Cook, note 23, p. 16.
30 Ascoli and Cook, note 23, p. 18.
31 E. Annoscia, letter to A.H. Cheetham, dated 24th June 1968.
32 N. Spjeldnæs, letter to members of the IBA advisory council ('board'), dated 17th March 1970.
33 T.J.M. Schopf, letter to A.H. Cheetham, dated 18th September 1970 (with attachment).
34 Smocovitis, note 6, p. 255.
35 W.O. Addicott, 'Proceedings of the sixty-sixth annual meeting, Miami Beach, Florida, November 18-20, 1974', *Journal of Paleontology*, 49 (1975), 578-579.
36 P.L. Cook, letter to A.H. Cheetham, dated 6th April 1971.
37 P.D. Taylor and A.H. Cheetham, 'Gilbert Powell Larwood (1930-1997)', in A. Herrera Cubilla and J. C. B. Jackson (editors), *Proceedings of the 11th International Bryozoology Association Conference* (Smithsonian Tropical Research Institute, Republic of Panama, 2000), vii-ix.
38 P.L. Cook, letter to A.H. Cheetham, dated 25th April 1974.
39 Anonymous, 'Abstract of minutes of business meeting of I. B. A., Geology Dept., Université Claude Bernard, 11 Sept. 1974'.
40 G. Lutaud, letter to A.H. Cheetham, dated 20th June 1968; Geneviève did attend later conferences, including those in 1971 and 1974.
41 G. Braga, 'Salvador Reguant, bryozoologist', in C. Blanch i Colat, P. Busquets i Buezo, and J. Serra-Kiel (editors), *Homenatge al Dr. Salvador Reguant i Serra* (Eumo Editorial, Barcelona, 1997), 63-70.
42 Y. Gautier, letter to A.H. Cheetham, dated 20th December 1967.

The Cincinnati Paleobryozoologists

Roger J. Cuffey,[*] Richard Arnold Davis[†] and John E. Utgaard[§]
[*]*Department of Geosciences (412 Deike Bldg.), Penn State University, University Park, PA 16802, USA*
[†]*Biology Department, College of Mount Saint Joseph, 5701 Delhi Road, Cincinnati, OH 45233-1670, USA*
[§]*Department of Geology, Southern Illinois University, Carbondale, IL 62901-4324, USA*

1. Introduction

Fossil bryozoans are extraordinarily abundant, diversified, well-preserved, and easily collected from the flat-lying Upper Ordovician limestones and shales exposed in the hills and ravines within a 150-km radius around Cincinnati, in southwestern Ohio, southeastern Indiana, and north-central Kentucky. These fossils have therefore been of great scientific interest during most of the history of paleontology; they have also attracted the attention of the general public in the region to an extent unusual in America.

The "Cincinnati paleobryozoologists" are those bryozoan paleontologists who have devoted significant effort toward studying Late Ordovician bryozoans specifically from the Cincinnati region, i.e., from the "type-Cincinnatian" strata. Especially in the 19th century, many (but not all) of these people grew up or were employed in that or nearby cities. Later, more lived or worked elsewhere, but all shared a common emphasis on analyzing these fossils.

The Cincinnati paleobryozoologists can be understood best by reference to their participation in the various phases in the history of type-Cincinnatian bryozoan studies.[1] These phases include the initial discovery of type-Cincinnatian bryozoans, preliminary studies based on external characters, introduction of thin-section methods, taxonomic description of many new genera and species, then relative quiescence, and finally modern proliferation of varied researches ranging from faunal monographs and taxonomic revisions to paleoecologic, paleobiologic, and evolutionary investigations.

Numerous publications deal primarily with type-Cincinnatian bryozoans. A complete listing of these is beyond the limits of the present paper, but most can be seen in three bibliographies.[2] We cite here only those contributions which had greatest impact, or which are still readily available and essential for current students. In addition, some modern papers include or mention only a few type-Cincinnatian forms among many more from elsewhere, but being collectively analyzed; such are generally omitted here, although would be necessary to include in a comprehensive bibliography.

The Cincinnati-area bryozoans are accompanied by rich assemblages of shelly invertebrates, especially brachiopods, crinoids, trilobites, pelecypods, gastropods, and nautiloids.[3] These fossils stimulated many other paleontologists in the same way as they have the Cincinnati paleobryozoologists, thereby leading some writers to label all those people as the "Cincinnati school of paleontology".[4] The term "school" implies a more coherent leadership and following than is true for the bryozoan scientists here, and hence seems less appropriate for our purpose. In addition to the bryozoan workers considered here, the Cincinnati "school" included a number of other renowned geologists of the 20th century, such as Charles Schuchert, W.H. Twenhofel, W.B. Scott, and N.S. Shaler. Other bryozoan paleontologists received their professional education at Cincinnati under Kenneth E. Caster (later with the cooperation of Richard S. Boardman at the Smithsonian) and went on to research bryozoans outside the Ordovician period or the Cincinnati area; Helen Duncan, Osborne Nye, and George Farmer may be mentioned among these.

Terminology relating to these fossils can be tricky. The strata around Cincinnati are the standard reference section for the Cincinnatian Series, the North American Upper or Late Ordovician, correlated with the Upper Caradoc plus Ashgill of the British type section recently adopted for international use. Consequently, "Cincinnatian" bryozoans mean animals of Late Ordovician age regardless of where in North America they come from - there are thus Cincinnatian fossils from Pennsylvania, Illinois, and Wyoming, as well as Ohio. In the present paper, we attempt a narrower focus, namely Cincinnatian-age fossils specifically from the Cincinnati region geographically, as the common shared link uniting all the Cincinnati paleobryozoologists.

Present-day readers also need to recall that the earliest works on Cincinnati-area bryozoans refer to them as being in the "Hudson River" and "Utica" rocks, and of "Silurian" or "Lower Silurian" age, because they pre-date introduction of the terms

"Cincinnati Group" and "Ordovician Period". Likewise, early Cincinnati organizations may be referred to as "western", from being near the early 19th century frontier of settlement; the Great Plains, Rocky Mountains, and California coast had not yet entered the consciousness of the eastern American public.

2. Discovery

Prior to European settlement, the wooded hills and ravines that were to become the Cincinnati region were sparsely inhabited by two Algonquian Indian tribes, the Miami and the Shawnee. Although natural outcrops scattered throughout the area must have yielded many weathered-out fossil bryozoans, no evidence exists that these Native Americans noticed them. In contrast, other Indians further west were well aware of crinoid columnals, Cambrian trilobites, and dinosaur footprints.

Losantiville had been founded about 1788 on the north bank of the Ohio River, and was designated to be the capital of the Northwest Territory, as well as the site for a frontier protective Fort Washington. By 1790, the settlement had been renamed Cincinnati, and began to serve as a base for launching military pacification expeditions. The town thereafter grew steadily in population right through establishment of the State of Ohio in 1803, that resulted in ever more exposure of bryozoan-rich bedrock, leading in turn to the first geologic inspections of the region.[5] Inevitably, the abundant fossils weathering out, including bryozoans, soon attracted scientific attention, as well as popular or amateur interest.

J.W. Van Cleve, a naturalist with the Navy-led Wilkes Expedition through the Mid-West, collected and wrote up the first bryozoan species (*Constellaria constellata*) from the type-Cincinnatian. However, death intervened, and his manuscript description was published later by James Dwight Dana (Figure1A), the father of the eminent mineralogy professor also at Yale University, in the report of that expedition.[6] Various problems since have plagued formal recognition of this species,[7] although Van Cleves' collections were re-studied just before adoption of thin-section methods.[8]

At roughly the same time, Cincinnati-area bryozoans had already been taken over to Europe, and in 1850 the great French paleontologist Alcide M. d'Orbigny published several new type-Cincinnatian species, including three of the most abundant forms still recognizable today and now known as *Monticulipora mammulata*, *Parvohallopora ramosa*, and *Heterotrypa frondosa*.[9] D'Orbigny's type specimens were thin-sectioned early in the 20th century.[10] A year later, Henri Milne-Edwards and Jules Haime, two other French scientists, described more, including *Parvohallopora rugosa* and *Dekayia aspera*.[11]

Initially, the Cincinnatian fossil bryozoans were considered to be small corals, especially the massive and robust-ramose trepostomes, frequently called

Figure 1. The most important Cincinnati paleobryozoologists: (A) Dana (collection of Southern Illinois University); (B) Nicholson (Hinde, 1899); (C) Ulrich (collection of Smithsonian Institution); (D) Bassler (Caster, 1965); (E) Nickles (Bassler, 1947); (F) Cumings (Shrock, 1970).

"monticuliporoids" after the best-known earliest-described genus. As the 19th century waned, opinion shifted to regarding them as bryozoans, a viewpoint finally firmly established by Cumings in 1912.[12]

3. External-character studies

Once fossil bryozoans were recognized around the Cincinnati area, paleontologists began to describe the various species and genera that could be distinguished among these abundant specimens. At that time, American professional higher education had not yet assumed the institutional form characteristic of the 20th century, and so these paleontologists came from varied, non-academic backgrounds; Caster refers to them as "publishing amateurs".[13] Among these, the four Cincinnati paleobryozoologists included here authored significant numbers of short bryozoan papers, far too many to cite, but listed elsewhere.[14] Many other Cincinnati writers mentioned bryozoans occasionally, or dealt with other phyla exclusively, and so are omitted here.

Most of the information derivable from examining bryozoan fossils in these early days concerned external surface characters; a little more came from observing broken or fractured specimens. Most of the genera and species in these papers would not be validly recognizable today except for re-study by numerous later workers who thin-sectioned the type specimens of these earlier papers. Even so, some early taxa have yet to be re-studied in this way, and so remain shrouded in nomenclatural fog.

Carl Ludwig Rominger was born in Germany, received his medical degree (but with a geological thesis) from Tubingen, but almost immediately fled the 1848 revolutions and came to Cincinnati, where the fossil "corals" attracted his attention. One of his earliest papers established one of the few Cincinnati trepostome species distinctive enough to be recognizable externally, now known as *Rhombotrypa quadrata* (Rominger, 1866), and whose original type specimens were much later sectioned and studied in the modern era.[15] At about the same time, he moved to Michigan, where - in addition to his medical practice - he became both Professor of Geology at the University of Michigan and the State Geologist of Michigan.[16]

The next Cincinnati paleobryozoologists of note were the prolific father and son pair Uriah Pierson James and Joseph Francis James, who published during 1871-1888 and 1881-1897 (respectively). Uriah James came to Cincinnati from upstate New York as a young printer, and by 1850 had become a publisher and proprietor of a major bookstore (still in business as of this writing). He was an active collector, especially of the local bryozoans (which he always thought were corals), displayed many in his bookstore, and not only authored papers himself,[17] but also published periodicals like *The Paleontologist*.[18] His son, Joseph James, was obviously exposed to Cincinnati-area bryozoans as a child and young clerk in his father's bookstore. He too wrote on bryozoans,[19] and also jointly with his father.[20] Joseph went on to become a Professor of Geology and Botany at Miami University of Ohio, and eventually joined the U.S. Department of Agriculture and also became a practicing physician.[21] The many type specimens designated by the Jameses were later sectioned and re-illustrated by Bassler in 1906.[22]

Samuel Almond Miller was a lawyer and publisher (particularly of the *Journal of the Cincinnati Society of Natural History*) who was massively involved with Cincinnati fossils - he reputedly named a thousand taxa, based on his collection of a million specimens, and published from 1874 until 1897. Most of his efforts involved other phyla, and lengthy, comprehensive, highly useful, but sometimes disparaged compilations.[23]

Figure 2. John M. Nickles and Raymond S. Bassler, Upper Eden exposure, Cincinnati, Ohio, 1900 (Photograph from the Bassler Photo Collection, Department of Paleobiology, National Museum of Natural History, Smithsonian Institution).

Figure 3. Bassler and E.O. Ulrich at Bassler's home, Cincinnati, Ohio, July 1900 (Photograph from the Bassler Photo Collection, Department of Paleobiology, National Museum of Natural History, Smithsonian Institution).

However, he did contribute to Cincinnati-area paleobryozoology both by himself and in cooperation with the local soap and candle maker Charles Brian Dyer.[24] Finally, two more workers began publishing on Cincinnatian-age bryozoans from the Cincinnati area in the 1870s - H.A. Nicholson in 1874[25] and E.O. Ulrich in 1878.[26] Both were destined to soon become two of the most eminent Cincinnati paleobryozoologists.

4. Introduction of thin-sections

Around 1880, the University of Aberdeen Professor H. Alleyne Nicholson (Figure 1B) introduced the preparatory technique of thin-sectioning to bryozoan studies, a contribution that may well rank as the most significant advance in investigations of the phylum. The sizeable, well-calcified type-Cincinnatian trepostomes were ideally suited to application of this method.

Nicholson's thin-section studies began with papers on the Paleozoic "corals" of Ohio published in 1876[27] and in general account of the group in 1879;[28] both works included some type-Cincinnatian bryozoans. Then, in 1881, he published a monumental work focused on trepostomes from Cincinnati (especially "*Monticulipora*"), which elevated understanding of these fossils to a whole new level and thus established Nicholson's role as the leading Cincinnati paleobryozoologist until that date.[29]

Type specimens of many of Nicholson's species have been re-located and restudied in recent years.[30] Although he died at a relatively young age, Nicholson produced a huge body of palaeontological work.[31]

5. Taxonomic descriptions

With the greatly expanded information made available by thin sections, the pace of description and illustration of Cincinnatian bryozoans quickened and remained high for the next 35 years. As soon as Nicholson's publications became available, Edgar Oscar Ulrich (Figures 1C, 3, 4A, 4B), a young bryozoan paleontologist just starting out, adopted thin-sectioning to illustrate his species' internal morphology.[32] He embarked on a series of papers beginning in 1882,[33] which established many Cincinnatian-age taxa from both the Cincinnati area and elsewhere. His studies culminated in two large monographs,[34] landmarks in Ordovician bryozoan studies, and confirming Ulrich's achievement as the second truly outstanding Cincinnati paleobryozoologist.

Ulrich grew up in Covington, across the river from Cincinnati proper. He was introduced (at the age of 7) to fossils by his minister. He worked briefly as a surveyor, took a year of medical college, became active in the Cincinnati Society of Natural History by 1877, and worked at various times for different state (and U.S.) geological surveys. Ulrich developed a laboratory business making and selling bryozoan thin-sections; he recruited

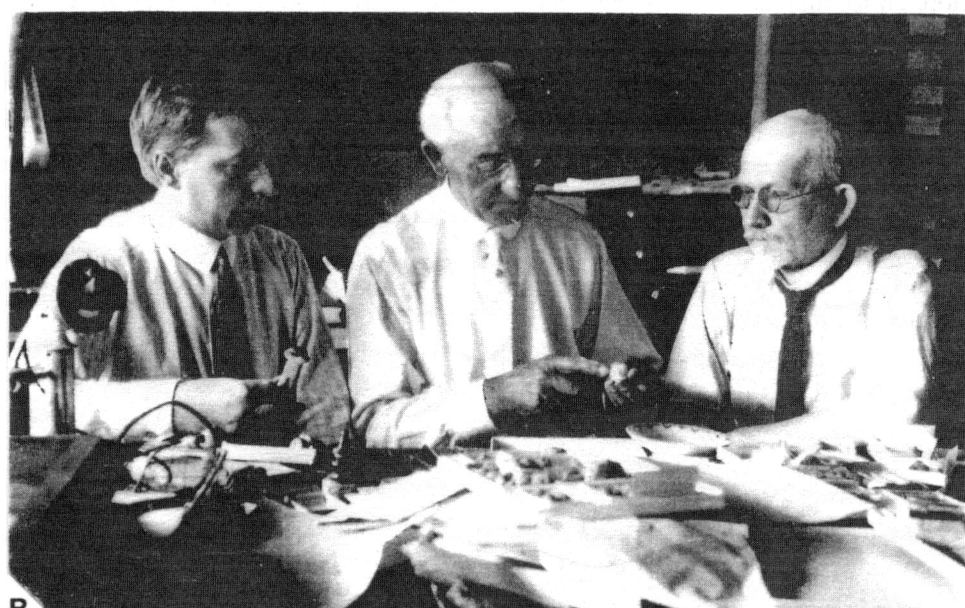

Figure 4. Cincinnati paleobryozoologists at work:
(A) E.O. Ulrich in the field near Cincinnati (Davis and Cuffey, 1998); (B) left to right R.S. Bassler, E.O. Ulrich, and their colleague A.F. Foerste in conference at the U.S. National Museum (collection of the Smithsonian Institution).

a number of local youngsters (including R.S. Bassler and J.M. Nickles) to collect fossils or grind sections for him. Eventually, in 1897, due to his increasing reputation, he was hired permanently by the U.S. Geological Survey and moved to Washington, D.C., where he spent the rest of his career, researching not only bryozoans, but also ostracods, conodonts, and Cambro-Ordovician stratigraphy.[35]

The third really outstanding Cincinnati paleobryozoologist was Ray Smith Bassler (Figures 1D, 2, 3, 4B), initially a high school student hired by Ulrich to help make bryozoan thin-sections. Bassler had already become interested in fossils by childhood collecting near Cincinnati. His high-school principal, George Harper, an amateur geologist himself,[36] introduced Bassler to Ulrich, and arranged Bassler's class schedule so that his afternoons were free to work in Ulrich's lab. During Bassler's final year in high school, he and Harper published a catalogue of local fossils.[37]

Bassler went on to undergraduate study at the University of Cincinnati, continuing to work for Ulrich. Near the end of his bachelor's program, he moved to Washington, D.C. and the Smithsonian when Ulrich was hired there. Bassler combined graduate study at Columbian University (now George Washington University) with paleontologic work at the museum. By the time Bassler received his Ph.D. in 1905, he and Ulrich had co-authored important papers on Cincinnati-area bryozoans.[38] By roughly ten years later, in 1915 Bassler had compiled a major bibliography still useful in studying type-Cincinnatian bryozoans.[39]

For the next half century, Bassler continued his major research interests in bryozoans, while adding other important fossil groups (notably corals, echinoderms, ostracods, and conodonts) to his repertoire. Simultaneously, he moved up through the ranks at the Smithsonian to become Head Curator of Geology, and across town eventually also Professor and Head of the Department of Geology at George Washington University. He is reported to have worked effectively with students, collaborators, and colleagues both here and abroad, often extending help beyond what might have been expected.[40]

From Nicholson to Ulrich to Bassler's early work, it had been a continuous 30 years of Cincinnati bryozoan research. While certainly cutting-edge for the time, and while still essential as a base for current investigation, their thin sections tend to be too thick and their microscopic magnifications too low, compared with modern practice. However, such qualifications should not detract from respecting the major contributions which these three important Cincinnati paleobryozoologists made.

Another local Cincinnati amateur acquainted with Ulrich and bryozoans was John Milton Nickles (Figures 1E, 2), about 20 years older than Bassler, but who worked with him to produce another major bryozoan bibliographic work.[41] Nickles soon thereafter also joined the U.S. Geological Survey in Washington, but then built his career compiling comprehensive bibliographies on geology in general.[42] Before leaving bryozoans, Nickles utilized them to biostratigraphically zone the type-Cincinnatian formations.[43]

Edgar Roscoe Cumings (Figure 1F) can be regarded as the fourth major Cincinnati paleobryozoologist, because of his important papers (some co-authored with his doctoral student J.J. Galloway) advancing fundamental understanding of Cincinnati bryozoans' colony development,[44] wall microstructure,[45] and affinities.[46] This last is especially significant because it settled the long-standing controversy in favour of the "stony" or "monticuliporoid" trepostomes being bryozoans rather than corals. In addition, he also carried out taxonomic studies,[47] and monographed faunas on the Indiana side of the Cincinnati region.[48] Cumings was Professor of Geology at Indiana University for many years,[49] and became well-known for pioneering studies of fossil reefs or bioherms.

Jesse James Galloway (1882-1962) was a graduate student co-author with Cumings on two significant Cincinnati bryozoan papers (cited above). Although he retained a life-long interest in those fossils, his research emphasis shifted to foraminiferans and stromatoporoids. He served on the geology faculty at Columbia University in New York, and then returned to Indiana University to finish out his long career.[50]

Emphasis during 1880-1915 focused mainly on basic descriptive taxonomy of the diverse species and genera among the type-Cincinnatian bryozoans, although the beginnings of other kinds of investigations were also evident. Ulrich's keen observations of anatomy especially, later led to greater paleobiological understanding of those fossils. Adoption of thin-section techniques came gradually, although by 1890 these were universal.

6. Quiescence

The advent of World War I shifted people's attention away from scientific pursuits viewed as being esoteric. The following decades of social change, economic depression, and more global conflict coincided with a low level of interest in type-Cincinnatian bryozoan studies, except for occasional papers including a few such materials as ancillary to faunas elsewhere (especially in Canada.[51]

The one Cincinnati paleobryozoologist remaining active during this quiescent phase was Bassler, who expanded his interests into Cenozoic and Recent bryozoans in cooperation with Ferdinand Canu in France.[52] Probably his most impactful contribution came toward the end of this time with the publication of his sole-authored bryozoan volume of the *Treatise on Invertebrate Paleontology*.[53] Not only did this publication draw together much information on type-Cincinnatian bryozoans, it also provided systematic context for these fossils by incorporating all the rest of their phylum. Moreover, historically, Bassler is remembered as the author of the first volume of the *Treatise* series, one of the most important paleontological publications in general, only now approaching completion after 50 years.

7. Modern diversified studies

Around 1960, interest in type-Cincinnatian bryozoans revived as part of a wide-spread renaissance in paleontology in general. In retrospect, the cause for this rebirth of the science appears simply to have been the post-World War II expansion of population and the economic growth which it stimulated. Young students grew up, pursued intellectual interests not practical in harder times, found opportunities for academic careers in universities and agencies devoted to serving the rapidly increasing numbers of people, and trained still more young people to carry on. (After a couple of decades, conditions reversed, beginning with collapse of the geologic sector of the oil industry; hence, growth has slowed or ceased, but the character of this renaissance continues to the present day.) A hallmark of this revival in paleontology has been its concern with biological as much as with geological aspects of the fossils which it studies.

These trends have resulted in a new generation of Cincinnati paleobryozoologists, distinguished from the earlier one in two major ways. First, the newer ones are employed professionally over a much wider range of locations and institutions. Second, while basic taxonomic research remains important in the modern phase, a great many additional types of investigations are being pursued by the new Cincinnati paleobryozoologists. Not every such publication is cited here, but enough can be given to indicate the diversity of current activity.

Enthusiastic local amateur collecting remains strong around Cincinnati. Some of this activity is individualistic; some is organized as the "Dry Dredgers", who have contributed significantly by bringing to light scientifically important bryozoan localities and specimens.

Cornerstones of continuing Cincinnati-area bryozoan studies are several faunal monographs detailing the bryozoan species found in particular stratigraphic formations at different horizons within the type-Cincinnatian: high, Whitewater Formation,[54] and Dillsboro Formation;[55] middle, Bellevue Limestone;[56] low, Eden Shale;[57] transitional, Clays Ferry Formation;[58] and immediately underlying, Lexington Limestone.[59] These are essential for the reliable or consistent species identifications on which numerous other investigations are based. Two monographs treat forms from everywhere in the type-Cincinnatian succession.[60] In addition, four monographs from the earlier phase of taxonomic descriptions are still needed in practice for identifying forms: Two deal with the Cincinnatian bryozoans of the northwestern part of the Cincinnati region;[61] two others include many type-Cincinnatian forms even though focused on other states.[62]

A number of recent studies coordinate and revise the species belonging to particular genera, especially their type species. Examples which include at least some Cincinnati-area materials are Boardman,[63] Boardman and Utgaard,[64] Utgaard,[65] Hickey,[66] Taylor and Wilson,[67] and Cutler.[68] Bibliographic compilations in the tradition of Bassler and Nickles have appeared.[69] Re-descriptions of critical type species, in addition to those already cited in previous sections of this paper, have been made.[70] Formal nomenclature adjustments have been proposed and accepted.[71] Many of these taxonomic investigations were undertaken in order to update, revise, and expand the original bryozoan volume of the *Treatise on Invertebrate Paleontology*.[72] The first part of the revised bryozoan *Treatise* has already appeared.[73]

Morphologic investigations involving new objectives and tools are reaching beyond basic taxonomy to questions of development and function. Certain studies analyzed soft-part and brown-body morphology.[74] Others explored new measurement techniques.[75] Some connected developmental anatomy to growth and function within a colony.[76] Still others examined the relationships among development, evolutionary processes, and environment.[77]

Other Cincinnati paleobryozoologists traced evolutionary trends in area lineages,[78] and implications for biostratigraphy.[79] Taphonomy has been considered on both regional,[80] and local scales,[81] including bioimmuration.[82] Some have studied bryozoan encrustation on unusual substrates like cephalopods,[83] and cobbles,[84] leading into biostromal[85] and biohermal[86] involvement, as well as participation in overall community succession[87] and sequence stratigraphy.[88] Assessment of particular paleoenvironmental factors was carried out for ancient salinity[89] and bathymetry.[90] Biogeographic aspects have been considered.[91] Finally, Tuckey and Anstey in 1992 discussed end-Ordovician extinctions affecting the Cincinnati-area bryozoans.[92]

8. Conclusion

The Cincinnati region of southwestern Ohio and adjacent southeastern Indiana and north-central Kentucky is world-renowned for its abundant and diversified bryozoans from the Upper Ordovician (type-Cincinnatian) Series. These and the many other associated fossils stimulated paleontologic careers for local residents, as well as for visitors from elsewhere. A number of both have made major contributions specifically to local bryozoan studies, and can thus be called "Cincinnati paleobryozoologists".

The abundant, diversified, well-preserved Cincinnati-region bryozoan faunas were discovered and initially described by Van Cleve, Dana, d'Orbigny, Milne-Edwards, and Haime (1846-1851).[93] Soon thereafter, many species of these fossils were described from external characters alone, by workers like Rominger, U. James, J. James, Miller, and C. Dyer.

About 1880, Nicholson introduced examination of bryozoans' internal structures via thin sections. Almost immediately, Ulrich applied this method extensively to Cincinnati and other bryozoans, and produced monumental works in the 1890s, which are still essential references. A younger local protégé, Bassler, began his lengthy career then, which culminated decades later in his authoring the first volume of the *Treatise on Invertebrate Paleontology*. Another protégé, Nickles, cooperated with Bassler to compile important bibliographies, and also used bryozoans to stratigraphically subdivide the type-Cincinnatian. Cumings demonstrated the bryozoan rather than coral affinities of the trepostomes, and he and his student Galloway published sizeable systematic monographs on bryozoans from the Indiana side of the Cincinnati region.

Then, from 1915 until 1960, specifically Cincinnati-area paleobryozoology was largely quiescent, until Perry,[94] Caster, Boardman, and their students and colleagues (Anstey, Brown, Cuffey, Horowitz, Karklins, Key, Pachut, Pohowsky, Singh, Utgaard, and Wilson) began modern investigations including these fossils, as well as bryozoans from other regions and horizons, investigations which continue to the present day.

9. Acknowledgements

We particularly thank the people who helped provide pictures of various paleobryozoologists: K. Bork, H. Henson , J. Sanner, P.D. Taylor, E. Yochelson. Susan Lauver prepared this manuscript for publication.

Notes

1 J.E. Utgaard and T.G. Perry, 'Trepostomatous bryozoan fauna of the upper part of the Whitewater Formation (Cincinnatian) of eastern Indiana and western Ohio', *Indiana Geological Survey Bulletin* 33 (1964), 111 pp., 23 pls.

2 Utgaard and Perry, note 1; K.E. Caster, 'Cincinnatian contributions to knowledge of the Lophophora', in *Lophophorates. Notes for a short course*, edited by T.W. Broadhead, *University of Tennessee, Department of Geological Sciences, Studies in Geology*, 5 (1981), 237-251; R.A. Davis and R.A. Spohn, 'Bibliography on the Type-Cincinnatian', in *Sampling the Layer Cake that Isn't: the Stratigraphy and Paleontology of the Type-Cincinnatian*, edited by R.A. Davis and R.J. Cuffey, *Ohio Division of Geological Survey Guidebook*, 13 (1998), 167-194.

3 K.E. Caster, E.A. Dalvé and J.K. Pope, *Elementary guide to the fossils and strata of the Ordovician in the Vicinity of Cincinnati, Ohio*. Cincinnati, Ohio, Cincinnati Museum of Natural History, 1955, 47 p., 8 pls; A. La Rocque and M.F. Marple, Ohio Fossils. *Ohio Division of Geological Survey Bulletin,* 54 (1956), v + 152 pp.; R.A. Davis, ed. *Cincinnati fossils. An elementary guide to the Ordovician rocks and fossils of the Cincinnati, Ohio, region.* Cincinnati, Ohio, Cincinnati Museum of Natural History (1992), ii + 61 p., 8 pls; R.L. Anstey and M.A. Wilson, 'Phylum Bryozoa', in *Fossils of Ohio,* edited by R.M. Feldmann and M. Hackathorn. *Ohio Division of Geological Survey, Bulletin* 70 (1996), 196-209; R.M. Feldmann and M. Hackathorn, eds. *Fossils of Ohio. Ohio Division of Geological Survey, Bulletin,* 70 (1996), xx + 577 pp.; R.A. Davis and R.J. Cuffey, eds. *Sampling the layer cake that isn't: the stratigraphy and paleontology of the type-Cincinnatian. Ohio Division of Geological Survey, Guidebook,* 13 (1998), iv + 194 pp.

4 K. Becker, 'Cincinnati area, the Mother of Geologists', *Compass*, 19(1) (1938), 188-196; K.E. Caster, 'Cincinnati, Spawning Ground of Geologists', *The Compass of Sigma Gamma Epsilon*, 28(2) (1951), 103-105; Caster 1981, note 2; K.E. Caster, 'The Cincinnati "School" of paleontology', *Earth Sciences History*, 1(1) (1982), 23-28.

5 D. Drake, 'Geological account of the Valley of the Ohio', *Transactions of the American Philosophical Society*, New Series, 2 (1825), 124-139; J. Locke, 'Prof. Locke's geological report [on southwestern Ohio]', *Ohio Geological Survey 2nd Annual Report*, (1838), 201-286, 15 pls; C. Lyell, *Travels in North America; with geological observations on the United States, Canada, and Nova Scotia.* London, John Murray, (1845), 316 pp (volume 1) and 272 pp (volume 2). [The geology of the Cincinnati area is discussed on p. 48-51 in volume 2, but see also the cross-section of the eastern United States on p. 92 of volume 1]

6 J.W. Van Cleve, (manuscript published in Dana, 1846); J.D. Dana, 1846 (text) and 1849 (plates), *Zoophytes: Reports of the United States Exploring Expedition, 1838-1842, under the Command of C. Wilkes, U. S. Navy*. Initially C. Sherman, later Lea & Blanchard, Philadelphia, volume 7, 740 p., 61 pls, p. 537-538; p. 52, fig. 6a, 6b; D.C. Haskell, *The United States Exploring Expedition, 1838-1842, and its publications, 1844-1874.* New York Public Library, New York (1942).

7 J.R.P.P. Ross, '*Constellaria* from the Chazyan (Ordovician), Isle La Motte, Vermont', *Journal of Paleontology*, 37(1) (1963), 51-56, pls. 5, 6; J.E. Utgaard, 'Paleobiology and taxonomy of the order Cystoporata', in *Bryozoa: Treatise on Invertebrate Paleontology*, Part G, revised, volume 1, edited by R.A. Robison. Geological Society of America and University of Kansas Press (1983), 327-439.

8 C.A. White, 'Van Cleve's fossil corals', *Indiana Department of Geology and Natural Resources, Eleventh Annual Report*, (1882), 376-401, pls. 44, 45; J. Hall, 'Van Cleve's fossil corals', *Indiana Department of Geology and Natural Resources, Twelfth Annual Report*, (1883), 239-318, 2 figs.

9 A.M. d'Orbigny, *Prodrôme de paléontologie stratigraphique universelle*. Volume 1; Victor Masson, Paris (1848-1852), lx + 394 pp.; R.J. Singh, 'Trepostomatous bryozoan fauna from the Bellevue Limestone, Upper Ordovician, in the tri-state area of Ohio, Indiana, and Kentucky', *Bulletins of American Paleontology*, 76(307), (1979), 157-288, pls. 16-48.

10 M. Boule and A. Thevenin, 1906, 'Types du prodrome de paléontologie stratigraphique universelle de d'Orbigny', *Annales de paléontologie*, 1 (1906), 97-100, pls. 8-9; 165-172, pls. 12-13.

11 Singh, note 9.

12 E.R. Cumings, 'Development and systematic position of the monticuliporoids', *Geological Society of America Bulletin*, 23 (1912), 357-370, pls. 19-22.

13 Caster, note 2, p. 237.

14 Davis and Spohn, note 2; Utgaard and Perry, note 1.

15 See C. Rominger, 'Observations on *Chaetetes* and some related genera, in regard to their systematic position, with an appended description of some new species', *Proceedings of the Academy of Natural Sciences of Philadelphia*, (1866), 113-123. Rominger's material has been redescribed by R.J. Cuffey and T.G. Perry, 'Redescription of the syntypes of the bryozoan species *Rhombotrypa quadrata* (Rominger)', *Contributions from the Museum of Paleontology, University of Michigan*, 19(4) (1964), 37-45, pls. 1-2; and by R.S. Boardman and F.K. McKinney, 'Skeletal architecture and preserved organs of four-sided zooids in convergent genera of Paleozoic Trepostomata (Bryozoa)', *Journal of Paleontology*, 50(1) (1976), 25-78,

16 pls.

16 H.M. Martin, 'Glimpses into the life of Carl Ludwig Rominger', *University of Michigan, Museum of Paleontology, Papers on Paleontology*, 5 (1973), 1-29; J.D. Wright, 'That remarkable man Carl Ludwig Rominger State Geologist', *University of Michigan, Museum of Paleontology, Papers on Paleontology*, 4 (1973), 1-148.

17 U.P. James, *Paleontology. Catalogue of Lower Silurian Fossils of the Cincinnati Group. Found at Cincinnati and Vicinity-within a circuit of 40 or 50 miles. New ed., much enl. With descriptions of some new species of corals and* Polyzoa. U.P. James, Cincinnati, Ohio (1875), 8 pp.; U.P. James, 'Descriptions of ten new species of *Monticulipora* from the Cincinnatian Group, Ohio, index, etc.', *The Paleontologist*, 6 (1882), 45-56, pl. 1.

18 A.J. Howe, W.H. Fisher and A.T. Keckeler, 'In Memoriam.—U. P. James', *Journal of the Cincinnati Society of Natural History*, 12(1) (1889), 5-7; J.F.J. (presumably Joseph F. James) 'Uriah Pierson James', *American Geologist*, (May 1889), 7 pp. + photograph of U.P. James.

19 J.F. James, '*Monticulipora*, a coral and not a polyzoon', *American Geologist*, 1(6) (1888), 386-392.

20 U.P. James and J.F. James, 'On the monticuliporoid corals of the Cincinnati Group, with critical revision of the species', *Journal of the Cincinnati Society of Natural History*, 10(3) (1887), 118-141; U.P. James and J.F. James, 'On the monticuliporoid corals of the Cincinnati Group, with critical revision of the species (continued)', *Journal of the Cincinnati Society of Natural History*, 10(4) (1888), 158-184, pl. 2; 11(1) (1888), 15-47, pl.1.

21 G.K. Gilbert, 'Joseph Francis James 1857-1897', *American Geologist*, 21(1) (1898), 1-11, pl. 1.

22 R.S. Bassler, 'A study of the James types of Ordovician and Silurian Bryozoa', *Proceedings of the U.S. National Museum*, 30 (1442) (1906), 1-66, pls. 1-6.

23 S.A. Miller, *The American Palaeozic Fossils: a Catalogue of the Genera and Species*, (Privately published by the author); printed by the Cincinnati Times Company, Cincinnati, Ohio (1877), xv + 253 pp.; S.A. Miller, *North American geology and palaeontology for the use of amateurs, students, and scientists*, (Privately published by the author, printed by the Western Methodist Book Concern), Cincinnati, Ohio (1889), 664 pp., 1194 text-figs; S.A. Miller, *Second appendix to North American Geology and Palaeontology*, (Privately published by the author), Cincinnati, Ohio (1897), 719-793; Caster, note 2, p. 238.

24 S.A. Miller, 'Observations upon *Stenopora fibrosa* and the genus *Chaetetes*', *Cincinnati Quarterly Journal of Science*, 1(4) (1874), 368-375; S.A. Miller and C.B. Dyer, 'Contributions to palaeontology', *Journal of the Cincinnati Society of Natural History*, 1(1) (1878*a*), 24-39, pl. 1-2. (There is a reprint paginated p. 1-16.); S.A. Miller and C.B. Dyer, *Contributions to Paleontology. No. 2*, (Privately published by the authors). Cincinnati, Ohio (1878*b*), 1-11, pls. 3-4; R.M. Byrnes, L.S. Cotton and F.W. Langdon, 'In Memoriam. Charles B. Dyer', *Journal of the Cincinnati Society of Natural History*, 6(3) (1883), 207-210. C.B. Dyer should not be confused with W.S. Dyer, who published over 40 years later on bryozoans of similar age but from Toronto, not Cincinnati.

25 H.A. Nicholson, 'Description of species of *Chaetetes* from the Lower Silurian rocks of North America', *Quarterly Journal of the Geological Society of London*, 30 (1874), 499-515, pls. 29-30.

26 E.O. Ulrich, 'Descriptions of some new species of fossils, from the Cincinnati Group', *Journal of the Cincinnati Society of Natural History*, 1(2) (1878), 92-100, pl. 4.

27 H.A. Nicholson, 'Notes on the Palaeozoic corals of the state of Ohio', *Annals and Magazine*

of Natural History, series. 4, 18 (1876), 85-95, pl. 5.
28 H.A. Nicholson, *On the Structure and Affinities of the "Tabulate Corals" of the Paleozoic Period, with critical descriptions of illustrative species*, William Blackwood and Sons, Edinburgh (1879), 342 pp., 15 pls., 44 text-figs.
29 H.A. Nicholson, *On the Structure and Affinities of the Genus* Monticulipora *and its Sub-genera, with critical descriptions of illustrative species*, William Blackwood and Sons, Edinburgh (1881), 240 pp., 6 pls., 50 text-figs.; Hinde, G.J. 'Prof. H.A. Nicholson, M.D., D.Sc., F.R.S., F.L.S., F.G.S.', *Geological Magazine*, 6 (1899), 138-144.
30 T.G. Perry, A.S. Horowitz and R.L. Anstey 'Restudy of some Nicholson's types of Palaeozoic bryozoans', in *Living and Fossil Bryozoa*, edited by G.P. Larwood. London, Academic Press (1973), 429-436, 1 pl; J.F. Pachut, R.L. Anstey and A.S. Horowitz, 'The H.A. Nicholson collection of Paleozoic stenolaemate bryozoans: Comparison of cladistic and phenetic classifications,' *Journal of Paleontology*, 68(5) (1994), 978-994.
31 See Perry, Horowitz and Anstey, note 30, p. 435-436; Singh, note 9, p. 246; Davis and Spohn, note 2, p. 185; M.J. Benton, 'H.A. Nicholson (1844-1899), invertebrate palaeontologist: bibliography and catalogue of his type and figured material', *Royal Scottish Museum Information Series. Geology* 7 (1979) Edinburgh, vii+94 pp.; M.J. Benton and N.H. Trewin, 'H.A. Nicholson Catalogue of type and figured material in the Palaeontological Collections, University of Aberdeen, with notes on the H.A. Nicholson Collection', *Publications of the Department of Geology and Mineralogy Aberdeen* 2 (1978), 1-28.
32 E.O. Ulrich, 'Descriptions of new genera and species of fossils from the Lower Silurian about Cincinnati', *Journal of the Cincinnati Society of Natural History*, 2(1) (1879*a*), 8-30, pl. 7; E.O. Ulrich, 'Description of a new genus and some new species of bryozoans from the Cincinnati Group', Journal of the Cincinnati Society of Natural History, 2(3) (1879*b*), 119-131, pl. 12.
33 E.O. Ulrich, 'American Paleozoic Bryozoa', *Journal of the Cincinnati Society of Natural History*, 5(3) (1882), 121-175, pls. 6-8; Utgaard and Perry, note 1, p. 111; Davis and Spohn, note 2, p. 192.
34 E.O. Ulrich, 'Paleozoic Bryozoa', *Illinois Geological Survey*, 8(2), section 6 (1890), 285-688, pls. 29-78; E.O. Ulrich, 'On Lower Silurian Bryozoa of Minnesota', *Minnesota Geological and Natural History, Final Report*, 3(1) (1893-1895), 96-332, 28 pls. (The title-page of the volume bears the date 1895, but page liii states that 100 copies of the bryozoan part were printed and distributed in early 1893.)
35 R.S. Bassler, 'Memorial to Edward Oscar Ulrich', *Geological Society of America, Proceedings for 1944* (1945), 331-352; L.E. Bradshaw, *E.O. Ulrich and the development of American stratigraphy*, University of Cincinnati (Department of History), Cincinnati, Ohio, unpublished Ph.D. dissertation (1989), v + 329 pp.
36 V. Kramer, 'Professor George W. Harper', *The Oracle*, 19(1) (1918), 12.
37 G.W. Harper and R.S. Bassler, *Catalogue of the fossils of the Trenton and Cincinnati Periods, occurring in the vicinity of Cincinnati, O.* Cincinnati, W. B. Carpenter Co. (1896), 34 pp.
38 E.O. Ulrich and R.S. Bassler, 'A revision of the Paleozoic Bryozoa, pt. 1, On genera and species of Ctenostomata', *Smithsonian Miscellaneous Collections*, 45 (1904*a*), 256-294, 4 pls; E.O. Ulrich and R.S. Bassler, 'A revision of the Paleozoic Bryozoa, pt. 2, On genera and species of Trepostomata', *Smithsonian Miscellaneous Collections*, 47(1470) (1904b), 15-55, pls. 6-14; see also Utgaard and Perry, note 1, p. 109; and Davis and Spohn, note 2, p. 168)
39 R.S. Bassler, 'Bibliographic index of American Ordovician and Silurian fossils', *U.S. National Museum Bulletin*, 92 (1915), 2 volumes, 1,521 pp., 4 pls.

40 K.E. Caster, 'Memorial to Ray S. Bassler (1878-1961)', *Geological Society of America, Proceedings for 1965*, (1965), 167-173.

41 J.M. Nickles and R.S. Bassler, 'A synopsis of American fossil Bryozoa, including bibliography and synonymy', *U.S. Geological Survey Bulletin*, 173 (1900), 663 pp.

42 R.S. Bassler, 'In Memoriam', *Geological Society of America. Bibliography and Index of Geology Exclusive of North America*. Volume 11 – 1945-1946 (1947), ii-v. (Obituary of John Milton Nickles, including photograph).

43 J.M. Nickles, 'The geology of Cincinnati', *Journal of the Cincinnati Society of Natural History*, 20(2) (1902), 49-100, pl. I [2-page map]; reprint edition, p. 1-53, 1 pl.

44 E.R. Cumings, 'Development of some Paleozoic Bryozoa', *American Journal of Science*, series 4, 17(97) (1904), 49-78.

45 E.R. Cumings and J.J. Galloway, 'Studies of the morphology and history of the Trepostomata or monticuliporoids', *Geological Society of America Bulletin*, 26 (1915), 349-374, pls. 10-15.

46 Cumings, note 12.

47 E.R. Cumings, 'A revision of the bryozoan genera *Dekayia*, *Dekayella*, and *Heterotrypa* of the Cincinnati Group', *American Geologist*, 29(4) (1902), 197-217, pls. 9-12.

48 E.R. Cumings, 'The stratigraphy and paleontology of the Cincinnati Series of Indiana', *Indiana Department of Geology and Natural Resources, Thirty-second Annual Report for 1907*, (1908), 605-1189, 55 pls; E.R. Cumings and J.J. Galloway, 'The stratigraphy and paleontology of the Tanner's Creek section of the Cincinnatian Series in Indiana', *Indiana Department of Geology and Natural Resources, Thirty-seventh Annual Report for 1912*, (1913), 353-479, 20 pls.

49 R.R. Shrock, 'Memorial to Edgar Roscoe Cummings', *Geological Society of America, Proceedings, P.* (1970), 177-183.

50 T.G. Perry, 'Memorial to Jesse James Galloway (1882-1962)', *Journal of Paleontology*, 36(4) (1962), 845-846.

51 W.S. Dyer, 'The stratigraphy and paleontology of Toronto and vicinity. Part 5. The paleontology of the Credit River section', *Ontario Department of Mines, Annual Report* 32, 32(7) (1925), 47-88, 7 pls; M.A. Fritz, 'Bryozoa (mainly Trepostomata) from the Ottawa Formation (Middle Ordovician) of the Ottawa-St. Lawrence Lowland', *Geological Survey of Canada, Bulletin*, 42 (1957), 75 pp., 30 pls.

52 F. Canu and R.S. Bassler, 'Fossil and Recent Bryozoa of the Gulf of Mexico region', *U.S. National Museum Proceedings*, 72(14) (1928), 1-199, pls. 1-34. See also J.-A. Sanner, this volume

53 R.S. Bassler, 'Part G. Bryozoa', in *Treatise on Invertebrate Paleontology*, edited by R.C. Moore. Geological Society of America and University of Kansas Press, (1953) 253 p.

54 Utgaard and Perry, note 1.

55 G.D. Brown Jr. and E.J. Daly, 'Trepostome Bryozoa from the Dillsboro Formation (Cincinnatian Series) of southeastern Indiana', *Indiana Geological Survey Special Report*, 33 (1985*b*), 95 pp., 12 pls.

56 Singh, note 9.

57 R.L. Anstey and T.G. Perry, 'Eden Shale bryozoans: a numerical study (Ordovician, Ohio Valley)', *Michigan State University, Publications of the Museum, Paleontological Series*, 1(1) (1972 [1973]), 80 pp., 22 pls. (title page bears the date 1972, but the verso of the title page states "Published 14 January 1973".).

58 O.L. Karklins, 'Trepostome and cystoporate bryozoans from the Lexington Limestone and the Clays Ferry Formation (Middle and Upper Ordovician) of Kentucky', *U.S. Geological Survey*

Professional Paper 1066-I (1984), 105 pp., 41 pls.
59 G.D. Brown Jr., 'Trepostomatous Bryozoa from the Logana and Jessamine Limestones (Middle Ordovician) of the Kentucky Bluegrass Region', *Journal of Paleontology*, 39(5) (1965), 974-1006, pls. 111-118.
60 R.A. Pohowsky, 'The boring ctenostomate Bryozoa: taxonomy and paleobiology based on cavities in calcareous substrata', *Bulletins of American Paleontology*, 73(301) (1978), 192 pp., 24 pls; O.L. Karklins, 'Ptilodictyoid Cryptostomata Bryozoa from the Middle and Upper Ordovician rocks of central Kentucky', *Journal of Paleontology*, 57(1) supplement (1983) (*Paleontological Society Memoir* 14), 31 p.
61 Cumings, note 48; Cumings and Galloway, note 48.
62 Ulrich, note 34.
63 R.S. Boardman, 'A revision of the Ordovician bryozoan genera *Batostoma*, *Anaphragma*, and *Amplexopora*', *Smithsonian Miscellaneous Collections*, 140(5) (1960), 28 pp., 7 pls.
64 R.S. Boardman and J.E. Utgaard, 'A revision of the Ordovician bryozoan genera *Monticulipora*, *Peronopora*, *Heterotrypa*, and *Dekayia*', *Journal of Paleontology*, 40(5) (1966), 1082-1108, pl. 133-142.
65 J.E. Utgaard, 'A revision of North American genera of ceramoporoid bryozoans (Ectoprocta): Part II; *Crepipora*, *Ceramoporella*, *Acanthoceramoporella*, and *Ceramophylla*', *Journal of Paleontology*, 42(6) (1968), 1444-1455, pls. 181-184.
66 D.R. Hickey, 'Bryozoan astogeny and evolutionary novelties: their role in the origin and systematics of the Ordovician monticuliporid trepostome genus *Peronopora*', *Journal of Paleontology*, 62(2) (1988), 180-203.
67 P.D. Taylor and M.A. Wilson, '*Corynotrypa* from the Ordovician of North America: colony growth in a primitive stenolaemate bryozoan', *Journal of Paleontology*, 68(2) (1994), 241-257; P.D. Taylor and M.A. Wilson, '*Cuffeyella*, a new bryozoan genus from the Late Ordovician of North America, and its bearing on the origin of the post-Paleozoic cyclostomates', in *Bryozoans In Space and Time*, edited by D.P. Gordon, A.M. Smith and J.A. Grant-Mackie. National Institute of Water and Atmospheric Research, Wellington, New Zealand (1996), 351-360.
68 J.F. Cutler, *Morphology, taxonomy, and evolution of the bryozoan* Constellaria *from the Cincinnati Arch*, Columbia University, New York, unpublished Ph.D. dissertation, (1968) 297 pp.
69 F.K. McKinney, 'Bibliography and list of the families Constellariidae and Dianulitidae (Ectoprocta, Order Cystoporata)', *Southeastern Geology*, 10(3) (1969), 175-184; F.K. McKinney, 'Bibliography and catalogue (1900-1969) of the Trepostomata (phylum Ectoprocta)', *Southeastern Geology, Special Publication*, 4 (1974), 147 pp.
70 J.R.P. Ross, 'Reevaluation of the type species of *Arthropora* Ulrich', *Journal of Paleontology*, 34(5) (1960), 859-861, pl. 108; N. Spjeldnaes, 'A redescription of the type species of the bryozoan genus *Crepipora*', *Journal of Paleontology*, 37(1) (1963), 64-68, pls. 9-10; R.L. Anstey and T.G. Perry, 'Redescription of cotypes of *Peronopora vera* Ulrich, a Cincinnatian (Late Ordovician) ectoproct species,' *Journal of Paleontology*, 43(2) (1969), 245-251, pls. 31-32.
71 J.E. Utgaard and R.S. Boardman, '*Heterotrypa* Nicholson, 1879, and *Peronopora* Nicholson, 1881, Bryozoa, Trepostomata: proposed designation of a type-species in conformity with generally accepted usage', *Bulletin of Zoological Nomenclature*, 22(2) (1965), 112-118; International Commission on Zoological Nomenclature, 'Opinion 838, Designation under the Plenary Powers of the type species in harmony with accustomed usage for genera *Heterotrypa*

Nicholson, 1879, and *Peronopora* Nicholson, 1881 (Class Bryozoa) (Ordovician)', *Opinions and Declarations rendered by the International Commission on Zoological Nomenclature*, 24(6) (1968), 835-841.

72 Bassler, note 53.

73 R.S. Boardman, A.H. Cheetham, D.B. Blake, J.E. Utgaard, O.L. Karlkins, P.L. Cook, P.A. Sandberg, G. Lutaud and T.S. Wood, 'Bryozoa, Part G, revised, volume 1: Introduction, Order Cystoporata, Order Cryptostomata', in *Treatise on Invertebrate Paleontology*, edited by R.A. Robison. Geological Society of America and University of Kansas Press (1983), 625 p.

74 R.S. Boardman, 'Mode of growth and functional morphology of autozooids in some Recent and Paleozoic tubular Bryozoa', *Smithsonian Contributions to Paleobiology*, 8 (1971), 51 pp., 11 pls; S.J. Morrison and R.L. Anstey 'Ultrastructure and composition of brown bodies in some Ordovician trepostome bryozoans', *Journal of Paleontology*, 53(4) (1979), 943-949; J.E. Utgaard, 'Mode of colony growth, autozooids, and polymorphism in the bryozoan order Cystoporata', in *Animal Colonies*, edited by R.S. Boardman, A.H. Cheetham and W.A. Oliver. Dowden, Hutchison, & Ross, Stroudsburg (1973), 317-360, 74 text-figs; Utgaard, note 7.

75 R.L. Anstey and J.F. Pachut, 'Fourier packing ordinate: A univariate size-independent measurement of the polygonal packing variation in Paleozoic bryozoans', *Mathematical Geology*, 12(2) (1980), 139-156; R.L. Anstey and J.W. Bartley, 'Quantitative stereology: an improved thin section biometry for bryozoans and other colonial organisms', *Journal of Paleontology*, 58(3) (1984), 612-625.

76 R.L. Anstey, J.F. Pachut and D.R. Prezbindowski, 'Morphogenic gradients in Paleozoic bryozoan colonies', *Paleobiology*, 2(2) (1976), 131-146; F.K. McKinney, 'Autozooecial budding patterns in dendroid Paleozoic bryozoans', *Journal of Paleontology*, 51(2) (1977), 303-329, 9 pls; M.E. Podell and R.L. Anstey, 'The interrelationship of early colony development, monticules and branches in Palaeozoic bryozoans', *Palaeontology*, 22(4) (1979), 965-982, pls. 126-130; R.L. Anstey, 'Zooid orientation structures and water flow patterns in Paleozoic bryozoan colonies', *Lethaia*, 14(4) (1981), 287-302.

77 J.F. Pachut and R.L. Anstey, 'A developmental explanation of stability-diversity-variation hypotheses: morphogenetic regulation in Ordovician bryozoan colonies', *Paleobiology*, 5(2) (1979), 168-187; R.L. Anstey, 'Astogeny and phylogeny: evolutionary heterochrony in Paleozoic bryozoans', *Paleobiology*, 13(1) (1987), 20-43; M.M. Key Jr. 'Partitioning of morphologic variation across stability gradients in Upper Ordovician trepostomes', in *Bryozoa: Present and Past*, edited by J.R.P. Ross. Western Washington University, Bellingham, Washington (1987), 145-152; J.F. Pachut, 'Population genetics of four species of Ordovician bryozoans: stereology and jackknifed analysis of variance', *Journal of Paleontology*, 61(5) (1987), 927-941; J.F. Pachut, 'Heritability and intraspecific heterochrony in Ordovician bryozoans from environments differing in diversity', *Journal of Paleontology*, 63(2) (1989), 182-194; J.F. Pachut, 'Morphological integration and covariance during astogeny of an Ordovician trepostome bryozoan from communities of different diversities', *Journal of Paleontology*, 66(5) (1992), 751-757.

78 G.D. Brown Jr. and E.J. Daly, 'Evolutionary trends in *Parvohallopora* Singh in the Dillsboro Formation (Upper Ordovician) of southeastern Indiana', in *Recent and Fossil Bryozoa*, edited by G.P. Larwood and C. Nielsen. Olsen and Olsen, Fredensborg (1981), 29-38; G.D. Brown Jr. and E.J. Daly, 'Analysis of evolution of structural characters of *Parvohallopora* Singh from the Dillsboro Formation, Cincinnatian Series, Indiana', in *Bryozoa, Ordovician to Recent*, edited by C. Nielsen and G.P. Larwood. Olsen & Olsen, Fredensborg (1985), 51-58.

79 O.L. Karklins, 'Bryozoan biostratigraphy of the upper Middle Ordovician and Upper Ordovician of Kentucky', in *Bryozoa, Ordovician to Recent*, edited by C. Nielsen and G.P. Larwood. Olsen & Olsen, Fredensborg (1985), 346-347.

80 R.L. Anstey and S.F. Rabbio, 'Regional bryozoan biostratigraphy and taphonomy of the Edenian stratotype (Kope Formation, Cincinnati area): graphic correlation and gradient analysis', *Palaios*, 4(6) (1989), 574-584.

81 S.C. Diekmeyer, *Quantitative analysis of cyclical faunal patterns in the Upper Ordovician Kope through Bellevue sequence, Cincinnati, Ohio.* University of Cincinnati, unpublished M.S. thesis (1990), 168 pp., 15 pls., 33 text-figs; D.H. Errett and R.J. Cuffey, 'Bryozoan mode of occurrence relative to individual beds in the type Cincinnatian (Upper Ordovician; Ohio, Indiana, Kentucky)', *Geological Society of America Abstracts with Programs*, 21(4) (1989), 11.

82 M.A. Wilson, T.J. Palmer and P.D. Taylor, 'Earliest preservation of soft-bodied fossils by epibiont bioimmuration; Upper Ordovician of Kentucky', *Lethaia* 27(3) (1994), 269-270.

83 G.C. Baird, C.E. Brett and R.C. Frey, '"Hitchhiking" epizoans on orthoconic cephalopods: preliminary review of the evidence and its implications', *Senckenbergiana Lethaea*, 69(5/6) (1989), 439-465, 4 pls.

84 M.A. Wilson, 'Disturbance and Ecologic Succession in an Upper Ordovician Cobble-Dwelling Hardground Fauna', *Science*, 228(4699) (1985), 575-577.

85 C.G. Dively and R.J. Cuffey, '*Monticulipora molesta* - Intraspecific? variability (Bryozoa, Upper Ordovician, southeastern Indiana)', *Geological Society of America Abstracts with Programs*, 26(3) (1994), 15.

86 R.J. Cuffey, 'The Maysville bryozoan reef mounds in the Grant Lake Limestone (Upper Ordovician) of north-central Kentucky', *Ohio Division of Geological Survey, Guidebook*, 13 (1998), 38-44.

87 F.W. Harris and W.D. Martin, 'Benthic community development in limestone beds of the Waynesville (upper Dillsboro) Formation (Cincinnatian Series, Upper Ordovician) of southeastern Indiana', *Journal of Sedimentary Petrology*, 49(4) (1979), 1295-1306.

88 J.R.P. Ross and C.A. Ross, 'Bryozoan evolution and dispersal and Paleozoic sea-level fluctuations', in *Bryozoans In Space and Time*, edited by D.P. Gordon, A.M. Smith and J.A. Grant-Mackie. National Institute of Water and Atmospheric Research, Wellington, New Zealand (1996), 243-258.

89 K.L. Butler and R.J. Cuffey, 'Reduced bryozoan diversity and paleoenvironmental stress in the Saluda Dolomite (uppermost Ordovician, southeastern Indiana)', in *Bryozoans In Space and Time*, edited by D.P. Gordon, A.M. Smith and J.A. Grant-Mackie. National Institute of Water and Atmospheric Research, Wellington, New Zealand (1996), 55-61.

90 R.L. Anstey, S.F. Rabbio and M.E. Tuckey, 'Bryozoan bathymetric gradients within a Late Ordovician epeiric sea', *Paleoceanography*, 2(2) (1987), 165-176; G.D. Brown Jr. and E.J. Daly, 'Trepostome bryozoan distributions in the Dillsboro Formation (Cincinnatian Series), Indiana: water depth controls?', in *Bryozoa: Present and Past*, edited by J.R.P. Ross. Bellingham, Washington, Western Washington University (1987), 41-47.

91 J.R.P. Ross 'Biogeography of Ordovician ectoproct (bryozoan) faunas', in *Bryozoa, Ordovician to Recent*, edited by C. Nielsen and G.P. Larwood. Olsen & Olsen, Fredensborg (1985), 265-271; R.L. Anstey, 'Bryozoan provinces and patterns of generic evolution and extinction in the Late Ordovician of North America', *Lethaia*, 19(1) (1986), 33-51; R.L. Anstey and T.L. Chase, 'Geographic diversity of Late Ordovician corals and bryozoans in North America', *Journal of*

Paleontology, 48(6) (1974), 1141-1148.

92 M.E. Tuckey and R.L. Anstey, 'Late Ordovician extinctions of bryozoans', *Lethaia*, 25(1) (1992), 111-117.

93 H. Milne-Edwards and J. Haime, 'Monographie des polypiers fossiles des Terrains Palaeozoïques, précédée d'un général de la classification des Polypes', *Archives du Muséum d'Histoire Naturelle, Paris*, 5 (1851), 1-504, 20 pls.

94 D.E. Hattin, 'Memorial to Thomas Gregory Perry, 1919-1972', *Geological Society of America Memorials*, 4 (1975), 154-159.

The French pre-Lamarckian bryozoologists

Jean-Loup d'Hondt
Laboratoire de Biologie des Invertébrés Marins et Malacologie, Muséum National d'Histoire Naturelle, 57, rue Cuvier, F-75231 Paris Cedex 05, France

1. Introduction
2. Guillaume Rondelet (1507-1566)
3. Jacques Barrelier (1606-1673)
4. Joseph Pitton de Tournefort (1635-1708)
5. Sébastien Vaillant (1689-1722)
6. René-Antoine Ferchault de Réaumur (1683-1757)
7. Bernard de Jussieu (1691-1777)
8. Jean-Etienne Guettard (1715-1786)
9. Louis-Augustin-Guillaume Bosc (1759-1828)
10. Marie-Jules-César Delorgne de Savigny (1772-1841)
11. Charles-Alexandre Lesueur (1778-1846)

1. Introduction

Rondelet discovered the bryozoans in 1555. The older authors, those working before the beginning of the 17th. century, considered the bryozoans to be animals. But for over a hundred years thereafter until the 18th century, most people except for Marsigli, considered these organisms to be algae, often corallines. The animal nature of the bryozoans was confirmed by Jussieu in 1742, and was then accepted by the majority of scientists, particularly Trembley, Rösel, Donati and Ellis. A total of 38 books by 34 authors were written about the bryozoans from Rondelet to Linnaeus, between 1555 and 1758. The publications on the Bryozoa, both books or in scientific papers, increased considerably after the work of Linnaeus. D'Hondt has prepared a critical study of the pre-linnean bryozoologists (nationalities, geographical origin, business, social background, lifespan, religions).[1]

The present paper concentrates on the French naturalists who worked on marine or fresh-water bryozoans between their discovery and the beginning of the nineteenth century. Thus, the study ends with the year (1816) in which the volume of Lamarck's *Histoire Naturelle des Animaux sans Vertêbres* was published (Lamarck's *Philosophie zoologique* was published in 1809).[2] Lamarck was the first scientist to publish a zoological monography on the invertebrates in France and in French. The systematics of the future Bryozoa (formerly a part of the "Zoophytes") was published after by Lamouroux in the same year in 1816[3] and was better than Lamarck's treatment. The main studies on the Bryozoa occurred infrequently;[4] for example, larvae of marine species were first observed by Lamarck in 1816[5] and again by Grant in 1827,[6] and the statoblasts of phylactolaemates by Réaumur in 1742[7] and by Rösel in 1746.[8] The polypide was first reported to be a parasitic or inquiline worm by Jussieu in 1742[9] and by Donati eight years later.[10] The complete functional anatomy of the polypide was established in 1826 by Henri Milne Edwards[11] and published in 1836 by Audouin and Milne Edwards.[12]

The 16 European authors who described, and sometimes illustrated, bryozoans at the time of the Renaissance, some of whom were included in the incomplete list published by Canu,[13] included only two Frenchmen. The work of the 8 other French workers on Bryozoology mentioned here was published between the beginning of the 18th. and the 19th centuries. The history of Bryozoology has been covered in reports by various authors including Pitt, Schopf and d'Hondt.[14]

2. Guillaume Rondelet (1507-1566)

Guillaume Rondelet (Figure 1) was born in Montpellier and died of food poisoning. He studied medicine in Paris and submitted his thesis in 1537. He travelled in the Netherlands and Italy where he visited Venice, Padua and Bologna. Rondelet was appointed professor of medicine at Montpellier in 1545 and became physician to the cardinal of Tournon (1489-1562), who was an important political and diplomatic figure. In this role, he travelled with him, and became familiar with the marine environment and marine life. He slept little and stayed up late into the night studying. He was very fond of cakes and fruit; but believed that alcohol was bad for the health.

Figure 1. Guillaume Rondelet (1507-1566).

Rondelet described and illustrated the first known bryozoan species in 1555.[15] He used the name "Giroflade de mer" that was given by the local fishermen to these organisms; comparing it to the beautiful flower "appelée en français oeillet" (*Dianthus*). This Mediterranean reteporiform species is commonly found by the fishermen along the Catalan and French coasts. He described its morphology, comparing it to a screen, fixed to the substratum by a tough "tail". Its red colour he assigned to a skin, although in fact the pigment is present throughout the zoarium. Rondelet was one of the rare older authors to

Figure 2. Jacques Barrelier [© Bibliothèque centrale M.N.H.N. Paris].

indicate the nature of the substratum, to which the animals were fixed: rocks and pieces of floating wood. Rondelet believed that the Bryozoa were animals and he included them in the Pisces (the general name for all the marine animals at that time), but placed them in the subdivision "Zoophytes". Rondelet's book, formerly in Latin, was published later in 1558 in French.[16]

Rondelet's species, according to its zoarial morphology and its colour, was assigned to *Sertella septentrionalis* (Harmer, 1934).[17] This later species was recognised by Reverter Gil and Fernandez Pulpeiro in 1999 to be the same as *Reteporella grimaldii* (Jullien, in Jullien and Calvet, 1903) which is its senior synonym.[18]

3. Jacques Barrelier (1606-1673)

Jacques Barrelier (Figure 2) was born in Paris and studied medicine, but he did not finish his thesis to become a Dominican priest. He was a highly cultured man and was under the protection of the Duke Gaston d'Orléans (1608-1660), the brother of king Louis XIII. He began to do research in theology and, as hobby, in botany and mycology. He was appointed assistant to the heads of the Dominican Order, Father Turco and Father Martin in 1846. In this position, he travelled with them in Provence, Languedoc and Italy, and founded a botanical garden in Rome. He came back to Paris in 1672, where he died of suffocation, after an asthma attack. After his death, his manuscripts were burned, but his engraved copperplates (334 plates, 1312 figures) were saved. They were published by Antoine de Jussieu (1686-1758) who was Professor of Botany in the King's botanical garden (the future Museum).

In 1714 Barrelier described an alga which he called *Muscus coralloides bifidus minor* (p. 121 and Figure 1275, VIII).[19] This is a zoarium of clearly a *Cellaria* species. He emphasized the bushy shape of the colony, its branching and the white colour of the branches. He also illustrated a second bryozoan (Figure 1275, VII) with an articulated zoarium, formed by a succession of cylindrical internodes, inflated in their central region, fixed to the substratum by rhizoids: this is a *Margaretta* (Ascophorina). There is a large species of *Margaretta* , *M. cereoides* (Ellis and Solander, 1786) in the Mediterranean, but its zoarial branching is not dichotomically terminal, but is lateral to the internode in a subterminal position, and both the branches on an internode are a little unwedged, giving the zoarium the appearance of a candelabra. One of Barrelier's drawings shows an arrangement that had been described previously. This species is very probably *M. cereoides*. Barrelier was, after the Italian naturalist Boccone (1633-1704), one of the first authors to have described and drawn a Cellariidae and a Margarettidae; although he believing them to be plants.

4. Joseph Pitton de Tournefort (1635-1708)

Tournefort was born in Aix-en-Provence into a comfortable, genteel family of the lords of Tournefort. He was destined to be entered into a religious order by his family, and studied in a Jesuit college, where he soon became interested in botany. However, he abandoned the Church and collected plants throughout his life in numerous European countries (Southwest France, Catalonia, the Pyrenees, Spain to Andalusia, Portugal and Great Britain). One of his best friends was Pierre Magnol (1638-1714), the botanist and creator of the taxonomic level the "Family". Tournefort reorganized the classification of plants and was appointed Professor in the "Jardin du Roy" and became a member of the French Academy of Sciences.[20] He was known to scientists throughout Europe. Unfortunately he was knocked down and killed by a cart outside the Jardin, when he was carrying a herbarium of plants back to study in his laboratory.

Tournefort in 1700 worked on two species of bryozoans, which he called *Eschara* and believed to be coralline algae.[21] This name "*Eschara*" was later given to many highly calcified organisms. He considered the first of them to be the same organism as the one illustrated by Rondelet in 1555,[22] Imperato in 1599,[23] and Jean Bauhin in 1651[24] (It is now considered to be a member of the Phidoloporidae). The second (now a *Frondipora*) he identified as the *Eschara* species described by Jean Bauhin in 1651,[25] the *Frondipora* and *Eschara* species from Boccone's work of 1674,[26] and Imperato's *Frondipora*.[27] Tournefort described another, unidentifiable species under the name of *Eschara maxima pilaeformis*. He believed that all three species were algae.

5. Sébastien Vaillant (1689-1722)

Sébastien Vaillant (Figure 4) was born at Vigny, near Paris, into a modest family. He was interested in botany and music from a very early age and was very gifted in both. He became organist at the Benedictine church of Pontoise, near Paris, when he was only eleven years old. While he was working in the charitable hospital in the city, he studied books of anatomy and medicine and became a surgeon, and started with this position in the army during foreign wars. On his return to Paris, he was appointed first as demonstrator, then as Professor, in the Royal Garden (the future Museum of Natural History), and was elected to the French Academy of Sciences in 1716.[28] He travelled widely, particularly in Britanny and Normandy, for 14 years. After his death, his herbarium was purchased by the French King Louis XV, and was later placed in the collection of the Royal Garden. Some of his scientific drawings were obtained by the Dutch naturalist Boerhave, in 1721, and were deposited in the Leiden Museum. The three great French botanists (Jussieu, Vaillant, Tournefort) of this period were all working independently of each other.

The oldest specimens of Bryozoa preserved in the French national collection are from herbarium of Vaillant. The Vaillant specimens were collected in Britanny and Normandy between 1703 and 1707. The material became detached from the plates and was later put in jars and included in the general collection of recent Bryozoa (now in the Laboratoire de Biologie des invertébrés marins et Malacologie, dry specimens). They belonged to this collection, according to Nyèvre's catalogue,[29] before 1867 (some recent identification was perhaps given by Nyèvre). This material is now registered as:

LBIMM-BRY-4428: *Bugula fulva* (Ryland, 1960), determined d'Hondt, 2001. Labelled *Umbella corallina punctata similior*, Lipp. 82, by Vaillant. As *Bugula flabellata* Busk, Determined Nyèvre (1867). 4 colonies.

LBIMM-BRY-4477: *Crisia denticulata* (Lamarck, 1816). "Je l'ai trouvée en Normandie en 1707" (Vaillant). Determined Nyèvre. 1867 and d'Hondt, 2001. 1 colony.

LBIMM-BRY-4497: *Vesicularia spinosa* (Linné, 1767) and some specimens of *Electra pilosa* (Linné, 1767) var. *eucrateiformis* Norman, 1894, determined d'Hondt, 2001. Picardie, 1706. Labelled *Vesicularia spinosa* Johnston, by Nyèvre, 1867. 2 colonies.

LBIMM-BRY-4650: *Scrupocellaria reptans* (Linné, 1767), determined d'Hondt, 2001. Labelled *Corallina alba geniculata creberine conflata quasi vermiculare ammolatere dinticulata* N° 79 by Vaillant). As *Canda reptans* Busk, Determined Nyèvre, 1867.

Figure 3. Sébastien Vaillant [© Bibliothèque centrale M.N.H.N. Paris].

1 colony.

LBIMM-BRY-4737: *Eucratea loricata* (Linné, 1758), determined d'Hondt, 2001. Labelled *Gemellipora loricata* Busk, by Nyèvre, 1867. "Elle adhérait à la Sertulaire Argentée" (Vaillant). 1 colony.

LBIMM-BRY-4856: *Bugula stolonifera* Ryland, 1960, determined d'Hondt, 2001. *Corallina mollior, albida conflata quasi vermicularior sed uno latere denticulat.* Lipp. 147. (Vaillant). Labelled "Sertulaire Bicellaria avicularia Blainv." by Nyèvre (1867). 1 colony.

LBIMM-BRY-5245: *Scrupocellaria reptans* (Linné, 1767), determined d'Hondt, 2001.

Figure 4. One of the specimens collected by Sébastien Vaillant (Fécamp, Normandy, 1703), the oldest bryozoan in the French national collections.

Det. as "Bicellaire de Geoffroy-Blainv. *Canda reptans* Busk" by Nyèvre (1867). Labelled by mistake "Savigny Pl. II, fig. 4", because in fact designed on Savigny's plate XI, fig. 4.[30] 3 colonies.

LBIMM-BRY-5320 (Figure 4): *Flustra foliacea* (Linné, 1758), determined Nyèvre, 1867 and d'Hondt, 2001. "Fucus telam livream, sericeamve texturâ sua amulans . Ray. Hist. 71. Je l'ay ramassé en Normandie en 1707; de Fécamp. 1703" (Vaillant). By mistake (Vaillant) in synonymy with *Porus cervicornis* Imperato. 2 colonies.

The book by Vaillant published posthumously in 1727,[31] contains an illustration of an animal that was found in the Paris area, named *Corallo-fungus-candissimus*. This is probably an arborescent *Alcyonidium*; this may have been due to a mistake (exchange of labels), attributing a freshwater origin to a marine specimen. The Bryozoa in the Vaillant herbarium are the oldest specimens of bryozoans in the French collections of the Museum of Paris and amongst the oldest specimens of Recent Bryozoa now preserved.

6. René-Antoine Ferchault de Réaumur (1683-1757)

René-Antoine Ferchault de Réaumur was born at La Rochelle into a very rich family. His father, the lord of Réaumur, des Angles et de la Bermandière, was a lawyer. He was educated in a Jesuit college and then began his legal studies. He was also interested in geometry and went to Paris when he was 19 years old. He became a member of the French Academy of Sciences when he was very young, 24 years old, and he remained an active member for almost 50 years. He first worked as a physicist[32] and invented a thermometer with reference positions for water freezing and boiling (0° and 100°C). He later studied applied science (metallurgy) and natural sciences. He was interested in the formation of pearls and fish scales and in the reproduction of crustaceans. He published the 6 volumes of a considerable natural history of insects between 1634 and 1642. His life style was aristocratic and he lived on his estates. He died after falling from his horse.

Réaumur is remembered for having been the first French zoologist to have described in 1742 the lophophore of the bryozoans,[33] and to have observed the statoblasts of phylactolaemates. These were also reported by Rösel in 1746 from material collected by Guettard (who discovered the phylactolaemates in France).[34] Réaumur was a friend of Abraham Trembley and was also the first author to consider Bryozoa to be animals, after Rondelet and similarily as Gesner (1516-1565). This was after a long period during which they were considered to be plants.

7. Bernard de Jussieu (1691-1777)

Jussieu was born in Paris and belonged to a very famous family of French naturalists.[35] He was a student in a Jesuit college in Lyon and came to Paris to join his brother who was Professor of Botany in the King Garden. He then travelled with him to collect plants in Spain and Portugal. On his return, he studied medicine in Lyon and succeeded Sébastien Vaillant as demonstrator in botany. He welcomed Linnaeus to the King's Garden in 1738 when he visited Paris. He was Professor of Buffon, and became a member of the French Academy of Sciences when he was 26 years old. At that time he managed the royal gardens of Trianon and Versailles for King Louis XV. He observed polypides in 1742,[36] but like the Italian marine biologist Vitaliano Donati (1713-1763) at the same period, he considered them to be inquiline worms living in the cavities of the zoarium.

8. Jean-Etienne Guettard (1715-1786)

Jean-Etienne Guettard was born in Etampes and was the son of an apothecary. He was a physician in the faculty of medicine in Paris. He was a very active naturalist, geologist, zoologist and botanist, and a friend of Bernard de Jussieu. In addition, he collected biological material for Réaumur, and became curator of the private collections of the Duke Louis d'Orléans and of his son Louis Philippe Joseph, the nephew of the French King Louis XIV. He was an active field botanist,[37] and discovered the first phylactolaemates collected in France, near Paris. He then gave his material to Réaumur, who (as indicated previously) had himself described the animals. Guettard was a very active man with a very high degree of scientific training, but he was only a collector and an observer of bryozoans; he has never published on these animals.

9. Louis-Augustin-Guillaume Bosc (1759-1828)

Our next scientist, Louis-Augustin-Guillaume Bosc was born at Paris, and was the son of the physician to the French King Louis XVI. He worked for the royal mail, but his hobbies were entomology and botany. As he was familiar with the literary fashionable circles of the day and a friend of many of the politicians implicated in the French Revolution, particularly Jean-Marie Roland, Home Secretary, he was a victim of the rivalities between the various political clans at that time.[38] He was dismissed by Maximilien Robespierre, the leader of the French Revolution, and went into hiding in a monastery, in a forest west of Paris, near Versailles, until the end of the socio-political confusion.[39] However he had new private and professional problems and went to the United States. There he was appointed the French Consul in New York in 1798 and 1799. He later returned to France, and became inspector of the French gardens. In 1825, he was appointed Professor of Botany in the Muséum national d'Histoire naturelle de Paris.

In his *Histoire Naturelle des Vers*, published in 1802,[40] Bosc described and illustrated 47 species from 6 genera of bryozoans, most of them belonging to known species and genera ("*Sertularia*", *Cellaria*, *Flustra*, *Cellepora*, *Alcyonidium*, *Cristatella*), but expanded and completed the descriptions of Ellis and Solander.[41] Another genus, *Eschara*, was added in 1803 in his *Nouveau Dictionnaire d'Histoire Naturelle*.[42] He described and illustrated *Flustra tuberculata*, now *Jellyella tuberculata* (Bosc, 1802), the first living bryozoan species described by a French author. He found this species on algae floating near the European and American coasts of the Atlantic.

10. Marie-Jules-César Delorgne de Savigny (1772-1841)

The ninth of our naturalists was born in Provins (90 km from Paris) into a family of provincial nobility. He was educated in religious colleges, where he read stories about the great travellers and first became interested in botany. He continued his studies, and he met some famous scientists, including Duméril , Lamarck, Cuvier and Etienne Geoffroy-Saint-Hilaire. He was appointed to act as zoologist (although, he was a botanist !), and as a member of the "Commission des Sciences et Arts", with Etienne Geoffroy-Saint-Hilaire. This was part of a scientific military campaign organized by Napoléon Bonaparte in Egypt between 1799 and 1802. There he built up very rich collections on various organisms, including about 80 species of Bryozoa.[43] He came back to France when he was 24 years old and prepared engravings of his Bryozoa between 1805 and 1812 (Figure 5). But in 1817 he rapidly developed an nervous illness and by 1822 he was extremely sensitive to light.[44] He was no longer able to study his biological material and consequently was unable to publish the descriptions of his new species. A young investigator, entomologist and later marine biologist,[45] Jean-Victor Audouin (1797-1841), was recommended by Cuvier to carry out a systematic study of the collection. Audouin had no help from Savigny nor did he have access to the specimens, and worked exclusively from

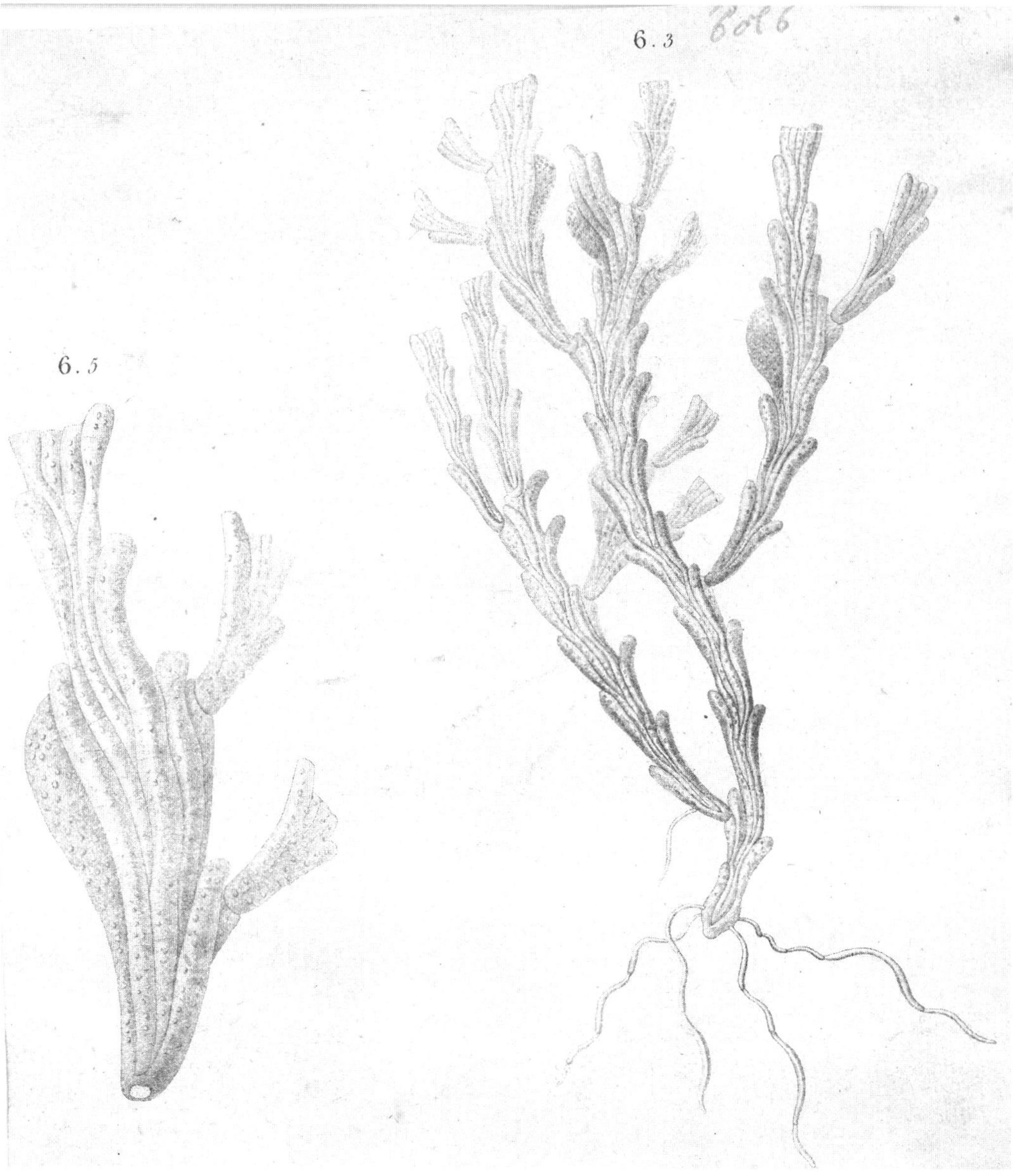

Figure 5. A plate by Savigny.

the engraved plates. Savigny lived in almost permanent darkness from 1822 until his death, and emerged into the light only with his face covered by a thick black cape. Thus Savigny was only, because his illness, the collector of bryozoological material that was later studied by a colleague. But he was the author of the original illustrations of the animals.[46]

Figure 6. An example of the engraved plates by Lesueur.

11. Charles-Alexandre Lesueur (1778-1846)

Charles-Alexandre Lesueur was born in Le Havre and was the son of a celebrated artist. He was appointed petty officer and gunner in the sea-going expedition of the *Géographe* and the *Naturaliste*.[47] This expedition was under the command of Baudin and was ordered by Napoléon Bonaparte to explore the southern regions (1800-1804). He became a friend of one of the zoologists on the campaign, François Péron (1775-1810) and became also the expedition's illustrator.[48] Péron became seriously ill before the end of the expedition, and he died early. In 1816, Lesueur went to the United States on various scientific positions,[49] coming back only in 1837, and then travelled in Great Britain and along the Mediterranean coasts. He was appointed curator of the Museum of Natural History at le Havre and held this post until his death. His own collection of specimens and the corresponding plates[50] were purchased by the city and given to the Museum. Now his collection of Bryozoa (93 specimens) is preserved in the Paris Muséum national d'Histoire naturelle, in the collection of the Laboratoire de Biologie des Invertébrés marins et Malacologie.[51] Péron and Lesueur never described their collections, and this material was studied and named by Lamarck in 1816 and Lamouroux in the same year.[52] Before his journey to United States, Lesueur engraved 14 plates of bryozoans (Figure 6), but only a few copies were made; many of the sets have been lost. Some of them are preserved in Muséum du Havre, the Muséum national d'Histoire naturelle de Paris (Central Library), the Natural History Museum (London), the Museum of Maastricht, and in the library of the author (personal gift of the Louis Calvet's family). Unfortunately for Lesueur, only specific names, without generic labels, were engraved, only near some figures. Consequently, these specific identifications must be considered to be *nomina nuda*, and most of the specimens have not been identified. This material was revised by d'Hondt in 1979, 1993 and 1997.[53] Lesueur was, therefore, only a collector and illustrator, but did not describe the species he collected.

Notes

1 Hondt, J.-L. d', 'La littérature pré-linnéenne sur les Bryozoaires : les auteurs et leurs apports scientifiques', *Bull. Hist. Epistém. Sc. Vie.* (in press, a)

2 J.B. Lamarck, *Philosophie zoologie*, I. Dentu, Paris (1809) (Reed. 1994, Flammarion: 718 pp); J.-B.P.A. de Monet de Lamarck, *Histoire naturelle des Animaux sans Vertêbres*. T. 2. Verdière, Paris (1816), 568 pp.

3 J.V.F. Lamouroux, *Histoire des Polypiers Coralligènes Flexibles, vulgairement nommés Zoophytes*. F. Poisson, Caen (1816), 560 pp.

4 J.-L. d'Hondt, 'Les phases de la connaissance des larves et de la métamorphose larvaire des Bryozoaires', in *L'observation dans les Sciences*. Editions du CTHS, Paris (2001), 223-245.

5 Lamarck, 1816, note 2. Larvae of phylactolaemates were first described by G.J. Allman in 1856. See G.J. Allman, - *A Monograph of the fresh-water Polyzoa.* Ray Society, London (1856), 119 pp., 11 pl.

6 R.E. Grant, Observation on the structure and nature of *Flustrae . Edinburgh New Philos. Journ.*, 3 (1827),107-118; 337-342.

7 A.F. de Réaumur, *Mémoires pour servir à l'Histoire des Insectes*. 6 volumes, Paris (1742).

8 A.J. Rösel von Rosenhof, *Documents entomologiques* (Reed. 1988, Ed. Mazenod-Citadelles, Paris) (1746-1761), 288 pl.

9 B. de Jussieu, Examen de quelques productions marines qui ont été mises au nombre des plantes et qui sont l'ouvrage'une sorte d'Insecte de mer. *Mém. Acad. Roy. Sc.*, (1742) 290-302.

10 V. Donati, *Essai sur l'Histoire Naturelle de la Mer Adriatique* (Original 1750; French translation, 1758). Pierre de Hondt, La Haye, 73 pp., 10 pl.

11 H. Milne-Edwards, 'Recherches anatomiques, physiologiques et zoologiques sur les Eschares', *Ann. Sc. zool.*, 2(6) (1836), 5-53.

12 J.V. Audouin and H. Milne Edwards, 'Résumé des recherches sur les animaux sans vertêbres, faites aux îles Chausay', *Ann. Sc. Nat.*, 15(14) (1836), 1-19.

13 F. Canu, 'Bibliographie primitive relative aux Bryozoaires', *Bull. Soc. géol. Fr.*, 4th series, 14(1916), 287-292.

14 L.J. Pitt, 'An Account of the Works on Polyzoa', *Proceedings of the Geologists' Association*, 72(2) (1961), 171-186; T.J.M. Schopf, 'The Literature of the Phylum Ectoprocta (1555-1963)', *Systematic Zoology*, 16(4) (1967), 318-327; J.-L. d'Hondt, 'Contributions des chercheurs français à la connaissance des Bryozoaires actuels', *Histoire et Nature*, 26-27 (1985 [1988]), 3-26; J.-L. d'Hondt, 'Origine des collections de Bryozoaires actuels du Muséum National d'Histoire Naturelle de Paris', *Histoire et Nature*, 30 (1993), 55-61; J.-L. d'Hondt, 'Histoire et chronologie des recherches sur les Bryozoaires', *Bull. Hist. Epistém. Sc. Vie*, 2(1) (1996), 98-111; J.-L. d'Hondt, 'Aux sources de l'invention taxinomique: le cas des Bryozoaires Eurystomes actuels', in *Les sciences et leurs langages*, Editions du CTHS, Paris (2000), 157-176; Hondt, d', note 1; J.-L. d'Hondt, 'Bryozoaires', in *Histoire naturelle.* Nouvelles explications des planches de la "Description de l'Egypte". Institut d'Egypte, Paris, Edition Impériale (in press, b).

15 G. Rondelet, *Universae aquatilum historiae*. Mathias Bonhomme, Lyon (1555), 240 pp.

16 G. Rondelet, *De Natura aquatilium carmen*. Mathias Bonhome, Lyon, 2e partie (1558*a*), 181 pp.; G. Rondelet, *L'histoire entière des poissons*. Macé Bonhome, Lyon, 2e partie (1558*b*), 181 pp.

17 Hondt, d' note 1.

18 O. Reverter Gil and E. Fernandez Pulpeiro, 'Some little-known species of Bryozoa described by J. Jullien', *Journal of Natural History*, 33 (1999), 1403-1418.

19 J. Barrelier, *Icones plantarum per Galliam, Hispaniam et Italiam observatae*. Stephanum Ganeau, via Jaccobea, Paris (1714), 148 pp., 1626 figs., index.

20 G. Becker, H. Bianchi, C. Carri ère, J.-P. Coste, P.J. Dufert, R. Duguy, G. Duprat, H. Gaussen, P. Guiral, H. Humbert, J.-F. Leroy, F. Meyer, J. Motte and G. Ranson, *Tournefort*. Muséum national d'Histoire naturelle, Paris (1957), 321 pp.

21 J.P. de Tournefort, *Institutiones Rei Herbariae*. I. Typographia Regia, Paris (1700), 695 pp.

22 Rondelet, note 15.

23 F. Imperato, *Dell'Historia Naturale*. Constantino Vitale, Porta Reale, Neapel (1599), 791 pp.

24 J. Bauhin, *Historia plantarum universalis*. III. Ebroduni (1651), 882 pp.

25 Bauhin, note 24.

26 P. Boccone, *Icones et descriptiones rariorum plantarum Siciliae, Melitae, Galliae, et Italiae. Quarum unaquæque proprio charactere signata, ab aliis ejusdem classis facile distinguitur.* (Oxford, 1674), 96 pp.

27 Imperato, note 23.

28 J. Rousseau, 'Sebastien Vaillant an oustanding 18th century botanist', *Regnum vegetabile*, 71 (1970), 195-228.

29 M. Nyèvre, *Catalogue détaillé des collections du Muséum National d'Histoire Naturelle, Bryozoaires*. (1867), 104 pp. (ms).

30 Hondt, d' (in press, b), note 14

31 S. Vaillant, *Botanica parisiense ou dénombrement par ordre aalphabétique des plantes qui se trouvent aux environs de Paris.* Jean and Herman Verbeek & Balthazar Lakeman, Leiden & Amsterdam (1727), 205 pp., 32 pl.
32 J. Torlais *et al.*, 'Réaumur (1683-1757)', *Rev. Hist. Sc. Appl.*, 11(1) (1958), 1-84.
33 Réaumur, note 7.
34 Rösel, note 8.
35 L. Figuier, 'Bernard de Jussieu', in *Les quatre Jussieu.* Hachette, Paris (1870), 327-339.
36 Jussieu, note 9.
37 A. de Soland, *Etude sur Guettard.* Imprimerie Laîné Frères, Angers (1873), 57 pp.
38 G. Cuvier, *Eloge historique de M. Bosc.* Institut Royal de France, Imp. Firmin Didot, Paris (1829), 28 pp.
39 A. Rey, *Le Naturaliste Bosc et les Girondins à Saint-Prix.* Imprimerie A. Pâris, Pontoise (1882), 45 pp; M. Creiner (coord.), (1989).- *L.A.G. Bosc d'Antic.* Sides, Paris (1989), 191 pp.
40 L. Bosc, *Histoire naturelle des Vers, III.* Guilleminet, Paris (1st edition (1802); 2nd edition (1830)), 212 pp.
41 J.-L. d'Hondt, 'Louis-Augustin-Guillaume Bosc, conventionnel et naturaliste, premier systématicien français des Bryozoaires actuels', in *Scientifiques et Sociétés.* Editions du CTHS, Paris (1989), 241-258.
42 L. Bosc, *Nouveau Dictionnaire d'Histoire Naturelle appliqué aux Arts.* I. Déterville, Paris (1803), 512 pp.
43 J.C. Savigny, *Description de l'Egypte, ou recueil des observations et des recherches qui ont été faites en Egypte pendant l'expédition de l'Armée française, publié par ordre du gouvernement. Histoire Naturelle, Planches. Tome deuxième. Polypes.* Imprimerie Royale, Paris (1817).
44 P. Pallary, 'Marie-Jules-César Savigny. Sa vie et son oeuvre. Première partie. La vie de Savigny', *Mém. Inst. d'Egypte* , Le Caire (1931). 1-111.
45 J. Théodoridès, 'Les débuts de la biologie marine en France: Jean-Victor Audouin et Henri-Milne Edwards, 1826-1829', *Bull. Inst. Océanogr. Monaco*, N° Special (1968), 417-437; J. Théodoridès, *Un zoologiste de l'époque romantique: Jean-Victor Audouin (1797-1841).* Mém. 6, Editions du CTHS, Section des Sciences, Paris (1978), 128 pp.
46 Hondt, d' (in press, b), note 15.
47 F. Baudin, *Mon voyage aux Terres Australes.* J. Bonnemains (coord.), Imprimerie Nationale (Paris) & Ville du Havre (2000), 467 pp.
48 Y. Laissus, 'Un naturaliste exemplaire: François Péron, 1775-1810', *Bull. Soc. Emulation Bourbonnais, Lettres, Sciences, Arts*, 58 (1976), 68-80; Y. Laissus, 'Un naturaliste exemplaire: François Péron, 1775-1810', *Annales du Muséum du Havre,* 12 (1978), 1-8.
49 G. Breton, 'Charles-Alexandre Lesueur (1778-1846). L'art au service des Sciences Naturelles', *Ann. Mus. Havre*, 8 (1977), 1-12.
50 G. Ord, 'A memoir of Charles-Alexandre Lesueur', *The Amer. Journ. Science and Arts*, 8 (1849), 189-216; J. Bonnemains *et al.*, 'Charles-Alexandre Lesueur (1778-1846), peintre, voyageur, naturaliste havrais', *Bulletin trimestriel de la Société géologique de Normandie et des Amis du Muséum du Havre*, Suppl. LXV (1) (1978), 1-23; J. Bonnemains, 'Les artistes du "Voyage aux Terres Australes" (1800-1804): Charles-Alexandre Lasueur et Nicolas-Martin Petit', *Bulletin trimestriel de la Société géologique de Normandie et des Amis du Muséum du Havre*, 76(1) (1989), 11-55; J. Goy, *Les méduses de François Péron et de Charles-Alexandre Lesueur.* Editions du CTHS , Paris (1995), 395 pp.
51 J.-L. d'Hondt, 'Révision des Bryozoaires de Lesueur et Péron conservés dans les collections

du Muséum National d'Histoire Naturelle de Paris', *Bulletin trimestriel de la Société géologique de Normandie et des Amis du Muséum du Havre*, 66(3) (1979), 9-24.

52 Lamarck, 1816, note 2.

53 Lamouroux, note 3.

54 Hondt, d', note 15; Hondt, d', 1993, note 14; J.-L. d'Hondt, 'Apports de Lamarck dans la connaissance des Bryozoaires', in *Jean-Baptiste Lamarck (1744-1829)* edited by G. Laurent, Editions du CTHS, Paris (1994 [1997]), 287-314.

History of investigations and current state of knowledge of bryozoan species diversity in the Bering Sea

Andrei V. Grischenko

Systematics and Evolution, Division of Biological Sciences, Graduate School of Science, Hokkaido University, Sapporo, 060–0810, Japan

1. Introduction

The Bryozoa are one of the most abundant and widely distributed groups of macrobenthos in the Bering Sea. Although investigations of the phylum have taken place over a century, knowledge of species diversity in this sea is still very incomplete. The coastal waters of the Bering Sea belong territorially to Russia and the United States of America and, accordingly, study of the bryofauna has been achieved generally by the efforts of the Russian and American bryozoological schools. For a number of reasons, their investigations were conducted independently and, because the investigators identified specimens collected within their "national" sea areas, species occurring in the eastern and southeastern shelves of the sea were generally studied by American scientists and those in western coastal waters by Russians. Therefore thc history of bryozoan investigations of the Bering Sea is most usefully presented according to the two lines of research.

2. Investigations of the American bryozoological school

The first reliable data about bryozoans in the Bering Sea were connected with biological investigations of the Alaskan shelf and reported by Alice Robertson.[1] She recorded three species – *Membranipora membranacea* (L.), *Bugula purpurotincta* (later changed to *B.*

pacifica, see below) and *Celleporina incrassata* (Lamarck), collected near the Pribilof Islands by the Harriman Alaska Expedition. Two years later a new ctenostome bryozoan, *Alcyonidium pedunculatum*, was described from the same area.[2] In further taxonomic work on Eastern Pacific cheilostomes Robertson[3] described two new species – *Bugula pacifica* and *Tricellaria pribilofi* (afterwards synonymised with *T. ternata* (Ellis and Solander),[4] based on new material from coastal waters off the Pribilof Islands. Two other species, *Dendrobeania murrayana* (Johnston) and *Terminoflustra membranaceotruncata* (Smitt), were newly recorded for the fauna of the Bering archipelago.[5] After reorganization of the Harriman Alaska Expedition into a division of Smithsonian Institution, the data on the Alaskan bryozoan fauna were formally repeated in the Harriman Alaska Series of the Smithsonian Institution.[6]

More than a decade later, Raymond Osburn[7] reported finding bryozoan colonies in the stomachs of eider ducks taken from nests in the Pribilof Islands. Fragments of *Crisia* sp. and *Celleporina incrassata* (Lamarck) were taken from the alimentary canals of king and pacific eiders (*Somateria spectabilis* and *S. nigra* respectively), caught near St Paul Island. *Tricellaria pribilofi* and *Myriapora subgracilis* (d'Orbigny) were included in an alimentary bolus of king eider from St George Island. Additionally, several colonies were found in the stomachs of the fishes *Tautoga onitis* and *Tautogolabrus adspersus* taken near the islands.

New data on Beringian bryozoans were published almost thirty years later. Nine anascans were indicated for the American sector of the Bering Sea in the monograph of western North American Bryozoa by Osburn.[8] Four of them, *Bidenkapia spitzbergensis* (Bidenkap), *Callopora craticula* (Alder), *Tegella arctica* (d'Orbigny), and *T. armifera* (Hincks), were registered near Punuk Island situated in the neighbourhood of St. Lawrence Island. *Callopora lineata* (L.) was found in shallow water off Nunivak Island; *B. pacifica* Robertson and *T. membranaceotruncata* (Smitt) were mentioned with reference to Robertson[9] for the Pribilof Islands. Disappointingly short inscriptions ("Bering Sea", without any more details) were on the labels accompanying specimens identified by Osburn as *Dendrobeania murrayana* (Johnston) and *Microporina articulata* (Fabricius).[10]

Seven ascophoran bryozoans were reported from the Bering Sea in the second part of his monograph.[11] Five species, *Arctonula arctica* (M. Sars), *Porella concinna* (Busk), *Stomachetosella cruenta* (Busk), *S. sinuosa* (Busk), and *Cylindroporella tubulosa* (Norman), recorded near Punuk Island, and two representatives of the genus *Celleporina* (*C. incrassata* (Lamarck) and *C. ventricosa* (Lorenz)) were recorded in the coastal waters of the Pribilof Islands and Nunivak Island accordingly. The concluding part of Osburn's monograph[12] contains records or indications of five cyclostomes and a single ctenostome bryozoan (*Alcyonidium pedunculatum* Robertson with reference to Robertson),[13] gathered in the Alaskan shelf of the Bering Sea. The cyclostomes included a new genus and species, *Bathysoecia hastingsae*, from Nunivak Island together with a couple of other cyclostomes,

Diplosolen obelium (Johnston) and *Diaperoecia johnstoni* (Heller), newly recorded for the region.

Taxonomic research on bryozoans of Beringian Alaskan waters was renewed in 1988 when ten species were described among the intertidal bryozoan fauna of the Narrow Strait, Kodiak Islands, Gulf of Alaska.[14] The additional material of *Callopora craticula* (Alder) from Punuk Island and of *Terminoflustra membranaceotruncata* (Smitt) and *Myriozoella crustacea* (Smitt) from the vicinity of the Pribilof Islands was examined. Material of seven other species was studied from the Beringian marine sector of Unalaska Island (Aleutian Islands). Six of them – *Stomachetosella sienna*, *Microporella germana*, *Celleporella reflexa*, *Hippothoa mawatarii*, *Hippoporidra truculenta*, and *Cryptosula okadai* (later recognized as a junior synonim of *C. zavjalovensis* Kubanin 1976, see Kubanin)[15,16] – were described as new to science.[17] These species, together with *Dendrobeania exilis*, supplemented the list of the previously known bryozoans of the Bering Sea.

A new species of *Microporella*, *M. speciosa*, has recently been described from the Pacific coast of Alaska, and some additional material of this species has also been examined from the shore of Nanvak Bay, Bering Sea coast of Alaska.[18]

Several cyclostome bryozoans, generally identified to genera or to family level and collected in Alaskan waters of the Bering Sea (mainly Unalaska to Cook Inlet) were illustrated in the work of Boardman.[19]

Some indications of the importance of bryozoans in the benthic communities of the Alaskan shelf of the Bering Sea are given in a few hydrobiological papers. Thus, in Lower Cook Inlet, Alaska, *Flustrellidra* sp. and *Dendrobeania* spp. were associated with the largest catches of the juvenile red king crab *Paralithodes camtschaticus* (Tilesius).[20] Four species common in Cook Inlet were mentioned by Kessler[21] – *Eucratea loricata* (L.), *Carbasea serrulata* (Busk), *Escharopsis lobata* (Lamouroux), and *Rhamphostomella costata* (Lorenz). A brief account of bryozoan diversity in the Beringian shelf of Alaska has also been given by MacIntosh.[22]

3. Investigations of the Russian bryozoological school

Perhaps the first investigator to note the abundance and appreciable role of bryozoans in the benthic communities in the Bering Sea was Evpraksiya F. Gurjanova, who gathered some rich collections from the intertidal zone of Bering Island in the Commander Islands.[23]

In the early 1950s, Herman Kluge (Figure 1) described six cheilostome species – *Tegella inermis*, *Callopora obesa*, *C. amissavicularis*, *Hippoponella fastigatoavicularis*, *Porella tumida*, *Smittina beringia* – and single cyclostome bryozoan, *Borgella tumulosa*,

Figure 1. Herman Kluge (photograph of 1949)

from the Bering Strait.[24] Shortly after, a representative of the abyssal fauna, *Kinetoskias* beringi, was described from 3812 m in the Commander Valley of the Bering Sea.[25] Twenty seven of 35 bryozoans included in the *Atlas of Invertebrates of the Far Eastern Seas of U.S.S.R.* were enumerated for the Bering Sea.[26]

The most detailed information about 135 Beringian bryozoans (comprising 13 cyclostomes, 14 ctenostomes, and 108 cheilostomes), accompanied by depth and locality data from more than twenty explored areas within the sea, was given in the *List of the species (Bryozoa) of the Far Eastern Seas of S.S.S.R.*, the posthumous work of Kluge.[27] In that publication, 74 species, together with newly described *Schizoporella stelloperforata* from the Avacha Gulf, were reported from the shelf zone of Eastern Kamchatka.

One hundred and six species, distributed mainly on the shelves of the Arctic Russian seas, are enumerated for the Bering Sea in the life-work of Kluge.[28] Taxonomic descriptions and the text-figures for two of them, the ctenostome bryozoans *Flustrellidra vegae* (Silén) and *F. cervicornis* (Robertson), were based on specimens from the Commander Islands region.

Androsova[29] cited *Kinetoskias beringi* Kluge (with reference to the early paper of Kluge)[30] for the Bering Sea basin in an account devoted to the abyssal fauna of the Pacific Ocean. In another generalization on cyclostomes of the family Heteroporidae Androsova[31] has indicated that the Bering sea is an area of distribution for two cerioporines, *Heteropora pelliculata* Waters and *Borgella tumulosa* Kluge.

Preliminary data on the intertidal bryozoan fauna of the western coast of the Bering Sea were published only in the late 1970s.[32] Eleven of the 16 species reported for the whole studied area came from the shores of the Commander Islands. Later, the intertidal bryozoan fauna of the Commander archipelago was chosen as the subject of my own investigations.[33,34] These publications gave ecological data about 65 species, including 59 species of Eurystomata. Results of studies on the intertidal bryozoan fauna of the coast of the Russian Far Eastern seas were summarized recently with more than 70 species recorded within the Bering Sea.[35] Eight of them were found along the tidal belt of the Chukchi Peninsula, 18 along Eastern Kamchatka, and about 70 species in the Commander Islands.

Gontar[36] provided information about nine rather common bryozoans as a small part of some preliminary results of floral, faunal, and hydrobiological investigations in the Avacha Inlet (Avacha Gulf of Eastern Kamchatka). Seven cheilostomes (including two species new to the fauna of the Bering Sea, *Schizobrachiella perforata* (Kluge) and *Stomachetosella hincksi* Powell), were reported for the Bering Strait in the contemporaneous paper of Gontar and Denisenko[37] describing the pattern of biogeographical distribution of Bryozoa in the Arctic basin.

Taxonomic work on Beringian bryozoans was continued by examination of three umbonulomorph species (i. e., having umbonuloid frontal shields), *Desmacystis sandalia* (Robertson), *Rhamphostomella scabra* (Fabricius), and *Umbonula arctica* (M. Sars) using material from the shallow water of the Commander Islands. Zooidal morphology and the distribution of zooidal polymorphs were given as reasons to transfer *U. arctica* to a new genus *Arctonula* gen. nov.[38]

The results of an inventory of different groups of macrobenthos of the Commander Islands has been published by Nauka Press in 1997 in the volume *Benthic Flora and Fauna of the Shelf Zone of the Commander Islands*. Nine ctenostomes and 132 cheilostome species (103 new to the fauna of the islands) were reported for the shelf zone of the archipelago.[39]

Some Bering Sea representatives of *Doryporella* – *D. spathulifera* (O'Donoghue and O'Donoghue) and *D. armata* Gontar – were recently used for taxonomic revision of this poorly known cheilostome genus.[40]

The visible role of bryozoans as permanent components of benthic communities of the Bering Sea has been mentioned in the hydrobiological literature. Thus, for example, Kuznetsov[41] has recorded the high frequency of bryozoans in many benthic communities of hard, soft, and mixed substrata of the Eastern Kamchatka shelf, though an overwhelming majority of the species indicated in his monograph were identified only provisionally.

Indications of the relatively high biomass and diversity of bryozoans in some intertidal biocoenoses of the Commander Islands were presented in the work of Tarakanova[42] for Bering Island and by Kussakin and Ivanova[43] for Mednyy Island. Bryozoans were recognized as subdominant forms in at least three benthic communities of hard substrata within the zone of encrusting calcareous red algae in the upper shelf of the Commander Islands.[44]

One or more species of *Alcyonidium* (Ctenostomata) have been cited among the most typical and frequent components of fouling communities on the ships cruising along the eastern coast of the Kamchatka Peninsula.[45, 46] According to Kubanin,[47] about 19 species of fouling bryozoans have been recorded on anthropogenical substrata within the Bering Sea. Also, at least eight cheilostome species have been introduced via shipping to ports and harbours in the region of Eastern Kamchatka.[48, 49] In some cases, *Alcyonidium* forms particularly dense encrustations in the ports of Avacha Inlet.[50] Some preliminary data on the role of bryozoans in some fouling shallow-water communities off Bering Island were presented by Oshurkov and Ivanjushina.[51, 52]

Associations between nudibranch molluscs and bryozoans in intertidal pools of rocky shores of Bering Island were described by Roginskaya.[53] The special importance of erect bryozoans as food items for the Beringian nudibranchs *Colga pacifica* (Bergh) and *Triopha catalinae* (Cooper) has been indicated in recent publications of Martynov[54] and Grischenko and Martynov,[55] and at least five species of decapod crustaceans that preferentially associate with bryozoans in shallow-water assemblages off the Commander Islands have been recorded.[56]

A history of investigations of marine bryozoans in the waters of the Kamchatka Peninsula and an account of taxonomic diversity in the regional fauna has been briefly described by Grischenko.[57]

4. Current knowledge on the bryozoans of the Bering Sea.

4.1 Total diversity

Historically and in recent years several attempts have been made to estimate the total bryozoan diversity of the Russian Far Eastern seas[58] and also the diversity of the separate seas of the region. Nevertheless, these inventories have only concerned the Sea of Okhotsk [59, 60, 61] and the Japan Sea.[62, 63] Although two accounts of the bryozoan fauna of the Bering Sea have been recently published discussing the shelves of both Eastern Kamchatka[64] and Alaska,[65] the total taxonomic diversity of the sea – coast, shelves, basin – has not previously been estimated. Based on published data, there are 223 species including 22 cyclostomes, 18 ctenostomes, and 183 cheilostomes (Table 1, Figure 2).

4.2 Regional diversity

It will be appreciated from the above chronicle of historical and recent investigations that bryozoan studies have been adhoc and piecemeal, so that some areas have been explored more intensively than others.

Eastern and Southeastern Bering Sea: It has been truly mentioned by Dick and Ross[66] that "bryozoans have been scantily studied in Alaska...", since only 46 species were reported in previously published papers dealing with the fauna of the Beringian shelf of Alaska (including 7 cyclostomes, 7 ctenostomes, and 32 cheilostomes). Twenty-four species were recorded near Unalaska Island, 24 species in shallow water off the Pribilof Islands, five species near Nunivak Island, and one species, *Carbasea serrulata* (Busk), off St Matthew Island.

Western and Northwestern Bering Sea: Our knowledge of bryozoans in the western and northwestern parts of the Bering Sea is also still rather incomplete. Fifty-eight species are known for the Bering Strait, 62 for the coast of the Chukchi Peninsula, and 89 for the Eastern Kamchatka Peninsula.[67] The most intensively studied region is the Commander Islands. Thus the known diversity of Bryozoa recorded from the western part of the Sea is about 90 species, from the north/northwestern parts of the Sea about 106 species, and 144 species for the Commander Islands.

5. Discussion

In spite of a long history of investigation, the Bryozoa of the Bering Sea have been studied only irregularly and our knowledge of the diversity and distribution of the total fauna is still incomplete. At the same time, there are several reasons to suspect a very rich bryofauna.

First, the Bering Sea is subdivided topographically into two highly contrasting areas – a shallow north/northeastern shelf (about 43.7% of the bottom area) with depths less than 200 m, and deep-water south/southwestern basins – the Aleutian and Commander Valleys (about 43.3% of the bottom area) – with depths over 2000 m. The boundary between these two main areas is the abrupt continental slope (about 13% of area of the seafloor) with depths 200–2000 m.[68]

Secondly, the Bering Sea is relatively open. The wide north/northeastern shelf of the sea narrowly connects at several places with the North Pacific Ocean via the straits of the Aleutian Islands in the South, with the Chukchi Sea via the Bering Strait in the North, and with the narrow shelf of Eastern Kamchatka in the West. The circulation, transport, and exchange of water mass in the Bering Sea[69-74] occurs mainly via the Bering Strait and also via the straits of the Aleutian ridge and along Eastern Kamchatka.

Thirdly, the Bering Sea is connected to some areas with high bryozoan diversity, i.e., to the Kuril Islands in the south-west (where more than 210 species have been recorded – see Gontar[75-85] and to the Gulf of Alaska in the south-east (with noticeably rich fauna described by Osburn;[86, 87, 88] Cuffey and Turner;[89] Dick and Ross.[90] The transitional setting of the sea and its general environmental conditions are conducive for an intensive exchange of varied biogeographical elements (of Pacific Asian, American, and Arctic origin) and hence for enrichment of the fauna. Thus, in accordance with some unpublished data (kindly supplied by Ms Nora R. Foster, the Aquatic Collection, University of Alaska, Fairbanks, pers. comm. 1993) there are at least 43 additional unrecorded species of cheilostomes distributed on the Alaskan shelf of the Bering Sea. Thirty-two of them occur along the Aleutian Islands chain, 19 in the Bristol Gulf, 12 in Norton Sound, and eight in other less well-known localities. Taking these figures into account, the total bryozoan diversity in the Beringian shelf of Alaska may be as high as 90 species.[91] Further, preliminary identification of specimens (deposited in the Kamchatka Institute of Ecology and Nature Management, Petropavlovsk-Kamchatsky) collected by RV *Nazarovsk* along the Eastern Kamchatka shelf in 1988 has demonstrated that a diverse bryofauna occurs in the area with a significant number of undescribed taxa. Moreover, 38 cheilostomes (14 anascans and 24 ascophorans) were found recently in the shelf and slope of the Commander Islands (unpublished data of the author). Hence, the cteno-cheilostome fauna of the archipelago comprises more than 175 species. Even taking into consideration both the published and unpublished data mentioned above, the total diversity of Bryozoa in the western part of the Bering Sea is likely to increase significantly, especially if more thorough sampling and detailed analysis are carried out.

Migration of Pacific temperate biogeographical elements of Asian and American origin through the Bering Sea may have been reasonably intensive, during the existence of the Beringia, until the Late Miocene[92] prior to the opening of the Bering Strait. Current knowledge is inadequate to reconstruct species distributions and the origin of the fauna in the region, though some paleontological data from both Asian[93, 94] and American[95] coasts of the North Pacific obliquely suggest a significant "crossroads" role for the Bering Sea in relation to bryozoan diversity. As has been mentioned by many authors,[96, 97] shelf and slope zones of the Commander-Aleutian Ridge constitute a "bridge" connecting the American and Asian continents, supporting the distribution of both American and Asian faunal elements (including bryozoans) at the present day. For example, at least 15 species previously considered to the Eastern Pacific[98] – *Callopora nuda* Dick and Ross, *C. decidua* Dick and Ross, *Dendrobeania exilis* (Hincks), *Porella columbiana* (O'Donoghue and O'Donoghue), *Hippoporina vulgaris* Dick and Ross, *Celleporina robertsoniae* (Canu and Bassler), etc. – were recently found in coastal waters of the Commander Islands.[99]

Inadequate sampling of the bryofauna in the Bering Sea does not allow objective comparisons among different parts of the sea. Further, many of the reported species require taxonomic revision and not enough ancillary data are reported in the relatively little literature to allow environmental and other comparisons. In short, we are still in the discovery phase where the Bering Sea bryofauna is concerned and more work needs to be done. Finally, as is well known, many parts of the Bering Sea are subject to intensive fisheries for teleosts and king crabs. As in many parts of the world, such activities are accompanied by impacts on seafloor habitats and benthic species abundance. It has been pointed out by Nelson and Gordon: "There is no doubt that seafloor habitats are being modified and their species negatively affected faster than we can inventory them".[100] Thorough inventories of coastal and shelf assemblages everywhere, including the Bering Sea, are desirable in order to make reliable within- and between-assemblage comparisons of bryozoans and other benthos – a worthy goal of the international bryozoological community.

6. Acknowledgements

I wish to express the appreciation to my supervisor, Professor Shunsuke F. Mawatari, for his valuable suggestions during preparation of the paper. I am also indebted to Dr Dennis P. Gordon (National Institute of Water and Atmospheric Research, Wellington) for critical reading of the original manuscript.

Notes

1 A. Robertson, 'Papers from the Harriman Alaska Expedition, 6, The Bryozoa', *Proceedings of the Washington Academy of Science*, 2 (1900), 315–40, pls 19–21.
2 A. Robertson, 'Some observations on *Ascorhiza occidentalis* Fewkes, and related Alcyonidia', *Proceedings of the California Academy of Science, Zoology*, 3 (1902), 99–108, pl. 14.
3 A. Robertson, 'Non-incrusting chilostomatous Bryozoa of the West Coast of North America', *University of California Publications in Zoology*, 2 (1905), 235–322, pls 4–16.
4 R.C. Osburn, 'Bryozoa of the Pacific Coast of America, Part 1, Cheilostomata-Anasca', *Allan Hancock Pacific Expeditions*, 14(1) (1950), 1–269.
5 Robertson, note 3.
6 A. Robertson, 'The Bryozoa of the Expedition', *Harriman Alaska Series*, 11 (1910), 221–43 (315–35), pls 23–5.
7 R.C. Osburn, 'Bryozoa as food for other animals', *Science*, n.s., 53 (1376) (1921), 451–3.
8 Osburn, note 4.
9 Robertson, note 3.
10 Osburn, note 4, pp. 106 and 165.
11 R.C. Osburn, 'Bryozoa of the Pacific Coast of America, Part 2, Cheilostomata-Ascophora', *Allan Hancock Pacific Expedition*, 14(2) (1952), 271–611.
12 R.C. Osburn, 'Bryozoa of the Pacific Coast of America, Part 3, Cyclostomata, Ctenostomata, Entoprocta and addenda', *Allan Hancock Pacific Expedition*, 14(3) (1953), 613–841.
13 Robertson, note 2.
14 M.H. Dick and J.R.P. Ross, *Intertidal Bryozoa (Cheilostomata) of the Kodiak vicinity, Alaska* (Center for Pacific Northwest Studies, Western Washington University Press, Bellingham, Occasional Paper No 23, 1988), 1–133.

15 A.A. Kubanin, 'Intertidal Bryozoa of Zav'yalov Island (Sea of Okhotsk)', *Biologiya Morya*, 1 (1976), 30–5. (In Russian with English summary)
16 A.A. Kubanin, 'Phylum Tentaculata, subphylum Bryozoa', in *A check-list of animals, plants and fungi from the intertidal zone of Far Eastern seas of Russia*, edited by O.G. Kussakin, M.B. Ivanova, A.P. Tsurpalo (Vladivostok, Dalnauka Press, 1997), 119–25. (In Russian)
17 Dick and Ross, note 14.
18 T. Suwa, M.H. Dick and S.F. Mawatari, 'A new species of *Microporella* (Bryozoa, Cheilostomata) from Alaska', *Zoological Science (Zoological Society of Japan)*, 15 (1998), 589–92.
19 R.S. Boardman, 'Reflections on the morphology, anatomy, evolution, and classification of the Class Stenolaemata (Bryozoa)', *Smithsonian Contributions to Paleobiology*, 86 (1998), 1–59.
20 K.A. Sundberg and D. Clausen, 'Post-larval king crab (*Paralithodes camtschatica*) distribution and abundance in Kachemak Bay, Lower Cook Inlet, Alaska, 1976', in *Environmental studies of Kachemak Bay and Lower Cook Inlet*, edited by L. Trasky, L. Flagg, D. Burbank (Alaska Department of Fish and Game, Anchorage, Alaska, 5, 1977), 36.
21 D.W. Kessler, 'Alaska's saltwater fishes and other sea life: a field guide', *Alaska Northwest Publications* (Anchorage, Alaska, 1985), 1–358.
22 R. MacIntosh, 'Bryozoans', in *Habitat Areas of Particular Concern Eastern Bering Sea Invertebrates* (Compiled by members of the Alaska Fishery Science Center's Resource Assessment and Conservation Engineering Division, 2000). http://www.afsc.noaa.gov/groundfish HAPC/Bryozoans_synopsis.htm
23 E.F. Gurjanova, 'The Commander Islands and its marine coastal flora and fauna', *Priroda*, 11 (1935), 64–72. (In Russian)
24 H.A. Kluge, 'Bryozoa of the Chukchi Sea and the Bering Strait', *Far North-East of the USSR, 2, The Fauna and Flora of the Chukchi Sea* (Moscow, Leningrad, Academy of Sciences of the USSR Press, 1952), 138–68. (In Russian)
25 H.A. Kluge, 'On the distribution of the bryozoan genus *Kinetoskias* Danielsson', *Trudy Zoologicheskogo Instituta Akademii Nauk SSSR*, 13 (1953a), 210–15. (In Russian)
26 H.A. Kluge, 'The Bryozoa', *Atlas of Invertebrates of the Far Eastern Seas of the USSR* (Moscow, Leningrad, Academy of Sciences of the USSR Press, 1955), 99–109. (In Russian)
27 H.A. Kluge, 'Species list of Bryozoa of the far eastern seas of the USSR', *Explorations of the Far Eastern Seas of the USSR*, 7 (1961), 118–43. (In Russian)
28 H.A. Kluge, 'Bryozoa of the northern seas of the USSR', *Opredeliteli po Faune SSSR*, 76 (1962), 1–584. (In Russian)
29 E.I. Androsova, 'Bryozoa', *Deep-water benthic fauna, Biology of the Pacific Ocean*, 2 (Moscow, Nauka Press, 1968), 90–1. (In Russian)
30 Kluge, note 25.
31 E.I. Androsova, 'A contribution to the biology and distribution of the ancient family Heteroporidae (Bryozoa, Cyclostomata)', *Zoologicheskii Zhurnal*, 50 (1971), 1341–46. (In Russian with English summary)
32 E.A. Izyumova and A.A. Kubanin, 'Phylum Podaxonia, Class Bryozoa', in A list of the animal species from the intertidal zone of the eastern Kamchatka and western coast of Bering Sea, compiled by O.G. Kussakin, *et al.*, *The Intertidal Zone of the Bering Sea and Southeastern Kamchatka* (Moscow, Nauka Press, 1978), 170–1. (In Russian)
33 A.V. Grischenko, 'Eurystomatous Bryozoa of the intertidal zone of the Commander Islands', Unpublished Diploma Paper, Department of Invertebrate Biology and Aquatic Ecology, Biological Faculty (Perm State University, Perm, 1993), 1–84. (In Russian)

34 A.V. Grischenko, 'Some ecology peculiarities of the Commodore Islands intertidal Eurystome Bryozoa (Ctenostomida, Cheilostomida)', Abstract of report, All-Russian and International Bryozoological Conference *Fossil and Living Bryozoa of the Glob*, edited by V.P. Ozhgibesov (Perm State University, Perm, 1994), 21–2.
35 Kubanin, note 16.
36 V.I. Gontar, 'Phylum Bryozoa, The list of macrophytes and invertebrates of macrobenthos of the Avacha Inlet', *Hydrobiological Explorations in the Avacha Inlet* (Vladivostok, Far Eastern Division of the Academy of Sciences of USSR, 1989), 113. (In Russian)
37 V.I. Gontar and N.V. Denisenko, 'Arctic Ocean Bryozoa', in *The Arctic Seas, Climatology, Oceanography, Geology, and Biology*, edited by Y. Herman (Van Nostrand Reinhold Company, New York, 1989), 341–71.
38 D.P. Gordon and A.V. Grischenko, 'Bryozoan frontal shields: the type species of *Desmacystis, Rhamphostomella, Rhamphosmittina, Rhamphostomellina*, and new genus *Arctonula*', *Zoologica Scripta*, 23 (1) (1994), 61–72.
39 A.V. Grischenko, 'Bryozoans (Ctenostomida, Cheilostomida) of the Commander Islands shelf zone'. *Benthic Flora and Fauna of the Shelf Zone of the Commander Islands* (Vladivostok, Dalnauka Press, 1997), 153–92. (In Russian with English summary)
40 A.V. Grischenko, S.F. Mawatari and P.D. Taylor, 'Systematics and phylogeny of the cheilostome bryozoan *Doryporella*', *Zoologica Scripta*, 29 (2000), 247–64.
41 A.P. Kuznetsov, *Fauna of benthic invertebrates of Kamchatkan waters of the Pacific Ocean and of Northern Kuril Islands* (Moscow, Academy of Sciences of the USSR Press, 1963), 1–271. (In Russian)
42 T.F. Tarakanova, 'The quantitative distribution of intertidal macrobenthos of the Bering Island', *The Intertidal Zone of the Bering Sea and South-Eastern Kamchatka* (Moscow, Nauka Press, 1978), 63–77. (in Russian with English summary)
43 O.G. Kussakin and M.B. Ivanova, 'Macrobenthos of the intertidal communities of Mednyy Island (the Commander Islands)', *Biologiya Morya*, 21 (1995), 99–107. (In Russian with English summary)
44 E.A. Ivanjushina, A.V. Rzhavsky, O.N. Selivanova and V.V. Oshurkov, 'Structure and distribution of benthic communities of the shallow-water zones of the Commander Islands', *Natural resourses of the Commander Islands (reserves, current state, problems of the protection and use)*, "The Commandors" programme, 2 (Moscow State University Press, 1991), 155–70. (In Russian with English summary)
45 N.A. Rudyakova, 'The fouling of ships cruising in Far Eastern seas north of the Japan Sea', *Trudy Instituta Okeanologii*, 85 (1967), 3–17. (In Russian)
46 N.A. Rudyakova, *Fouling in the northwestern part of the Pacific Ocean* (Moscow, Nauka Press, 1981), 1–68. (In Russian)
47 A.A. Kubanin, 'Geographical distribution of marine-fouling bryozoans', *Ecology of Fouling in the North-western Part of the Pacific Ocean* (Vladivostok, Far Eastern Science Centre, Academy of Sciences of the USSR, 1980), 109–40. (In Russian)
48 A.A. Kubanin, 'Marine Bryozoa (Class Gymnolaemata) and their participation in ship-fouling in the Northwestern Part of the Japan Sea', Synopsis of thesis, Candidate in Biological Sciences (Vladivostok, 1984a), 1–23. (In Russian)
49 A.A. Kubanin, 'Marine Bryozoa (Class Gymnolaemata) and their participation in ship-fouling in the Northwestern Part of the Japan Sea', Unpublished Ph.D. thesis (Vladivostok, 1984b), 1–240. (In Russian)

50 A.A. Kubanin, 'Marine bryozoans in the fouling communities of Far Eastern seas of USSR, Abstract of report, *7th All-Russian Colloquium on Fossil and Modern Bryozoa* (Moscow, 1986), 24–6. (In Russian)
51 V.V. Oshurkov and E.A. Ivanyushina, 'Effect of experimental harvesting on kelp regrowth and on the structure of the shallow-water communities of Bering Island (Commander Islands)', *Asian Marine Biology*, 10 (1993), 95–108.
52 V.V. Oshurkov and E.A. Ivanyushina, 'Structure and distribution of some fouling communities off Bering Island (Commander Islands)', *Biofouling*, 8 (1994), 35–45.
53 I.S. Roginskaya, 'Phases in the distribution of nudibranch molluscs in the *Thalassiophyllum clathrus* biocenosis in intertidal pools of Bering Island (The Commander Islands), Abstract of report', *14th Pacific Science Congress*, Khabarovsk, Committee F (Marine Science), Section F2 (Marine Biology), Subsection F2a (Biology of the Shelf) (Moscow, Nauka Press, 1979), 65–6. (In Russian)
54 A.V. Martynov, 'Opisthbranch gastropods at the Commander Islands with remarks on the fauna of the Russian Far Eastern Seas', *Benthic Flora and Fauna of the Shelf Zone of the Commander Islands* (Vladivostok, Dalnauka Press, 1997), 230–41. (In Russian with English summary)
55 A.V. Grischenko and A.V. Martynov, 'Bryozoa as food items for genera *Colga* and *Triopha* (Nudibranchia, Doridacea), Abstract of report', Russian and International Bryozoan Conference *Bryozoa of the World*, edited by A.A. Dobrovolsky, P.D. Taylor (Zoological Institute Russian Academy of Science and Saint Petersburg State University, Saint Petersburg, 1997), 13.
56 E.A. Ivanjushina, 'Decapods of the upper shelf zone of the Commander Islands', *Benthic Flora and Fauna of the Shelf Zone of the Commander Islands* (Vladivostok, Dalnauka Press, 1997), 193–206. (In Russian with English summary)
57 A.V. Grischenko, 'History of studies and current state of knowledge of Bryozoan species diversity in the coastal waters of Kamchatka Peninsula', *Conservation of biodiversity of Kamchatka and coastal waters*, edited by R.S. Moiseev, Section 1, History of scientific studies on biodiversity of Kamchatka and its current state, Materials of regional scientific conference (Petropavlovsk-Kamchatsky, 2000), 23–6. (In Russian)
58 Kluge, note 27.
59 H.A. Kluge, 'Class Bryozoa'. *The Fauna of the Sea of Okhotsk and the Conditions of its Existence*, edited by P.V. Uschakov (Leningrad, Academy of Sciences of the USSR Press, 1953b), 169–79. (In Russian)
60 E.I. Androsova, 'Bryozoa of the northeastern part of the Pacific Ocean, Abstract of report', *Conference on Fauna and Flora Combined Researches* (Leningrad, Zoological Institute, Academy of Sciences of the USSR, 1962), 1–6. (In Russian)
61 A.V. Grischenko, D.P. Gordon and P.D. Taylor, 'A unique new genus of cheilostomate bryozoan with reversed-polarity zooidal budding', *Asian Marine Biology*, 15 (1999), 105–17 (for 1998)
62 E.I. Androsova, Bryozoans of the order Cheilostomata of the northern part of the Japan Sea', *Explorations of the Far Eastern Seas of the USSR*, 5 (1958), 90–205. (In Russian)
63 E.I. Androsova, Bryozoans of the orders Cyclostomata and Ctenostomata of the northern part of the Sea of Japan', *Fauna of the seas of north-west Pacific*, *Explorations of the Fauna of the Seas*, 3 (1965), 72–114. (In Russian)
64 Grischenko, note 57.
65 MacIntosh, note 22.
66 M.H. Dick and J.R.P. Ross, 'Intertidal cheilostome bryozoans in rock-pile habitat at Narrow

Strait, Kodiak, Alaska', in *Bryozoa: Ordovician to Recent*, edited by C. Nielsen and G.P. Larwood (Fredensborg, Olsen and Olsen, 1985), 87–93.
67 Grischenko, note 57.
68 A.K. Leonov, *The Regional Oceanography, Part 1, Bering, Okhotsk, Japan, Caspian and Black Seas* (Leningrad, Hydrometeorology Press, 1960). (In Russian)
69 F. Favorite, 'Flow into the Bering Sea through Aleutian island passes', *Oceanography of the Bering Sea*, edited by D.W. Hood and E.J. Kelley (University of Alaska, Fairbanks, 1974), 3–37.
70 D. Hood, 'The Bering Sea', *Estuaries and Enclosed Seas*, edited by B.H. Ketchum (Amsterdam e. a., Elsevier, 1983), 337–73.
71 F.M. Hughes, L.K. Coachman and K. Aagaard, 'Circulation, transport and water exchange in the western Bering Sea', *Oceanography of the Bering Sea*, edited by D.W. Hood, E.J. Kelley (University of Alaska, Fairbanks, 1974), 59–98.
72 K. Ohtani, 'Oceanographic structure in the Bering Sea', *Memoires of the Faculty of Fishery of the Hokkaido University*, 21 (1973), 65–105.
73 K. Ohtani, Y. Akibi and A.Y. Takenouti, 'Formation of Western Subarctic water in the Bering Sea', *Biological oceanography of the northern North Pacific Ocean*, edited by A.Y. Takenouti, *et al.*, (Tokyo, Idemitsu-Shoten, 1972), 31–44.
74 A.Y. Takenouti and K. Ohtani, 'Currents and water masses in the Bering Sea: a review of Japanese work', *Oceanography of the Bering Sea*, edited by D.W. Hood, E.J. Kelley (University of Alaska, Fairbanks, 1974), 39–57.
75 V.I. Gontar, 'Bryozoa of the Kuril Islands upper shelf', *Conformities of Distribution and Ecology of Coastal Water Biocoenoses* (Leningrad, 1978a), 60–2. (In Russian)
76 V.I. Gontar, 'Bryozoa of the Iturup Island coastal water', *Biologiya Morya*, 1 (1978b), 10–16. (In Russian)
77 V.I. Gontar, 'Cheilostomates of the upper shelf of the Kuril Islands, Abstract of report', *5th All-Russian Colloquium on Fossil and Modern Bryozoa* (Khabarovsk, 1979a), 17–19. (In Russian)
78 V.I. Gontar, 'Bryozoa of the order Cheilostomata in shallow water off Simushir Island', *Biology of the Kuril Islands Shelf* (Moscow, Nauka Press, 1979b), 234–48. (In Russian)
79 V.I. Gontar, 'Bryozoan fauna of the order Cheilostomata of the Kuril Islands coastal water', Synopsis of thesis, Candidate in Biological Sciences (Leningrad, 1980), 1–23. (In Russian)
80 V.I. Gontar, 'On the Cheilostomata (Bryozoa) of the Kuril Islands', in *Recent and Fossil Bryozoa*, edited by G.P. Larwood, C. Nielsen (Fredensborg, Olsen and Olsen, 1981), 101–3.
81 V.I. Gontar, 'New species of the order Cheilostomata (Bryozoa) from the region of the Kuril Islands', *Zoologicheskii Zhurnal*, 61 (1982), 543–53. (In Russian, with English summary)
82 V.I. Gontar, 'New species of Cheilostomida from the Kuril Islands (Bryozoa)', *Zoosystematica Rossica*, 1 (1992), 9–12.
83 V.I. Gontar, 'Phylum Bryozoa. List of invertebrate species of the continental slope of the Kuril Islands', *The Fauna of the Continental Slope of the Kuril Islands, Exploration of the Fauna of the Seas*, 46 (1993a), 200–3. (In Russian)
84 V.I. Gontar, 'New deepwater species of Cheilostomatida from the Kuril Islands and the Pacific Ocean (Bryozoa)', *Zoosystematica Rossica*, 2 (1993b), 41–5.
85 V.I. Gontar, 'Cyclostomida fauna of the upper shelf zone of the Kuril Islands', Abstract of report, Russian and International Bryozoan Conference *Bryozoa of the World*, edited by A.A. Dobrovolsky and P.D. Taylor (Zoological Institute Russian Academy of Science and Saint Petersburg State University, Saint Petersburg, 1997), 42.

86 Osburn, note 4.
87 Osburn, note 11.
88 Osburn, note 12.
89 R.J. Cuffey and R.F. Turner, 'Modern bryozoans on the Kodiak shelf, Southern Alaska', in *Bryozoa: Present and Past*, edited by J.R.P. Ross (Western Washington University, Bellingham, 1987), 65–72.
90 Dick and Ross, note 14.
91 MacIntosh, note 22.
92 G.A. Evseev and E.V. Krasnov, 'The role of Beringia in the formation and migrations of marine invertebrate faunas', *Beringia in the Cenozoic*, Materials of All-Russian symposium (Vladivostok, 1976), 43–53. (In Russian with English summary)
93 T. Hayami, 'Miocene Bryozoa from southwest Hokkaido, Japan', *Transactions Proc. Palaeontological Society of Japan, n.s.*, 7 (1970), 316–36.
94 T. Hayami, 'Neogene Bryozoa from northern Japan', *Scientific Reports Tohoku University, ser. 2 (Geology)*, 45 (2) (1975), 83–126.
95 J.D. Soule and M.M. Duff, 'Fossil Bryozoa from the Pleistocene of southern California', *Proceedings of the California Academy of Science ser.*, 4 (29) (1957), 87–146.
96 V.O. Mokievsky, 'Marine invertebrates and economic development of the shallow-water zone of the Commander Islands', *Rational resources management on the Commander Islands (status and protection of natural communities, problems of economic and ethno-cultural development)* (Moscow State University Press, 1987), 108–13. (In Russian)
97 Ivanjushina, Rzhavsky, Selivanova and Oshurkov, note 44.
98 Dick and Ross, note 14.
99 Grischenko, note 39.
100 W.A. Nelson and D.P. Gordon, 'Assessing New Zealand's marine biological diversity – a challenge for policy makers and systematists', *New Zealand Science Review*, 54 (1997), 58–66.

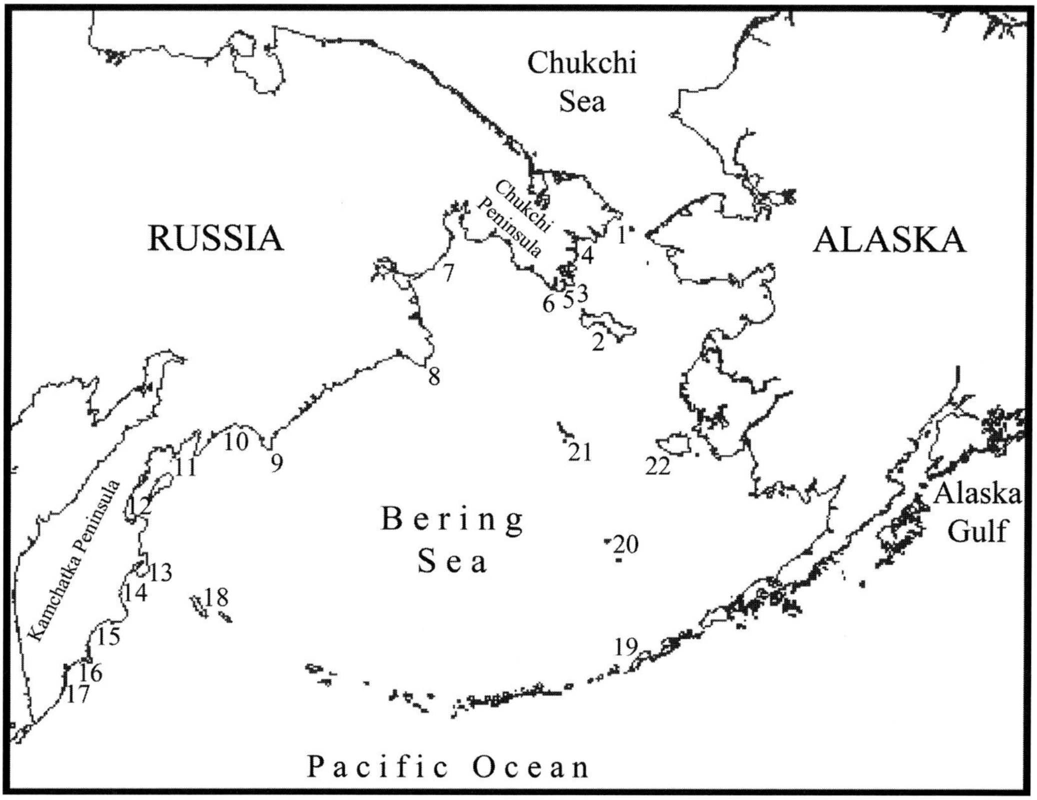

Figure 2. Map showing examined localities within the Bering Sea. The numbers of localities corresponds to Table 1.

Table1. Species composition and distribution of Bryozoa in the Bering Sea (compiled on the basis of literature sources).

*Numbers for examined localities: 1. Bering Strait; 2. St Lawrence Island; 3. Between St Lawrence Island and Chukotskiy Cape; 4. Lavrentiya Bay; 5. Chukotskiy Cape; 6. Chaplin Cape; 7. Gulf of Anadyr; 8. Navarin Cape; 9. Olyutorskiy Cape; 10. Olyutorskiy Gulf; 11. Korfa Gulf; 12. Karaginskiy Gulf; 13. Africa Cape; 14. Kamchatskiy Gulf; 15. Kronotskiy Gulf; 16. Avacha Gulf; 17. Avacha Inlet; 18. Commander Islands; 19. Unalaska Island; 20. Pribilof Islands; 21. St Matthew Island; 22. Nunivak Island; 23. Unknown localization. * The Kronotsky and Avacha Gulfs are accepted here as belonging to the Bering Sea.*

Taxa	**Examined localities within the Bering Sea**																						
	1	2	3	4	5	6	7	8	9	10	11	12	13	14	15	16	17	18	19	20	21	22	23
Order Cyclostomata																							
Crisia eburnea (L.)																		•					
Crisia eburneodenticulata Smitt																•							
Crisia cornuta (L.)						•																	
Crisia sp.																		•		•			
Oncousoecia polygonalis (Kluge)	•		•																				
Tubulipora liliacea (Pallas)																		•					
Bathysoecia hastingsae Osburn																						•	
Idmidronea fenestrata (Busk)		•						•															
Diplosolen obelium (Johnston)																						•	
Diplosolen obelium arctica Waters																•			•				
Entalophoroecia capitata (Rob)																		•	•	•			
Mesenteripora meandrina (S. Wood)			•																				
Diaperoecia johnstoni (Heller)																						•	
Fasciculiporoides americana (d'O)							•																
Heteropora pelliculata (Waters)																							•
Borgella tumulosa Kluge	•																						
Borgiola sp.																	•						
Discocytis sp.																			•				
Disporella hispida (Fabricius)																							•
Disporella alaskensis Osburn																							•
Disporella crassiuscula Smitt	•															•							
Patinella verrucaria (Fabricius)															•	•		•	•	•			
Order Ctenostomata																							
Alcyonidium gelatinosum (L.)	•																•	•					
Alcyonidium mytili Dalyell	•						•			•	•	•		•	•	•	•	•	•				
Alcyonidium vermiculare Okada	•																						
Alcyonidium mamillatum Alder	•													•	•	•		•					
Alcyonidium erectum Andersson							•																
Alcyonidium hirsutum (Fleming)														•	•	•							
Alcyonidium disciforme Smitt	•																						
Alcyonidium pedunculatum Rob.																				•			
Flustrellidra hispida (Fabricius)															•			•	•	•			
Flustrellidra corniculata (Smitt)																	•	•	•				
Flustrellidra cervicornis (Robertson)	•															•		•		•			
Flustrellidra gigantea (Silén)	•			•			•							•	•	•		•					
Flustrellidra vegae (Silén)														•	•	•		•	•				
Flustrellidra kurilensis (Mawatari)																		•					

	1	2	3	4	5	6	7	8	9	10	11	12	13	14	15	16	17	18	19	20	21	22	23
Nolella dilatata (Hincks)																							•
Bowerbankia arctica Busk							•																
Bowerbankia composita Kluge	•																						
Bowerbankia imbricata (Adams)																			•				
Order Cheilostomata																							
Membranipora membranacea (L.)																				•			
Eucratea loricata (L.)	•	•		•		•	•	•		•	•	•			•	•							
Electra arctica (Borg)	•																						
Electra baltica (Borg)														•	•	•							
Electra crustulenta (Pallas)															•			•					
Bidenkapia spitzbergensis (Bid.)	•	•	•	•	•	•	•	•								•	•	•					
Callopora craticula (Alder)		•	•													•		•					
Callopora decidua Dick and Ross																		•					
Callopora lamellata Androsova																		•					
Callopora lineata (L.)		•				•	•							•	•	•		•				•	
Callopora longispinosa Androsova																		•					
Callopora minuscula (Hincks)																		•					
Callopora nuda Dick and Ross																		•					
Callopora obesa Kluge	•																						
Callopora sp.																		•					
Callopora whiteavesi Norman																		•					
Cauloramphus cymbaeformis (Hin.)		•				•								•		•		•		•			
Cauloramphus disjunctus (C & B)																		•					
Cauloramphus echinus (Hincks)										•	•	•			•	•							
Cauloramphus intermedius Kluge														•	•	•		•					
Cauloramphus magnus Dick & Ross																		•					
Cauloramphus spiniferum (Johnst')				•		•	•	•				•			•	•		•					
Doryporella spathulifera (Smitt)	•															•							
Doryporella armata Gontar												•						•					
Doryporella alcicornis (O'D)																		•					
Ellisina marginulata Gontar																		•					
Tegella amissavicularis Kluge	•																						
Tegella anguloavicularis Kluge	•	•					•	•															
Tegella aquilirostris (O'Donoghue)																		•					
Tegella arctica (d'Orbigny)	•	•								•	•	•	•		•	•		•					
Tegella armifera (Hincks)	•	•		•			•		•	•	•	•			•	•	•	•					
Tegella horrida (Hincks)																		•					
Tegella inermis Kluge	•												•					•					
Tegella japonica (Canu and Bassler)																		•					
Tegella nigrans (Hincks)														•	•	•							
Tegella retroversa (Kluge)	•		•		•		•	•	•						•		•	•					
Tegella unicornis (Fleming)							•																
Amphiblestrum quadratum H	•																						
Flustra foliacea (L.)																		•					
Flustra securifrons (Pallas)	•	•													•								
Terminoflustra membranaceotruncata (Smitt)							•											•		•			
Carbasea carbasea (Ellis & S)	•	•			•	•	•	•															
Carbasea serrulata (Busk)	•	•		•			•	•									•				•		
Carbasea nordenskjoldi (Kluge)							•	•															
Hincksina longiavicularia Gontar																		•					
Bugula pacifica (Robertson)																		•		•			
Corynoporella tenuis Hincks								•	•							•		•					
Dendrobeania lichenoides (Robert')																		•					
Dendrobeania exilis (Hincks)																		•	•				
Dendrobeania murrayana (Johnston)	•								•									•		•			
Dendrobeania pseudomurrayana K																		•					

	1	2	3	4	5	6	7	8	9	10	11	12	13	14	15	16	17	18	19	20	21	22	23
Dendrobeania tenuis (Kluge)																		•					
Dendrobeania fessa (Kluge)	•		•					•	•							•		•					
Dendrobeania fruticosa (Packard)	•																	•					
Dendrobeania quadridentata (Loven)	•					•																	
Dendrobeania curvirostrata (Robertson)																		•					
Dendrobeania flustroides (Levinsen)							•	•															
Dendrobeania levinseni (Kluge)		•					•																
Dendrobeania pseudolevinseni Kluge																		•					
Caberea ellisi (Fleming)								•					•					•					
Bicellarina alderi (Busk)						•																	
Kinetoskias beringi Kluge																							•
Scrupocellaria elongata (Smitt)		•			•		•	•	•						•			•		•			
Scrupocellaria scabra (Van Beneden)								•								•		•					
Scrupocellaria praenulata (Norman)	•		•		•								•					•					
Scrupocellaria orientalis (Kluge)																							•
Semibugula birulai Kluge	•																	•					
Tricellaria ternata (Ellis and Solander)		•					•	•	•					•	•	•		•		•			
Tricellaria gracilis (Van Beneden)					•			•								•				•			
Tricellaria peachi (Busk)							•											•					
Tricellaria beringia (Kluge)	•							•					•					•					
Beania vegae Silén																		•					
Beania columbiana O'Donoghue													•					•					
Microporina articulata (Fabricius)																		•					•
Micropora coriacea (Johnston)																		•					
Euritina arctica Osburn																		•					
Cribrilina annulata (Fabricius)		•												•	•	•		•					
Cribrilina spitzbergensis Norman	•																						
Membraniporella solida Kluge																		•					
Celleporella hyalina (L.)	•	•	•	•	•	•	•		•	•	•	•	•	•	•	•	•	•	•	•			
Celleporella reflexa Dick and Ross																		•	•				
Hippothoa mawatarii Dick and Ross																		•	•				
Hippothoa expansa Dawson																		•					
Hippothoa arctica (Kluge)			•										•					•					
Lepraliella contigua (Smitt)	•																	•					
Harmeria scutulata (Busk)	•	•																					
Arctonula arctica (M. Sars)		•	•			•	•	•						•	•	•		•					
Arctonula kunashiri Gontar																		•					
Escharopsis lobata (Lamouroux)					•			•	•				•			•		•					
Ragionula rosacea (Busk)																		•					
Desmacystis sandalia (Robertson)																		•					
Desmacystis sandalia concinna Gontar																		•					
Porella alba Nordgaard														•	•	•		•					
?*Porella minuta* (Norman)														•	•	•							
Porella acutirostris Smitt		•	•	•		•	•		•				•			•		•					
Porella concinna (Busk)	•	•	•										•			•		•		•			
Porella belli (Dawson)	•						•									•		•					
Porella immersa Mawatari																		•					
Porella tumida Kluge	•																						
Porella obesa (Waters)	•												•			•		•					
Porella fragilis Levinsen							•	•										•					
Porella columbiana (O'Donoghue)																		•					
Porella compressa (Sowerby)			•		•	•		•										•					
Cystisella saccata (Busk)	•							•	•														
Cystisella beringia (Kluge)		•	•		•		•	•	•							•							
Rhamphostomella scabra (Fabricius)														•	•	•		•					
Rhamphostomella scabra orientalis Kluge															•			•					

	1	2	3	4	5	6	7	8	9	10	11	12	13	14	15	16	17	18	19	20	21	22	23
Rhamphostomella costata (von Lorenz)		•			•	•	•	•					•			•		•					
Rhamphostomella cristata (Hincks)																		•					
Rhamphostomella spinigera von Lorenz																		•					
Rhamphostomella ovata (Smitt)							•							•	•	•							
Rhamphostomella radiatula (Hincks)													•					•					
Rhamphostomella plicata (Smitt)			•																				
Rhamphostomella bilaminata (Hincks)																	•						
Rhamphostomella sibirica (Kluge)		•					•	•					•			•		•					
Rhamphostomella curvirostrata (O'Donoghue)																		•					
Rhamphostomella sp.																		•					
Codonellina argentea (Hincks)																		•					
Codonellina acuta (Ortmann)																		•					
Codonellina sp.																		•					
Phylactella pacifica O'Donoghue	•		•															•					
Escharella immersa Fleming			•																	•			
Escharella ventricosa (Hassall)	•						•												•				
Escharoides jacksoni rostrata Kluge		•																					
Smittina rigida von Lorenz	•						•									•		•					
Smittina mucronata (Smitt)	•					•										•		•					
Smittina majuscula (Smitt)																		•					•
Smittina minuscula (Smitt)	•															•		•					
Smittina tuberosa Kluge																		•					
Smittina beringia Kluge	•																						
Parasmittina jeffreysii (Norman)																•		•					
Parasmittina trispinosa (Johnston)									•									•		•			
Hippoporina pertusa (Esper)														•	•	•							
Hippoporina reticulatopunctata (Hincks)						•	•	•										•					
Hippoporina propinqua (Smitt)				•			•																•
Hippoporina ussowi (Kluge)																		•					•
Hippoporina vulgaris Dick and Ross																		•					
Schizobrachiella stylifera (Levinsen)																•							
Schizobrachiella perforata (Kluge)	•															•							
Schizoporella ortmani Kluge													•										
Schizoporella biaperta (Michelin)																		•	•				
Schizoporella stelloperforata Kluge																•							
Schizomavella lineata (Nordgaard)																		•	•				
Schizomavella porifera (Smitt)													•					•					
Arthropoma inarmata Gontar																		•					
Pachyegis producta (Packard)																•		•					
Pachyegis groenlandica (Norman)						•																	
Stomachetosella cruenta (Busk)		•	•															•					
Stomachetosella sinuosa (Busk)		•																•	•				
Stomachetosella limbata (von Lorenz)													•					•					
Stomachetosella sienna Dick and Ross																		•	•				
Stomachetosella magniporata (Nordgaard)																		•					
Stomachetosella hincksi Powell	•																						
Stomachetosella pachystega (Kluge)			•																				
Myriapora coarctata (M. Sars)								•	•				•					•					
Myriapora subgracilis (d'Orbigny)	•	•				•		•					•				•	•	•	•			
Myriapora subgracilis variabilis Androsova																		•					
Myriapora orientalis (Kluge)	•	•	•		•			•	•				•		•	•	•	•					
Myriozoella crustacea (Smitt)	•	•	•			•	•	•	•	•	•	•	•		•	•		•	•	•			
?*Cryptosula pallasiana* (Moll)															•			•					
Cryptosula zavjalovensis (Kubanin)				•			•			•	•	•		•	•	•		•	•				
Cylindroporella tubulosa (Norman)	•	•					•						•			•		•					
Microporella ciliata (Pallas)																		•		•			

	1	2	3	4	5	6	7	8	9	10	11	12	13	14	15	16	17	18	19	20	21	22	23
Microporella neocribroides D & R																		•					
Microporella germana D & R																			•				
Microporella speciosa Suwa, D&M																							•
Fenestrulina malusii (Audouin)																		•					
Cheilopora sincera (Smitt)								•	•														
Cheilopora praelucida (Hincks)	•																	•					
Cheilopora orientalis (Kluge)																	•						
Cheilopora inermis (Busk)																				•			
Hippoporidra truculenta D & R																		•	•				
Hippoponella hippopus (Smitt)	•	•																		•			
Hippoponella parva Androsova																		•					
Hippoponella fastigatoavicularis K'	•																	•					
Retepora imperatii tumescens Ort'																							•
Retepora cellulosa (L.)																							•
Phidolopora elongata (Smitt)																		•					
Cellepora nodulosa (von Lorenz)																		•					
Celleporina nordenskjoldi (Kluge)		•				•	•	•	•				•		•	•		•					
Celleporina robertsoniae (C & B)																		•					
Celleporina incrassata (Lamarck)	•		•		•			•	•						•			•	•	•			
Celleporina ventricosa (von L)			•	•	•		•	•								•		•				•	
Cyclostomata: 22	3	1	2	0	0	1	1	1	0	0	0	0	0	0	1	4	1	5	4	3	0	3	3
Ctenostomata: 18	8	0	0	1	0	0	4	0	0	1	1	1	0	5	6	6	3	9	5	3	0	0	1
Cheilostomata: 183	47	32	21	10	14	19	35	34	19	7	7	9	24	15	29	48	9	130	15	18	1	2	9
Total diversity: 223	58	33	23	11	14	20	40	35	19	8	8	10	24	20	36	58	13	144	24	24	1	5	13

Micropalaeontology in Vienna at the turn of the 19th century: foraminiferan and bryozoan studies of Leopold von Fichtel and Johann Paul Carl von Moll

Miklós Kázmér[*] and Norbert Vávra[†]
[*]*Department of Palaeontology, Eötvös University, H-1518 Budapest, P.O. Box 120, Hungary*
[†]*Institute of Palaeontology, University of Vienna, Althanstrasse 14, A-1010 Wien, Austria*

1. Introduction
2. Leopold von Fichtel (1770-1810)
3. Johann Paul Carl von Moll (1735-1812)
4. Fichtel and Moll: *Testacea microscopica* (1798)
5. Moll: *Eschara zoophytorum* (1803)
6. Discussion

1. Introduction

The Habsburg emperors of Austria had a centuries-old tradition of interest in natural history. Ferdinand I (1503-1564) invited Pietro Mattioli (1500-1577), the greatest botanist of his age, to be his 'physician in ordinary' in 1555, and supported the publication of his comments on *Dioscorides* in 1565. His son, Maximilian II (1527-1576), supported Carolus Clusius (1526-1609), *princeps botanicorum*, for two decades during his stay in the Austrian empire. Both rulers spent great sums on establishing and maintaining rich gardens and menageries, and Maximilian also owned a collection of natural and art objects.[1] Rudolf II (ruled 1576-1612) founded the collection of physical and astronomical instruments, and spared no costs to stock the garden and the menagerie with the most exotic plants and animals.[2] He was probably the first to attempt to establish a systematic collection of the mineral wealth of his empire by commanding the mining towns to send mineral specimens to the natural history collection.[3] Anselmus Boetius de Boodt (1550-1632) published the most authoritative account on minerals and gems in the seventeenth century based on Rudolf's collection in Prague.[4]

Into this family married Franz Stephan von Lothringen (1708-1765). Having no real political power since his wife, the Empress Maria Theresa was the ruler of Austria, Bohemia, and Hungary from 1740 to 1780, he cultivated his private interests.[5] Besides improving agricultural methods on his property[6], he generously supported scientific research[7]. He had a deep knowledge of mineralogy, and assembled a significant collection of minerals, rocks, and fossils, and another one of coins and antiques in the imperial court next to the existing physical intruments museum in Vienna.[8] For this purpose in 1748 he bought one of the largest natural history collections of the time from Johann Baillou (1679-1758) of Florenz, which was one of the best in Europe, containing more than 30,000 mineralogical, palaeontological and marine zoological specimens.[9] Additionally, the Emperor invited Baillou to serve as director of the new museum in Vienna[10]. The collection contained mostly minerals and fossils, and crabs, conchs and zoophytes among the animals. The zoological objects were collected to serve as an explanation for the fossil ones.[11] The emperor liked the collection so much that almost no day passed without him spending time there. He transferred his interest to one of his daughters, the Princess Maria Anna (1738-1789), who developed her own mineral and fossil collection.[12]

Upon the early death of Emperor Franz Stephan von Lothringen in 1765, the Empress transferred the natural history collections from the family's private property to the care of the state. New rooms were built and new personnel hired to work scientifically on the collections. Baillou's post was taken by Born as director in 1777.

Ignaz von Born (1742-1791) (Figure 1), a native of Hungary, who had long been known as a scientist and organiser of exceptional capabilities,[13] brought new life into the Imperial Natural History Cabinet. He enriched it with specimens of his own and gifts from his acquaintances and friends in several countries. Born himself published the catalogue of the shell collection.[14] A new exhibition was installed in the years of 1778-1780 with the enthusiastic help of curators and volunteers, among them a certain Karl von Moll.[15] The mineralogical and fossil collection was ordered following the schemes of Wallerius and Cronstedt, the crabs, shells, and radiolarians according to Linnaeus, while the system of Pallas was used to arrange the zoophytes.[16]

In 1780 Born – together with his friend Benedict Franz Herrmann – founded a freemasons' lodge in Vienna[17]. Unusual in its activities, the lodge *Zur wahren Eintracht* was a kind of academy of sciences, with regular sessions, a mineral collection,[18] and most importantly, with a scientific journal titled *Physicalische Arbeiten der einträchtigen Freunde in Wien* edited by Born himself (1783-1788).[19] The lodge, with which even the would-be emperor Joseph was associated, helped Vienna become a centre of German enlightenment.[20]

Masonic activities were first put under state control by Emperor Joseph II in 1785, then

Figure 1. Ignaz von Born (1742-1791).

forbidden in the wake of the French revolution from 1794 by the new emperor Franz II. Born lived to see the demise of his journal and the dissolving of his academy, the lodge. However, his other great achievement, the Natural History Cabinet-turned-scientific-museum, survived him as an early symbol of imperial power and wealth[21] in the time when free thinking was severely limited in countries opposing the stormy events in France.

From 1788 Abbot Andreas Stütz (1747-1806), author of works on mineralogy, served as vice-director of the Natural History Cabinet, and following Born's early death in 1791 took full responsibility for the growth, arrangement and study of the collection. Stütz was followed by Carl Schreibers, a zoologist in 1806, who rearranged the exhibitions and encouraged scientific publishing.[22]

This was the historical and spiritual background behind the activities of two associates of the Imperial Natural History Museum in Vienna, Leopold von Fichtel and Johann Paul Carl von Moll. The two friends established the science of micropalaeontology in the Austro-Hungarian Monarchy, and are still considered as major figures in the history of micropalaeontology.[23] Their works are highly valued even today, but the life of the authors and the circumstances in which they worked are largely unknown. This paper is a small addition to their respective biographies.

2. Leopold von Fichtel (1770-1810)

We are unaware about most of the details in the life of Leopold von Fichtel.[24] Often he is confused with his father, Johann Ehrenreich von Fichtel (1732-1795), a significant personality of Austro-Hungarian geology, palaeontology, and mining science.[25] A jurist by education, serving the governments of Transylvania and Austria in several capacities, the elder Fichtel was a celebrated scientist of his age, member of several scientific societies, and prolific author of geological and palaeontological monographs on the minerals and fossils of the Carpathians. He was an ardent vulcanist, and wrote a vitriolic pamphlet on contemporary mineralogical practices.[26] His biography is routinely included in lexicons and biographical compilations even today.[27] The son, probably a less colourful character and certainly a less prolific author, did not make his way into the favours of lexicographers. However, his lasting contribution to science certainly exceeds that of the elder Fichtel.

Leopold von Fichtel, the only son of his father, was born in Hermannstadt,[28] Transylvania in 1770.[29] We are unaware of the details of his education. His father was financial councillor in Hermannstadt up to the son's age of 15, then spent two years in Vienna as customs officer before returning to take the position of councillor to the Gubernium in Hermannstadt again.[30] Leopold probably never married and never took a paid position, at least not for any considerable time. His death record (he died in 1810 in Vienna) recorded him as *'Practicant beim k.k. Directorium'*, i.e. an unpaid assistant in the bureaucracy.[31]

Leopold possibly inherited the love for travel from his father. The elder Fichtel started travelling widely within Hungary while working for a law firm, and continued it during his occupations as mining supervisor and customs inspector. Leopold followed his father on many of his offical journeys.[32]

The scattered records tell that the younger Fichtel joined the work on the new exhibition in the Imperial Natural History Cabinet as volunteer.[33]

Probably the first scientific publication of the younger Fichtel was an annotated translation of a treatise on the lithology of Vesuvius.[34] His interest was probably started by his father, who was an ardent advocate of volcanist ideas.[35] Giuseppe Gioeni[36] (1743-1822) was professor of natural sciences at the University of Catania. He visited Vesuvius upon an invitation from Sir William Hamilton, the ambassador to the Bourbon court in Naples, author of the *Observations on Mount Vesuvius*.[37] Gioeni's work comprises an

Figure 2. The Imperial Collection

extensive geological description of Vesuvius and a catalogue of all volcanic products together with their mineralogical and chemical analyses. It is one of several contemporary treatises on the active volcanoes and their products in Italy.[38]

In 1798 Leopold von Fichtel published – together with J.P.C. von Moll – the *Testacea microscopica*, a lavishly illustrated treatise on Foraminifera at his own cost. The figures were drawn with the utmost care. A recent revision[39] accompanied by scanning electron photography of the original specimens proved the exceptional accuracy of the original illustrations arranged on twenty-four copper plates.

In 1804 10,000 gulden was paid for a rich collection of insects, which were collected by Fichtel during his stay in the East Indies.[40]

In 1806 Abbot Stütz died; he had been acting director of the Imperial Natural History Cabinet. Three scientists applied for the now vacant position: Leopold Fichtel, Moll, and Carl Schreibers. The last – an assistant professor, zoologist at the University of Vienna, besides being a promising scientist, had friends and relatives in higher positions than the other two. He was appointed to the position.[41] Under Schreibers's years as director, the Cabinet grew to a full-fledged, majestic natural history museum, with systematically arranged collections, where the scientists – employed and volunteers – published major monographs in their chosen fields.[42]

Schreibers, supported by Count Wrbna, convinced the Emperor to buy animals to fill the gaps in the systematic collections. Fichtel, who was an insider in the Cabinet for a long time and enjoyed the trust of the Emperor, undertook the task and travelled to England to supervise the buying. His aim was to buy all necessary and available specimens for the collection. He spent, through the Emperor's donation, 18,000 guldens in total for this project.[43] He obtained not only rare species but also a number of unique specimens for the Vienna Cabinet. It was during his visit to England that Fichtel laid the foundations for an ethnographic museum by obtaining some of the ethnographic collections of Captain James Cook.[44]

Leopold Fichtel enjoyed a considerable, although irregular income by gathering natural objects and selling whole collections to museums. He sold a large amount of the most rare and valuable conchs to the Natural History Cabinet in 1797 for a considerable sum.[45] The mineral collection he inherited from his father, rich especially in mineralized gold ores of Hungary and Transylvania,[46] was acquired by the School of Chemistry and Metallurgy in Kolozsvár[47] in 1799 for 5000 gold florins.[48]

The director of the Imperial Natural History Cabinet repeatedly sent Leopold Fichtel on buying trips, e.g. in 1806 to acquire the Lever and the Parkinson collection in England.[49] His further travels are less well known. We know that he spent considerable time in East India, since his collection was bought for 10,000 guldens by the Cabinet.[50] Other travels brought him to France, Spain, and Portugal, from where he returned with numerous mammal and bird specimens, a rich insect collection and other specimens, either collected by himself or acquired by purchase.[51]

The Emperor frequently added to the sum set aside for regular buying of natural objects from mineral traders and private persons. In the years between 1806 and 1809, 14,000 guldens was spent on buying besides what was allowed by the annual budget. Seemingly Fichtel sold the largest collection of minerals for the sum of 5132 guldens to the Cabinet during this period.

The younger Fichtel was well-known in scientific circles. He was a member of the

Linnean Society in London and of the Asiatic Society of Bengal in Calcutta.[52] He died on 18th March 1810 in Vienna from tuberculosis.

In 1812 the department of shells (the 'Conchylien-Abtheilung') was enriched with a major collection of foraminifers. This was assembled by Leopold von Fichtel and Johann Paul Carl von Moll, and contained the original material published in their *Testacea microscopica* of 1798. This collection, unique of its kind, exists even today.[53] It was bought – together with the original manuscript and figures, and 503 further plates and drawings of the rest of the collection besides *Argonauta* and *Nautilus* – for the high sum of 150 guldens from Fichtel's mother.[54]

3. Johann Paul Karl von Moll (1735-1812)

Johann Paul Karl von Moll[55] was born in Öttingen, Bavaria, on 30th October 1735. His father was a financial clerk in the court of the Count of Öttingen. He had at least an uncle and a godfather in Vienna, who had a keen interest in natural history. One of them held the fossil collection of the famous naturalist Nicolaus Lang.[56] Living probably with these relatives he acquired his knowledge in fossils.[57]

Moll was an ambitious and idealistic helper of Ignaz von Born in arranging the systematic exhibition and description of the Imperial Natural History Cabinet, while enriching the collection of the freemason lodge *Zur wahren Eintracht* with gifts,[58] and maintaining his own collection of shells.[59] Moll died poor in the city hospice on 20th February 1812.[60]

Both Fichtel and Moll were pioneers of micropalaeontology in Austria-Hungary. Educating no disciples, a generation passed until August E. Reuss started his work on microscopic fossils in the 1840s.

4. Fichtel and Moll: *Testacea microscopica* (1798)

The first well-illustrated major monograph on foraminiferans is the *Testacea microscopica* of Fichtel and Moll,[61] considered by Alcide d'Orbigny (1802-1857), the great French palaeontologist, professor in the Jardin des Plantes of Paris, to be one of the fundamental works in micropalaeontology.[62] It is the most important of the earliest works on the Foraminifera. The still earlier works on which Linné based his few species of Foraminifera were not well illustrated. The work of Fichtel and Moll is for the most part excellently illustrated. Cushman also remarked that in his own copy the colours of the plates are beautifully preserved in spite of their age.[63]

The work of Fichtel and Moll is important for the specific names used. On this basis several genera were erected by numerous later authors.[64] Fichtel and Moll supplied a general designation of the origin of the described specimens. Precise localities are given only in the case of fossil material.

The significance of this early major work on foraminifers is enhanced by the fact that the original material is preserved in the Natural History Museum in Vienna. A revision, accompanied with scanning electron micrographs, and a colour facsimile of the original plates has been made recently.[65] A further major complete monograph exists in manuscript titled *Microscopische Conchylien,* ready for publication, complete with drawings. Even nine plates have been prepared. This material was obtained from Fichtel's mother after the death of his son.

5. Moll: *Eschara zoophytorum* (1803)

This monograph has both a frequently cited Latin[66] and a rarely mentioned German[67] edition published in the same year. The description below follows the page numbering of the Latin edition.

The volume starts with a list of illustrations on six unnumbered pages (left column in Latin, right column in German). Then follow preliminary contemplations on zoophytes or phytozoa in general (*Preliminares contemplationes de zoophytis seu phytozois in genere*) (pp. 3-24, with abundant notices in the footnotes). Moll describes the anatomy, nutrition, and systematics of the bryozoans, with extensive references.

The systematic subdivision of the Bryozoa (i.e. *Eschara*) (pp. 25-30) follows, accepting Pallas's terminology and nomenclature. Descriptions are grouped under a first section: Escharae described by Pallas in his *Elenchus Zoophytorum* (pp. 30-50), and under second section: Escharae, undescribed either by Pallas in his *Elenchus Zoophytorum* or by others.

A full revision of Moll's species is still to be undertaken. Here a few remarks are given to emphasize the importance of the work. Pallas's species are the following:

Eschara fascialis (pp. 30-34) = *Pentapora fascialis* (Pallas, 1766)
  variety α proprie *fascialis* (Pl. I, figs 1A-F)
  variety β *lamellosa* (Pl. I, fig. 2)
Eschara spongites (pp. 34-35; Pl. I, figs 3A-B) = *Stylopoma spongites* (Pallas, 1766)
Eschara annularis (pp. 36-37; Pl. I, figs 4A-B)
Eschara pilosa (pp. 37-44) = *Electra pilosa* (Linnaeus, 1766)
  variety α *loeflingiana*
  variety β *ellisiana* (Pl. I, figs 5A-D)
  variety γ *reaumuriana* (Pl. II, figs 6A-H)
Eschara foliacea (pp. 44-48; Pl. II, figs 7A-E) = *Pentapora foliacea* (Ellis & Solander)
Eschara papyrea (pp. 48-50; Pl. II, figs 8A-B) = *Charbasea papyrea* (Pallas, 1766)

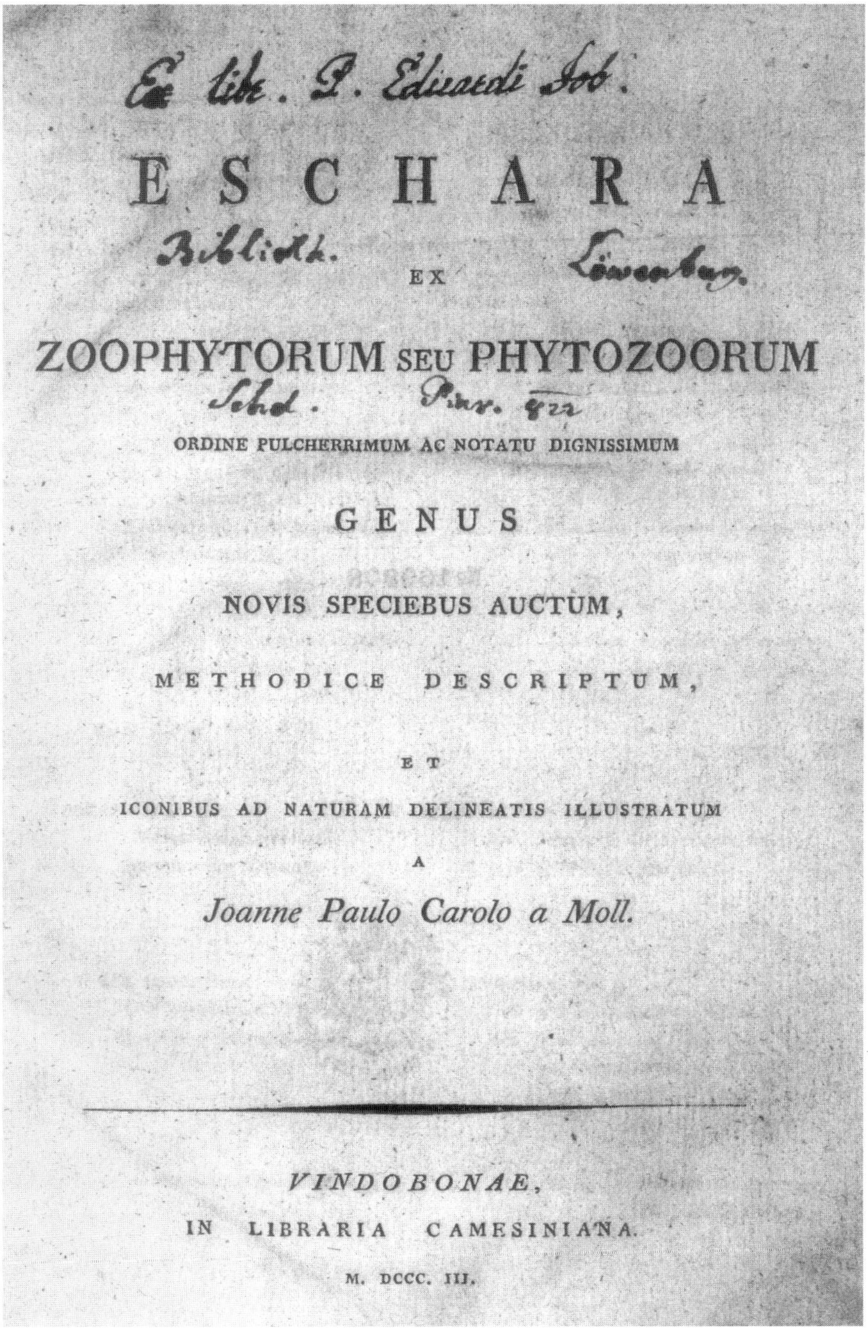

ESCHARA

EX

ZOOPHYTORUM SEU PHYTOZOORUM

ORDINE PULCHERRIMUM AC NOTATU DIGNISSIMUM

GENUS

NOVIS SPECIEBUS AUCTUM,

METHODICE DESCRIPTUM,

ET

ICONIBUS AD NATURAM DELINEATIS ILLUSTRATUM

A

Joanne Paulo Carolo a Moll.

VINDOBONAE,

IN LIBRARIA CAMESINIANA.

M. DCCC. III.

Figure 3. Title-page of Eschara zoophytorum.

The new species are as follows:

Eschara impressa, nova species (pp. 51-54 Pl. II, figs 9A-I) = *Calpensia nobilis* (Esper, 1796)
Eschara vulgaris (pp. 55-56). Type species of the genus *Escharina* Milne-Edwards, 1836.
variety α, nova species (with fixed lower labium) (Pl. III, figs 10A-B)
variety β, nova with integrated lower labium (Pl. III, figs 11A-C)
Eschara cyclostoma, nova species (pp. 56-57. Pl. III, figs 12A-D)
Eschara Pallasiana, nova species (pp. 57-58; Pl. III, figs 13A-B). Type species of the genus *Cryptosula* Canu & Bassler, 1929.
Eschara Borniana, nova species (pp. 58-60; Pl. III, figs 14A-C)
Eschara Otto-Müelleriana, nova species (pp. 60-62; Pl. III, 15A-C) = *Pentapora ottomulleriana* (Moll, 1803)
Eschara sedecimdentata, nova species (pp. 62-63; Pl. III, figs 16A-C) = ? *Electra* sp.
Eschara radiata, nova species (pp. 63-64; Pl. IV, figs 17A-I) = *Puellina radiata* (Moll, 1803)
Eschara bimucronata, nova species (pp. 65-66; Pl. IV, figs 18A-C) = *Haplopoma bimucronata* (Moll, 1803)
Eschara planata (p. 67; Pl. IV, fig. 19)
Eschara patellaria (pp. 68-69; Pl. IV, figs 20A-B) = *Mollia patellaria* (Moll, 1803)
Eschara depressa (pp. 69-70; Pl. IV, fig. 21) = *Calpensia nobilis* (Esper, 1796)

Genus, species and variety names are invariably given both in Latin and in German in the title line. A second line with smaller letter gives the name translated into the French.

Moll named three new species after major personalities in contemporary palaeontology:

Peter Simon Pallas (1741-1811) German naturalist and explorer. Professor of natural history at the St. Petersburg Academy of Sciences. Widely travelled in Russia and described the fauna, flora and minerals of Siberia. His first major study was the *Elenchus zoophytorum*, a major study in bryozoology, which founded his fame and brought him the invitation of Catherine II to St. Petersburg.[68]

Ignaz von Born (1742-1791) Hungarian naturalist, organiser of science, freemason, director of the Natural History Cabinet in Vienna. He published several volumes on mining, on mineral and fossil collections and on shells.[69]

Otto Frideric Müller (1730-1784), Danish naturalist, was born in Copenhagen. He studied theology, law and natural history in Denmark. After graduation he visited several European countries. He published a number of monographs and a large number of scientific papers, first in theology, and later in botany (*Flora Danica*) and zoology (insects mostly). He was a member of several European academies. He made a living from being lawyer and the archivist of the Norwegian Chamber. Müller published probably only a single work on bryozoans, titled *Memorie sur un nouveau Genre de Zoophytes* in the

Gazette Litteraire de Berlin.[70]

The bryozoan samples derive from a multitude of localities. They arrived from all over the world, collected during hundreds of years of donations and specialized gathering expeditions to South and Central America, southern Africa, India, and of course Europe.[71] A brief mention in the paragraph 'patria' indicates the Mediterranean Sea, Adriatic Sea, Indian Ocean, Atlantic Ocean, the Cape of Good Hope, American seas, and the North Sea between England and Belgium. The collectors are mostly unnamed; probably were unknown to the authors as well. Reaumur and Baillou are named as persons from whom specimens were derived.

All species are recent, but a few fossil localities are mentioned, too: the famous Kroisbach[72] quarry next to Lake Neusiedl[73] in Hungary, close to the Austrian border,[74] an unidentified locality near Vienna, and at Monte S. Petri next to Trajectum ad Mosam (Maastricht) in The Netherlands. The new species came invariably from the Mediterranean, and one from the Adriatic Sea (one specimen is of unknown origin).

The illustrations are strongly magnified,[75] exactly drawn and not stylised like in several later works of palaeontology.[76] A microscope was used for drawing them. Experimenting with a microscope from a slightly older age, Rögl and Hansen have shown that magnification and resolution was satisfactory for the proper and exact drawing of such minute features such as the ornament of foraminiferans a few tens of micrometres in size.[77] Drawing microscopes were available at that time. Mirrors were applied to view the object and the drawing surface simultaneously.[78] The highly faithful illustrations suggest that the artist – probably the author himself – used such drawing equipment.

Moll put in a few remarks on the ecology of the species. In the North Sea between England and Belgium, *Eschara pilosa* var. *loeflingiana* abundantly encrusts algae. Specimens on the alga *Ulva lactuca* from Flandria are in the collection of Moll.[79] He has observed that *Eschara pilosa* var. *ellisiana* is an encrusting species. Pallas found it on the alga *Sertularia longissima,* while, Moll's specimen is on *Sertularia cupressina.*

6. Discussion

Both Fichtel and Moll are considered as 'Klassiker der Mikropaläontologie in Österreich'.[80] Lamouroux named the genus *Mollia*[81] in 1816. The greatest personalities of foraminifer research have paid visits to the Fichtel and Moll collections in the Natural History Museum in Vienna throughout the twentieth century.[82]

Neither Fichtel nor Moll taught students and there were no immediate successors to the promising beginning in micropalaeontology that they made. Almost half a century passed until the first monographic descriptions on Austrian material appeared: Alcide d'Orbigny wrote on foraminiferans in 1846[83] and A.E. Reuss on bryozoans in 1848.[84] With the latter's monograph there began a tradition of high-quality micropalaeontological work in Austria-Hungary, which persists even today.[85]

The words '*methodice descriptum*' on the title page of the *Eschara* is a character of the age. Several works claim – justifiedly – in their title that they contain a rational and methodical description, so fitting to the rationalism of the Enlightenment.[86]

Notes

1 L.J. Fitzinger, 'Geschichte des kais. kön. Hof-Naturalien-Cabinetes zu Wien. I. Abtheilung. Älteste Periode bis zum Tode Kaiser Leopold II. 1792', *Sitzungsberichte der Kaiserlichen Akademie der Wissenschaften, mathematisch-naturwissenschaftliche Classe,* 21 (1856), 433-479, Wien, 468.

2 L. Hendrix, G. Bocskay, J. Hoefnagel and T. Vignau-Wilberg (editors*), Nature Illuminated: Flora and Fauna from the Court of Emperor Rudolf II* (Los Angeles, 1997). Robert John Weston Evans, *Rudolf II and His World. A Study in Intellectual History, 1576-1612* (Oxford, 1984).

3 W.E. Wilson, 'The History of Mineral Collecting', *Mineralogical Record* 25/6 (1994), 1-243, Tucson, Arizona, 38.

4 *Gemmarum et Lapidum Historia* (Lugduni Batavorum, 1609). See M.-C. Maselis, A. Balis and R.H. Marijnissen, *De albums van Anselmus de Boodt (1550-1632), geschilderte natuurobservatie aan het Hof van Rudolf II te Praag* (Tielt, 1989). C. Gysel, 'Anselme Boece De Boodt (1550-1632), lapidaire et médecin de Rodolphe II', *Vesalius* 3 (1997), 33-41.

5 G. Schreiber, *Franz I. Stephan: an der Seite einer grossen Frau,* (Graz, 1986)

6 R. Zedinger, 'Franz Stephan von Lothringen und sein Kreis unter besonderer Berücksichtigung der Tätigkeiten im Gebiet der heutigen Slowakei', in *Die goldene und silberne Reise des Kaisers Franz Stephan von Lothringen in die mittelslowakischen Bergstädte*, (Banská Stiavnica, 2001), 18-28.

7 C. Riedl-Dorn, 'Franz Stephan als Förderer naturwissenschaftlicher Aktivitäten', in *Die goldene und silberne Reise des Kaisers Franz Stephan von Lothringen in die mittelslowakischen Bergstädte*, (Banská Stiavnica, 2001), 29-38.

8 G. Niedermayr, 'The mineral collection of the Museum of Natural History', *Mineralogical Record,* 20 (1989), 347-354, Tucson, Arizona.

9 Wilson, note 3, p. 159.

10 Riedl-Dorn, note 7, pp. 30-32, and C. Riedl-Dorn, Chevalier de Baillou und das Naturalienkabinett, in *Lothringern Erbe. Franz Stephan von Lothringen (1708-1765) und sein Wirken in Wirtschaft, Wissenschaft und Kunst der Habsburgmonarchie.* (= Katalog des Niederösterreichischen Landesmuseums, Neue Folge 429 (St. Pölten, 2000), 111-115.

11 Fitzinger, note 1, p. 439.

12 This collection was bought by the University of Pest upon the retirement of the Princess from court activities, where it still exists as the core of the collection today (G. Papp, Tamás Weiszburg, 'The mineral collection of Archduchess Maria Anna', in *Museum and Collections in the History of Mineralogy, Petrology and Paleontology in Hungary,* edited by G. Vitális and T. Kecskeméti (Budapest, 1991). Annals of the History of Hungarian Geology, Special Issue

3, 135-143).

13 C. Riedl-Dorn, 'Ignaz von Born (1742-1791) – ein siebenbürgischer Naturforscher', *Stapfia* 45 (Linz, 1996), 345-355.

14 I. Born, *Index rerum naturalium Musei Caesarei Vindobonensis. Pars I. Testacea,* (Vindobonae, 1778) and I. Born, *Testacea Musei Caesarei Vindobonensis, quae jussu M. Theresiae Augustae disposuit et descripsit Ig. a Born* (Vindobonae, 1780). Earlier Born published a catalogue of his own collection, the *Lithophylacium Bornianum* (Pragae, 1772). It certainly has contributed to his invitation to the directorship of the Natural History Cabinet in Vienna. Catalogues published during previous centuries and also in Born's time served much more than an inventory of the accumulated specimens. Starting at least with Ulisses Aldrovandi, these were systematic treatments of all knowledge related to objects collected, and offered a system of classification. One of Born's catalogues (*Catalogue méthodique et raisonné de la Collection des Fossiles de Mlle Eleonore de Raab* [Viennae, 1790]) provided data on the chemical composition and reactions of several hundred mineral species.

15 Whether this person is identical with Johann Paul Carl von Moll, see F. Rögl, ´L.v. Fichtel und J.P.C. von Moll und ihre wissenschaftliche Bedeutung', *Annalen des Naturhistorischen Museums in Wien,* 84/A (1982), 63-77. Moll was Protestant, hence a use of his last given name is justified.

16 Fitzinger, note 1, p. 450.

17 D. Lindner, *Ignaz von Born, Meister der Wahren Eintracht. Wiener Freimauerei im 18. Jh* (Wien, 1986), 243.

18 Fitzinger, note 1, pp. 477-478 and his endnote 2.

19 It was not Born's only attempt to improve the scientific organisation of the Habsburg Empire. He founded an academy of science in Bohemia, with a journal (*Abhandlungen einer Privatgesellschaft in Böhmen, zur Aufnahme der Mathematik, der vaterländischen Geschichte, und der Naturgeschichte,* 1775-84). Born was the convenor of the first international scientific conference held at Skleno, Hungary, in 1786. On this occasion the first international scientific society, the 'Societät für Bergbaukunde' (again with a journal) was founded, of which Born was a president. See M. Teich, 'Born's amalgamation process and the international metallurgical gathering at Skleno in 1786', *Annals of Science,* (1975); and L. Molnár and A. Weiss, *Ignaz Edler von Born und die Societät der Bergbaukunde* (Wien, 1986).

20 E. Weigl, *Schauplätze der deutschen Aufklärung. Eine Städterundgang*, Rowohlts Enzyklopädie, (Reinbek bei Hamburg, 1997), 207-229.

21 S. Sheets-Pyenson, *Cathedrals of Science: The development of colonial natural history museums during the late nineteenth century* (Kingston, Ontario, 1988), 144. C. Riedl-Dorn, *Das Haus der Wunder. Zur Geschichte des Naturhistorischen Museums in Wien* (Wien, 1998).

22 Fitzinger, note 1, p. 460.

23 H. Zapfe, *Catalogus Fossilium Austriae. Heft XV: Index Palaeontologicorum Austriae* (Wien, 1971), 78.

24 A number of scattered data are supplied by Fitzinger, note 1, pp. 449, 463, 479, and L.J. Fitzinger, 'Geschichte des kais. kön. Hof-Naturalien-Cabinetes zu Wien. II. Abtheilung. Periode unter Franz II. (Franz I. Kaiser von Österreich) bis zu ende des Jahres 1815', *Sitzungsberichte der Kaiserlichen Akademie der Wissenschaften, mathematisch-naturwissenschaftliche Classe,* 57 (1868), 1013-1092 (1019, 1036, 1038, 1040-1, 1044, 1050, 1064, 1078.

25 H. Küpper, 'Österreichs Beiträge zur Entwicklung der Mikropaläontologie', *Erdöl-Zeitschrift*

75/5 (1959), 114-121 (115)

26 G. Papp, 'An ardent vulcanist from Hungary. Sketches to the scientific portrait of Johann Ehrenreich von Fichtel (1732-1795)', in *Volcanoes and History,* edited by Nicoletta Morello (Genova, 1998), 505-522.

27 The *Index bio-bibliographicus notorum hominum,* edited by Jean-Pierre Lobies (Osnabrück, 1973-) registers 27 encyclopedias and biographic collections up to 1961 recording data about the elder Fichtel.

28 Nagyszeben in Hungarian, Sibiu in Romanian, the seat of the autonomous government of Transylvania.

29 F. Schlichtegroll, *Nekrolog auf das Jahr 1795.* Vol. 6/2 (Gotha, 1798), 346, Zapfe (note 18), 29-30.

30 *Slovensky biograficky slovnik*, 2 (Bratislava, 1987), 85.

31 Zapfe, note 23, pp. 29-30; F. Rögl and H.J. Hansen, 'Foraminifera described by Fichtel & Moll in 1798. A revision of *Testacea microscopica*', *Neue Denkschriften des Naturhistorischen Museums in Wien* 3, (1984), 1-143 (15).

32 Sclichtegroll, note 29.

33 M. Kirchmayer, 'Johann Fridvaldszky, Johann Ehrenreich von Fichtel und dessen Sohn Leopold von Fichtel. Forscher der Periode 1750-1840 der Entwicklung der Naturwissenschaften in Österreich (Beitrag zur Geschichte der Geologie in Österreich)', *Verhandlungen der Geologischen Bundesanstalt* (1961) 113-115.

34 G. Gioeni, *Saggio di litologia vesuviana* (Napoli, 1790), 208. Fichtel's German translation is Joseph Ritter von Gioeni, *Versuch einer Lithologie des Vesuvs.* Aus dem italieneischer übersetzt, und mit einigen Anmerkungen begleitet von Leopold von Fichtel (Wien, 1793), 392. On Gioeni, see F. Aradas, *Elogio del Cav. Giuseppe Gioeni.* (Catania, 1869), 28.

35 Papp, note 26.

36 G. Buccieri, 'Gioeni, Giuseppe', in *Dizionario Biografico degli Italiani*, (2000, Roma), 115-118

37 W. Hamilton, *Observations on Mount Vesuvius, Mount Etna, and other volcanos* (London, 1772), 179.

38 E. Vaccari, 'Lazzaro Spallanzani and his geological travels to the "Due Sicilie": the volcanology of the Aeolian Islands', in *Volcanoes and History,* edited by Nicoletta Morello (Genova, 1998). 621-651.

39 Rögl and Hansen, note 31.

40 Fitzinger (note 24), p. 1036. At this time the director of the cabinet received 1500 guldens annually.

41 Rögl, note 11, casts doubts on it, since no trace of their applications was found in the archives, while other five applicants' documentation is still available.

42 Fitzinger, note 24, p. 1039.

43 Fitzinger, note 24, p. 1040.

44 Fitzinger, note 24, p. 1078 and endnote 2.

45 Fitzinger, note 24, p. 1019. Other parts of Cook's collections were given to Trinity College, Dublin and now are in the National Museum of Ireland in Dublin (see Wyse Jackson, P.N. 'The geological collections of Trinity College, Dublin', *The Geological Curator*, 5(7) (1992), 263-274).

46 Robert Townson, *Travels in Hungary with a short account of Vienna in the year 1793* (London, 1797), 506.

47 Klausenburg in German, Cluj-Napoca in Romanian.
48 Wilson note 3, p. 170.
49 Fitzinger, note 24, pp. 1040 and 1078, endnote 2.
50 Fitzinger, note 1, p. 479, also records his insect collection in Vienna.
51 Fitzinger, note 24, pp. 1040-1041.
52 Rögl, note 15, p. 73.
53 Rögl, note 15; Rögl and Hansen, note 31.
54 Fitzinger, note 24, p. 1064.
55 J.P.C. Moll has been repeatedly confused with another, however unrelated, scientist of the same family name. The two scientists bore no relation whatsoever. Karl Ehrenbert Freiherr von Moll (1760-1838) was a mineralogist, founding editor of a scientific journal (*Annalen der Berg- un Hüttenkunde,* 1797-1826), high official in the court at Salzburg, and politician. See O. Kühn, 'Die Autoren der *Testacea microscopica,* Wien, 1798', *Anzeiger der Österreichischen Akademie der Wissenschaften, Mathematisch-naturwissenschaftliche Klasse* 106/5 (1969), 89-93.
56 Nicolaus Lang (1670-1741), medical doctor in Luzern, Switzerland, author of *Historia lapidum figuratorum Helvetiae* (Venetiis, 1708), one of the last books proposing inorganic origin of fossils.
57 Rögl, note 15, p. 71.
58 Fitzinger, note 1, pp. 477-478, endnote 2; Kühn, note 55, pp. 90-91; Lindner note 17, p. 243.
59 Fitzinger, note 1, p. 479.
60 Zapfe, note 23, p. 78.
61 *Testacea microscopica aliaque minuta ex generibus* Argonauta *et* Nautilus *ad naturam delineata et descripta a Leopoldo a Fichtel et Jo. Paolo Carlo a Moll.* XII + 123 p., 24 plates. Anton Pichler, Wien. The second edition is identical, except the publisher was Camesianische Buchhandlung, Wien.
62 A. d'Orbigny, *Foraminiféres fossiles du bassin tertiaire de Vienne (Autriche)* (Paris, 1846), 312.
63 J.A. Cushman, 'The work of Fichtel and Moll and of Montfort', *Contributions from the Cushman Laboratory for Foraminiferal Research* 3 (1927), 168-171 (168)
64 Cushman, note 63, p. 168.
65 Rögl and Hansen, note 31.
66 *Eschara ex zoophytorum seu phytozoorum ordine pulcherrimum ac notatu dignissimum genus novis speciebus auctum, methodice descriptum, et iconibus ad naturam delineatis illustratum a Joanne Paulo Carolo a Moll. Vindobonae, in libraria Camesiniana. M. DCCC. III.* 9, 70 p., 4 plates, 4to.
67 Moll, Johann Paul Karl von, *Die Seerinde, aus der ordnung der Pflanzenthiere das schönste und merkwürdogste Geschlecht, mit neuen Arten vermehrt, methodisch beschrieben, und durch nach der Natur gezeichnete Abbildungen erläutert.* Camesinaiche Buchhandlung, Wien, 1803. 77 p., 4 plates, 4to.
68 F. Wendland, 'Peter Simon Pallas (1741-1811): Materialen einer Biographie', *Veröffentlichungen der Historischen Kommission zu Berlin*, 80 (Berlin, 1992)
69 Lindner, note 17.
70 C. Jonas in Bricka, C.F.: *Dansk Biografisk Lexikon*, 19 vols, Copenhagen, 1887-1905, fide Scandinavian Biographical Archive, fiche A-222, pages 231-259, Knaur, München.
71 Fitzinger, notes 1 and 24 faithfully records the most important collecting trips and their results enriching the museum's collections.

72 Kroisbach: now Fertörákos (Hungary).
73 Lake Neusiedl: Fertö-tó (Austria and Hungary).
74 Moll, note 67, 35.
75 Magnification is an obvious requirement today. However, in the time of Moll there were works where natural objects were illustrated in their original size, which makes comparisons hard, if not even impossible. The figures of insects in Robert Townson's *Travels in Hungary* (1797) have been draught in natural size, severely hindering the use of the illustrations. See O. Merkl, 'Robert Townson's "Entomologia"', in *Robert Townson's Travels in Hungary*, edited by P. Rózsa (Debrecen, 1999), 99-116.
76 Rögl, note 15, p. 69, emphasizes the extraordinary faithfullness of the original drawings of Fichtel and Moll, now preserved in the Natural History Museum in Vienna.
77 Rögl and Hansen, note 31, pp. 20-24.
78 Rögl, note 15, p. 66.
79 Moll, note 67, p. 40.
80 H. Hiltermann, 'Zur Geschichte der angewandten Mikropaläontologie', *Bericht der Naturhistorischen Gesellschaft zu Hannover* 109 (1965), 23-47.
81 J.V.F. Lamouroux, *Histoire des polypiers coralligenes flexibles, vulgairement nommés zoophytes* (Caen, 1816), 560.
82 Rögl, note 15, pp. 67-68; Rögl and Hansen, note 27, p. 40.
83 A. d'Orbigny, *Foraminiferes fossiles du bassin tertiaire de Vienne (Autriche)* (Paris, 1846), 312.
84 A.E. Reuss, 'Die fossile Polyparien des Wiener Tertiärbeckens', *Haidinger's Naturwissenschaftliche Abhandlungen* 2 (1848), 1-109.
85 N. Vávra, *Catalogus Fossilium Austriae, Heft Vb/3: Bryozoa tertiaria* (Wien, 1977), 210.
86 Born, note 14.

A survey on the history of freshwater bryozoology in Belgium and Luxembourg

Jos. A. Massard and Gaby Geimer

Centre Universitaire de Luxembourg, Département des Sciences, Séminaire d'Histoire des Sciences, Luxembourg, 162A, Avenue de la Faïencerie, L-1511 Luxembourg, Grand Duchy of Luxembourg

and Musée national d'histoire naturelle, Luxembourg, 25, Rue Münster, L-2160 Luxembourg, Grand Duchy of Luxembourg

1. B.C. Dumortier, a pioneer of bryozoology

Belgian freshwater bryozoology began in 1835 when Barthélemy Charles Joseph Dumortier (1797-1878) (Figure 1) published in the 'Bulletin de l'Académie royale des Sciences et Belles-Lettres de Bruxelles' his '*Recherches sur l'anatomie et la physiologie des polypiers composés d'eau douce*'.[1]

According to G.J. Allman's[2] formulation in his *Monograph of the fresh-water Polyzoa*

Figure 1. Barthélemy Charles Dumortier (1797-1878) (from G. Lebrun, Grandes figures de la Belgique indépendante, *Bruxelles, 1936).*

(1856) it was 'a memoir on the "Polype à Panache" of Trembley'.[3] Dumortier had found this animal in 1834 near Brussels in the plain of Mon-Plaisir,[4] in the canal between Brussels and Ruppel.[5] And he had created a new genus and a new specific name for it: *Lophopus crystallinus*.

The reasons were summed up by Allman:

> 'This Polyzoon, which had been previously confounded with *Alcyonella*[6] and *Plumatella*, was believed by Dumortier to be sufficiently distinct to render it the type of a new genus, which he accordingly constituted, under the name of *Lophopus*. In consequence of using lenses of too low a power, Dumortier persuaded himself of the absence of cilia on the tentacula, and made this supposed fact the principal character in his new genus. Notwithstanding, however, the erroneous observation on which Dumortier thus relied, the separation of the "Polype à Panache" from the other fresh-water Polyzoa was an important step, and is fully borne out by its general structure.'

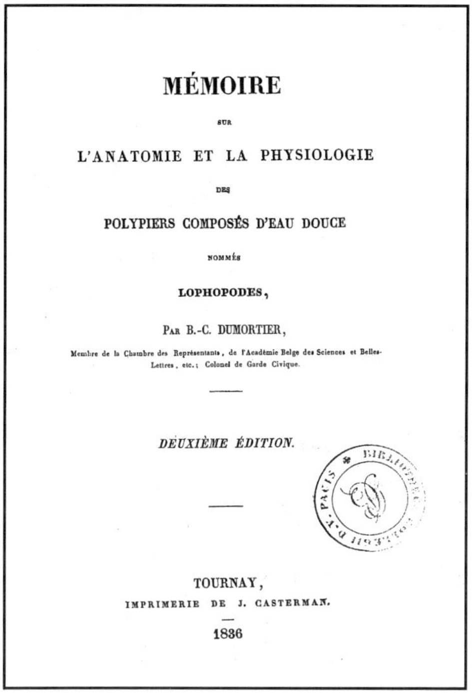

MÉMOIRE

SUR

L'ANATOMIE ET LA PHYSIOLOGIE

DES

POLYPIERS COMPOSÉS D'EAU DOUCE

NOMMÉS

LOPHOPODES,

PAR B.-C. DUMORTIER,

Membre de la Chambre des Représentants, de l'Académie Belge des Sciences et Belles-Lettres, etc.; Colonel de Garde Civique.

DEUXIÈME ÉDITION.

TOURNAY,

IMPRIMERIE DE J. CASTERMAN.

1836

Figure 2. Second edition of Dumortier's memoir on the anatomy and physiology of the composed freshwater polyps called lophopods (1836).

Later on Dumortier tried to justify his assertion concerning the alleged absence of tentacular cilia: due to the low power of his lenses he would not have seen the cilia themselves but however the effect they produced in the water.[7] In fact, in 1835 he had stated clearly that he had never seen any cilia: '*je n'ai jamais vu l'apparence de cils vibratiles*'.[8]

Eventually Allman emphasizes the general significance of Dumortier's work:

> " The memoir of Dumortier is chiefly valuable as giving us the most complete account of the anatomy of the Polyzoa which had up to his time been published.[9] To him is due the honour of having been the first to demonstrate a distinct nervous system in these animals; and he describes the cutaneous, circulatory, respiratory, manducatory, digestive, muscular, and reproductive systems, with much detail, and with a correctness which makes us the more surprised that he should have committed so grave an error respecting the tentacular cilia."

In the 'Bulletin Zoologique' of the same year,[10] Paul Gervais[11] made an analysis of the memoir of Dumortier, and contended against the right of the 'Polype à Panache' to assume the position of a distinct genus, insisting on its representing nothing more than a *Plumatella*. Dumortier tried to refute Gervais' arguments in the second and enlarged edition of the memoir he published in 1836 (*Mémoire sur l'anatomie et la physiologie des polypiers composés d'eau douce nommés Lophopodes*) (Figure 2)[12] and to demonstrate

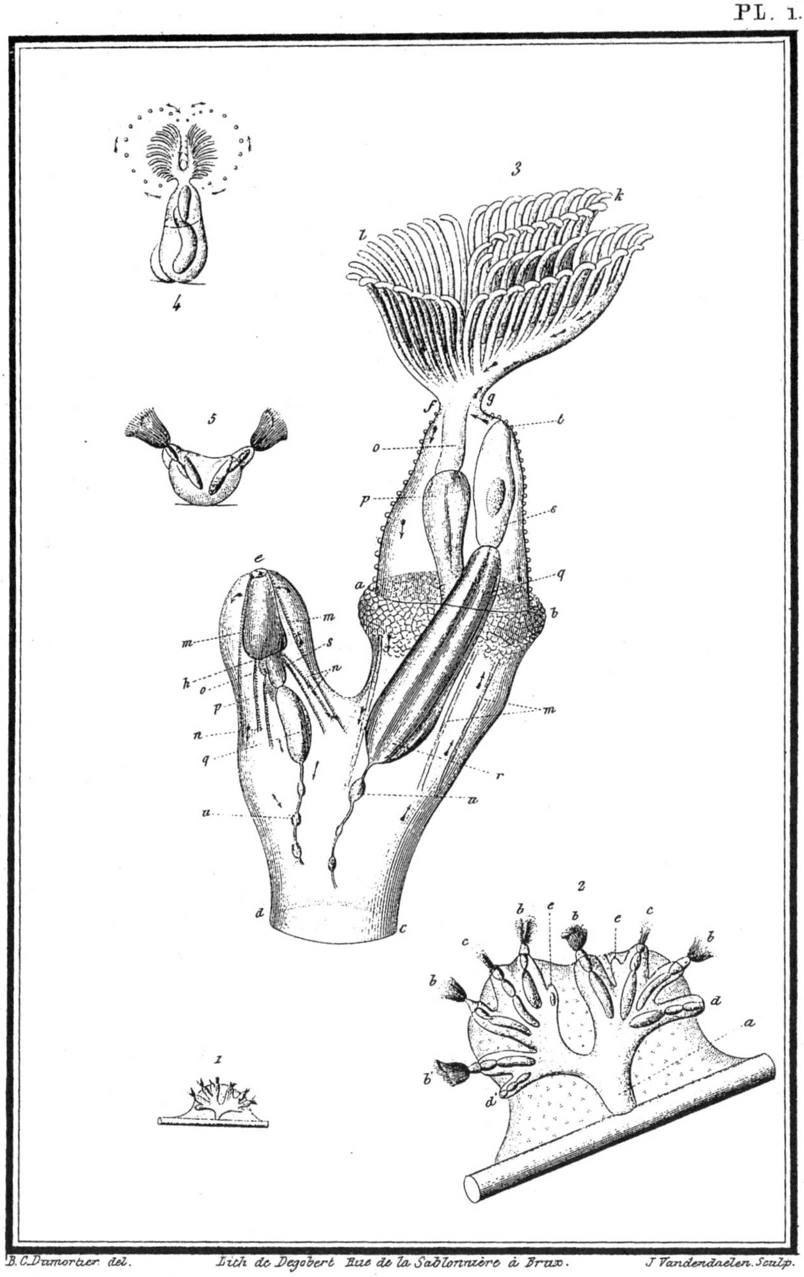

Lophopus crystallinus. Dmrt.

Figure 3. Lophopus crystallinus*: plate from Dumortier (1835).*

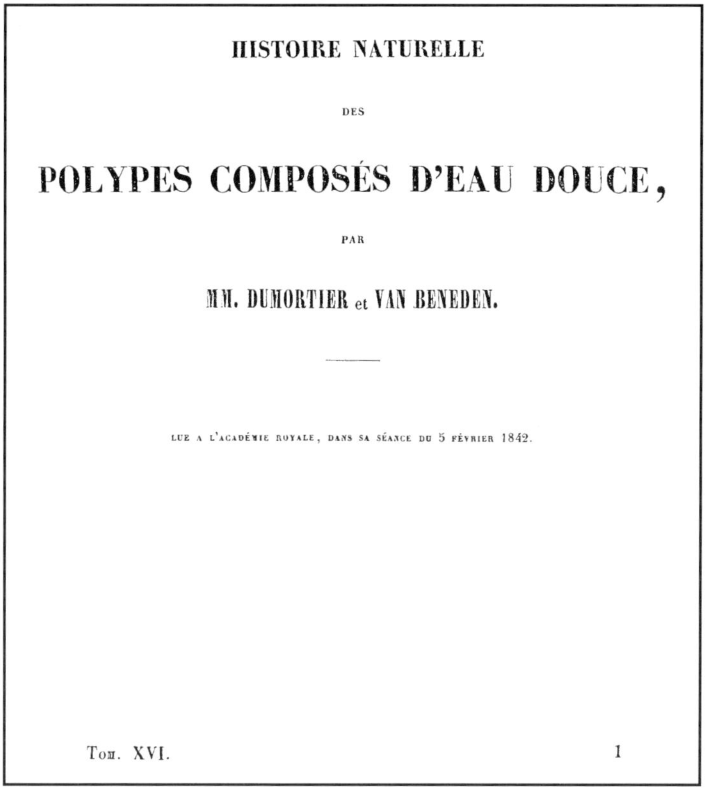
HISTOIRE NATURELLE

DES

POLYPES COMPOSÉS D'EAU DOUCE,

PAR

MM. DUMORTIER et VAN BENEDEN.

LUE A L'ACADÉMIE ROYALE, DANS SA SÉANCE DU 5 FÉVRIER 1842.

Tom. XVI. 1

Figure 4. Dumortier and Van Beneden's Natural History of the composed freshwater polyps published in 1843 in the Nouveaux Mémoires de l'Académie royale de Belgique *(Volume 16).*

that Gervais was wrong.[13]

Dumortier assembled in his newly created 'Lophopodes' (Lophopoda) all the bryozoan species with a U-shaped tentacular crown known at that time, including Blumenbach's *Tubularia sultana*,[14] which he placed in the genus *Plumatella* (*P. sultana*) in spite of its circular tentacular crown. A complete key to the genera and the species (with synonymy) was included in the second edition of his 'Mémoire'.[15]

Dumortier made a distinction between two *Lophopus* species, *L. cristallinus* (sic) (= Trembley's Polype à Panache) and *L. campanulatus*, the last one allegedly known from standing waters in Germany. Dumortier had only seen and studied the first one.

In 1843 the first part of the *Histoire naturelle des polypes composés d'eau douce* by Dumortier with P.J. Van Beneden as joint author came out (Figure 4).

The second part followed in 1850, illustrated with six plates.[16] In this work it was stated that *Lophopus crystallinus* ('lophopier de Trembley') had 'eggs' (i.e. statoblasts) that were oval and lacking spines, whereas *Lophopus campanulatus* ('lophopier couronné') had orbicular 'eggs' with spines.[17] It was even mentioned that Turpin had taken these 'eggs' for those of *Cristatella mucedo*,[18] but obviously Dumortier and Van Beneden did not share this opinion.

It is one of these spiny eggs that is shown in plate VI, with a caption referring to *L. crystallinus*. But it clearly is a statoblast of *Cristatella mucedo*. This was stressed by Allman in 1856 and Jullien in 1885,[19] however, who did not discuss the question of *L. campanulatus*, which on the ground of its 'eggs' should thus correspond to *Cristatella mucedo*.

2. The publications of Pierre-Joseph Van Beneden

In 1839 Pierre-Joseph Van Beneden (1809-1894) published in the 'Bulletin' of the Royal Academy of Sciences in Brussels some observations on the freshwater 'polyps', his first note on freshwater bryozoans.[20] This preliminary note to a larger research Van Beneden was busy with, was discussed later on by Allman[21] who did not agree with Van Beneden's statement that the sexes are separated in *Alcyonella*, which was in contradiction to his own observations. As far as the rest of the article was concerned, Allman made the following summary:

> 'He describes also a circulation of fluid in various parts of the body, and he supposes it due to the action of cilia, which he affirms to exist on the exterior of the alimentary canal as well as on the skin. This motion of the fluid in the interior had been, (...) already noticed by Trembley; and M. Van Beneden now for the first time refers it to its true cause by showing its dependence on vibratile cilia, though he incorrectly describes the external surface of the alimentary canal as ciliated. He believes he has seen at the base of each tentacle an aperture, which he regards as an aquiferous mouth ("bouche aquifère"), destined to give admission to the external water; though in a subsequent memoir he admits that this appearance is deceptive. He describes the great supraoesophageal ganglion, and mentions the existence of locomotive ciliated embryos in *Alcyonella*. He mentions having found, along with M. Gervais, the *Fredericella* and *Paludicella*; and *in fine* he shows how the form of *Alcyonella* is varied by accidental circumstances influencing its growth.'

In the following years Van Beneden who wrote in 1840[22] that during the last six years he had been gathering material on the genera *Alcyonella*, *Plumatella*, *Cristatella* and *Fredericella*, published a series of minor notes with observations on freshwater bryozoans such as observations on *Alcyonella*[23] and considerations on its 'so-called mobile eggs',[24] the development of *Tubularia* (1844),[25] the classification of bryozoans (1850).[26]

In 1848, Van Beneden published his *Recherches sur les Bryozoaires fluviatiles de Belgique*.[27] Allman (1856) called it 'a very valuable memoir on the fresh-water Polyzoa of Belgium', adding the following appreciation: 'The author here enters into important details of the anatomy and classification of the fresh-water Polyzoa, and gives excellent descriptions and copious synonyms of the several species. After the lapse of nearly a century, he restores to the *Alcyonella stagnarum* of Lamarck the specific name *fungosa* originally bestowed on it by Pallas. This is but an act of justice to its discoverer, and ought to be followed by subsequent systematists. Van Beneden also makes the addition of a second species of *Alcyonella*, which he describes under the name of *A. flabellum*.[28] K. Kraepelin considered it as an important new work for its time, with bibliography, synonymy, and discussion of the various systems of organs.[29] For Nelson Annandale (1876-1924), superintendant and trustee of the Indian Museum of Calcutta, founder-director of the Zoological Survey of India, one of the contributors to *The Fauna of British India*, it was 'the first really important work of comprehensive nature', even though it was 'superseded by Allman's *Monograph of the Fresh-Water Polyzoa*' (1856).[30]

Van Beneden studied also the marine bryozoans, and his publications gave quite a good survey on the bryozoan fauna of the Belgian coast,[31] including some entoprocts.[32]

3. Dumortier and Van Beneden as joint authors

As previously mentioned Dumortier, the elder and the more influential of both, had had Van Beneden joining him in his work on freshwater bryozoans. On 5th February 1842 they read before the Royal Academy of Brussels a first paper on the freshwater 'compound polyps'. This *Histoire naturelle des Polypes composés d'eau douce* was published in 1843 and consisted entirely of a historical introduction to the subject.[33] Allman (1856) considered it as 'most elaborate and learned'.[34]

The second part of this work was submitted to the Royal Academy of Sciences of Brussels on 9th May 1848; it was printed in 1850 in Brussels by Hayez, the official printer of the Academy.[35] Pointing out its scientific and iconographic qualities and discussing its main results Allman (1856) was quite enthusiastic about this 'very important memoir':

> 'This memoir is occupied with the anatomy of the genera *Paludicella*, *Fredericella*, *Alcyonella*, and *Lophopus*, and accompanied by numerous well-executed figures. It contains extensive and careful anatomical details of all the genera treated of in the memoir. It describes and figures with much minuteness the development of the bud in *Paludicella*, and mentions the occurrence of a peculiar winter bud in this Polyzoon, occupying the position of the ordinary buds, but destined to remain during the winter months in an undeveloped state. The structure of the testicle in *Alcyonella* is examined, and the spermatozoa, with their vesicles of evolution, demonstrated; but these cells are not sufficiently distinguished from the contained spermatozoon. The ciliated embryos of *Alcyonella* are also described and figured, but the authors do not pursue their development into much detail, while they consider them as identical with statoblasts in a particular stage of evolution, and deprived of their external shell. The statoblasts themselves in *Alcyonella* and *Fredericella* are described, but the essential structure of an ovum is attributed

to them, while some confusion has arisen with regard to the statoblast of *Lophopus crystallinus*, the body described as such being manifestly the statoblast of *Cristatella*.[36] A statement formerly made by M. Dumortier, that the tentacula of *Lophopus crystallinus* are deprived of cilia, is repeated here; it is asserted that, instead of ciliary vibrations, the tentacula of this Polyzoon present a moniliform current, which ascends one side and descends the other of each tentacle; the appearance of these currents is compared to that of an endless chain in uninterrupted motion, and attention is drawn to the analogy of this phenomenon with that of the decomposition of water by the galvanic battery. I have no doubt, however, that the phenomena thus described is truly a case of ciliary vibration, and that the cilia have merely escaped the observer in consequence of some defect in the microscope employed in their investigation. I have repeatedly had under my own observation a species of *Lophopus*, which I do not hesitate to refer to M. Dumortier's species, and yet I found the cilia in all cases perfectly distinct. An opinion previously expressed by M. Van Beneden, when he thought he had seen apertures ("bouches aquifères") for the admission of water into the perigastric space, is here given up, and the source of the error pointed out. Further, M. Van Beneden, now finding a testicle in the same cell with the statoblasts, modifies his previous views as to the unisexualism of *Alcyonella*, and, comparing this Polyzoon to the plants belonging to the twenty-third class of Linnaeus, he suggests that male, female, and hermaphrodite individuals may all coexist in the same coenoecium. On the whole, this memoir of the learned Belgian naturalists, though in some respects incorrect, must be regarded as the most important, in an anatomical point of view of any which had as yet appeared.'[37]

Kraepelin shares this opinion some twenty years later when he writes that 'in respect to bibliography, anatomy and systematics' this second part of the *Histoire naturelle des Polypes composés d'eau douce* 'is the most important earlier work on freshwater bryozoans, together with Allman's monograph'. He stresses that the fidelity of the figures outstrips by far those of Allman's work.[38]

So it may be concluded that in the first half of the 19th century Belgian bryozoology enjoyed an excellent reputation worldwide and has made an outstanding contribution to the research on freshwater bryozoans. Hence the high opinion they were held in by fellow researchers such as Allman who made the following statement in 1850: 'Among those who now most materially advanced our knowledge of the freshwater Polyzoa, must be especially mentioned MM. Gervais, Dumortier and Van Beneden. To Gervais we are indebted for the first complete zoographic view of the subject, the determination and diagnosis of the genera and their systematic distribution, while Dumortier and Van Beneden have both contributed most important information on the anatomical structure of certain species. Van Beneden moreover has given us a complete memoir on the whole of the species inhabiting the freshwaters of Belgium, a memoir, which, both in a zoographical and zootomical point of view, is certainly the most valuable we possess.'[39]

4. Biographical sketch of B.C. Dumortier

Barthélemy Charles Dumortier (Du Mortier)[40] was born on 3rd April 1797 in Tournai where he died on 9th July 1878. He studied the classics in Paris under the direction of a priest. Once returned to Tournai he married Philippine Rutteau in 1816. The couple had eight children.

Very active as a scientist and as a politician, Dumortier is known above all as a botanist. As a child he began collecting plants and from 1816 on he made numerous botanizing tours through the various regions of Belgium, also visiting Holland and the Rhine Provinces. In 1823 he was in Luxembourg where he discovered in the sandstone region of Mullerthal the filmy-fern *Hymenophyllum tunbrigense* an Atlantic species whose presence in this part of Europe was quite sensational and linked to the special microclimate of the sandstone gorges of the Mullerthal.[41] In 1822 his first publication came out: *Commentationes botanicae*; herein he attacked Linné's sexual system insisting on the necessity of introducing a more natural classification, rejecting at the same tome Jussieu's system which in his opinion lacked 'symmetry'. A next step in developing an own system was made in his *Analyse des familles des plantes* in 1829.

Dumortier was a renowned specialist of liverworts. In 1822 he had published a monograph of the Jungermanniae (Hepaticae); a much more complete work on the subject was to be published in 1874: *Hepaticae Europae: Jungermannideae Europae post semiseculum recensitae*, often considered as his most important contribution to botany.

His *Florula belgica* was published in 1827. His other botanical publications were devoted to the grasses (*Observations sur les Graminées*, 1823; *Etude agrostographique sur le genre Michelaria et la classification des Graminées*, 1868), to the willows (1825, 1862), and to a new classification of the fruits (1835).

In 1832 he published his *Recherches sur la structure comparée et le développement des animaux et des végétaux*. Although the conclusions of this comparative study of the structure and the development of animals and plants rapidly appeared quite inaccurate, the paper remains interesting from a historical point of view because it contains the description of the multiplication of cells by formation of a partitioning wall which Dumortier had observed in filamentous freshwater algae. In their biographical note Wildeman and Hauman emphasize that this was prior to the description of the plant cell division made in 1835 by Hugo von Mohl (1805-1872)[42] and that the fatherhood of the discovery of cell division by formation of a partition should be attributed to Dumortier rather than to von Mohl.[43]

It is striking that after 1838 there was a long interruption in Dumortier's botanical publishing which he resumed only when he was nominated chairman in 1862 of the newly founded Belgian botanical society ('Société royale de Botanique').

In the meantime he had turned to zoology, where he distinguished himself above all by his research on bryozoans, the object of his first zoological publication, *Recherches sur l'anatomie et la physiologie des Polypiers composés d'eau douce* (1835), soon followed, as we have seen, by the *Histoire naturelle des Polypiers composés d'eau douce* (1843, 1850) written in collaboration with P.J. Van Beneden. The subjects of his other zoological publications were the development of the embryo of gastropods (1837), some remarks on the '*merle roselin*' (1837) (i.e. *Sturnus roseus*, rose-coloured starling, a migrant bird not often invading our regions), on the modifications of the skull of the orang-utan (*Pongo pygmaeus*) (1838), on a North Sea beaked whale,[44] stranded at Ostend (1839).

Dumortier was also interested in history and archaeology. He wrote papers on two ancient manuscripts discovered in the library of Tournai (1830), on the cathedral of Tournai (1837, 1841, 1845, 1851), on the birthplace of Pierre-Paul Rubens (1861, 1862), on Constantine the Great whom he thought born in Tournai (1866), and on the identity of the holy image called Notre-Dame-la-Brune (1873).

Finally Dumortier wrote quite a number of political pamphlets, articles, speeches and reports as a result of the political activity he began with at the end of the 1820s. As an ardent patriot he was one of the leaders of the anti-Dutch movement in Tournai, and in 1830 he participated actively in the Belgian Revolution. In September 1831 he entered the parliament of the newly independent Belgium. He was to remain a member of this assembly for the rest of his life, and he played an important role in the politics of his country. He was even nominated minister of state.

The Belgian Academy of Sciences admitted Dumortier in 1829. From 1870 to 1878 he was the chairman of the board of trustees of the National Botanical Garden, whose creation was greatly due to him. He was a member of more than forty Belgian and foreign scientific societies. King Leopold II (1835-1909) rewarded him with the title of a count, but Dumortier had concealed this distinction from his family, so that his son didn't know anything about it in his father's lifetime.

His biographers insist that Dumortier was a very committed catholic and that his creationist convictions were never shaken by any Darwinistic ideas.

5. Biographical sketch of P.-J. Van Beneden

Pierre-Joseph Van Beneden (Figure 5) was born at Mechlin, Belgium, on 19th December 1809; he died at Louvain on 8th January 1894.[45] Having finished his secondary studies, Van Beneden was apprenticed to Louis Stoffels (1764-1853), a well-known Mechlin pharmacist and a great collector, especially of minerals and fossils, who had transformed his house into a kind of natural history museum. He aroused Van Beneden's interest in natural sciences. In 1830 Van Beneden took part in the Belgian Revolution, which broke out during his apprenticeship. Later on he recalled that while fighting the Dutch in front

Figure 5. Pierre-Joseph Van Beneden (1809-1894) (from G. Lebrun, Grandes figures de la Belgique indépendante*, Bruxelles, 1936).*

of the walls of Antwerp, he more than once was holding a fossil shell in one hand and a cartridge in the other.

Encouraged by Stoffels, Van Beneden decided to study medicine at the University of Louvain and he got his M.D. in 1832. Then, more interested in zoology than in practicing medicine, he left Belgium in order to complete his scientific education at the Muséum d'Histoire Naturelle in Paris where his remarkable dissecting skills made a great impression.

In 1835 he tried without any success to get a professorship at the newly reopened official universities of Gand or Liège, so that he was quite glad to accept an invitation by the 'rector magnificus' of the Catholic University of Louvain. On 10th April 1836 he was nominated professor of zoology and comparative anatomy. Later on palaeontology was added to these matters. His first lecture, which took place on 23rd April 1836, was a big success. Very popular among his students, Van Beneden continued teaching in Louvain during the next 57 years, until his death.

Throughout his life he was a most diligent worker, and the list of his contributions to scientific periodicals, which starts in 1835, amounts to some 350 papers. A good deal of these were published in the Bulletins or Memoirs of the Academy of Sciences of Brussels which he had belonged to since his nomination as a corresponding member in 1836; he became a full member in 1842.

In the earlier part of his career he directed his attention especially to marine molluscs, then to other marine invertebrates, which he studied during many vacations spent at Ostend, where he established in 1843, at his own expense, a marine laboratory and an aquarium for the further prosecution of these studies. This institution was one of the earliest, if not actually the first example of a place of study of marine zoology in any part of the world. Van Beneden's ambition was to make a complete survey of the animals of the Belgian coast. The results were outstanding publications on Coelenterata, various groups of Worms, Bryozoa, Crustacea, Tunicata, Fishes and Cetacea of the North Sea. Besides the purely faunistic aspects, Van Beneden was keen on considering the anatomy and embryology of his material.

Associated with this part of his work were his classical studies in connection with parasitic worm-like animals, the linguatulids and the so-called tetrarhynchids. Starting in 1847, Van Beneden investigated their development, transformation, and life histories. The linguatulids (Pentastomida, tongue worms) are endoparasites of the respiratory organs of Amniota especially reptiles (lungs), more rarely mammals (nasopharynx) and marine aquatic birds (air sacs). At the beginning of 1848 Van Beneden had the opportunity of studying fresh linguatulids (including eggs and embryos) he found in the lung of a boa received from the Antwerp Zoo. He recognised that these parasites were not worms and did not belong to the trematodes or to the cestodes. His conclusion was that in the light of their anatomical and embryological characteristics they had to be arthropods. He ranged them into the Lernean Crustaceans; later on the linguatulids or Pentastomida were considered as a subphylum of the Arthropoda[46] or an arthropod group of uncertain position (Arthropoda incertae sedis),[47] or as a separated 'pararthropod' or 'proarthropod' phylum. Recently they have been again replaced in the Crustacea by some authors,[48] but the discussion on their systematic position seems not yet closed.[49]

The so-called tetrarhynchids were known as parasites of teleosts; their body shows no annulation and its anterior part bears four retractile proboscides. Van Beneden noticed that typical tetrarhynchids were only hosted by teleosts and that they had no sexual organs. In skates and dogfishes, however, he found cestodes resembling to tetrarhynchids but differing from them by the presence of segments in the posterior part of their body and by the presence of sexual organs revealing that these were adult individuals. Van Beneden's intuition was that possibly the tetrarhynchids of the teleosts represented an immature developmental stage that could continue its development in the alimentary canal of the elasmobranchs feeding on their hosts. Thus the tetrarhynchids were no longer independent species, but an element of the life cycle of cestodes. The zoological group 'cystic worms' had to be dropped. Van Beneden first mentioned his results in May 1847 in a foot note in the article *Un mot sur la reproduction des animaux inférieurs*, then in January 1849 in a more explicit report on the development of the tetrarhynchids (*Note sur le développement des Tétrarhynches*) and finally in October 1849 in his article *Les Helminthes cestoïdes, considérés sous le rapport de leurs métamorphoses, de leur composition anatomique et de leur classification.* [50]

Van Beneden did not hesitate to generalize his results proclaiming that the 'cystica' are in fact 'incomplete tenioids'. This was quite a revolutionary explanation of the cysticerci, coenuri and echinococci of higher vertebrates and especially mammals, no longer considered as independent species but as larval stages of tapeworms. Van Beneden's views were soon experimentally confirmed by the German physician Friedrich Küchenmeister (1821-1890).[51] In order to convince some French opponents to his theory Van Beneden infested with cysticerci two dogs belonging to a lot of four, and demonstrated in April 1858 in Paris that *Taenia serrata*, the dog-tapeworm, was only present in the experimentally infested animals and not in the two control dogs.[52]

The elucidation of the life cycle of the intestinal worms stroke a blow at the idea still alive at that time that cestoid worms could be the result of a spontaneous generation. It was now up to Pasteur to give the ultimate demonstration that abiogenesis did not exist in bacteria either.

In 1853 Van Beneden was awarded the "Grand prix des sciences physiques" of the Institute of France for his work on the cestodes. In 1875 he published his famous vulgarising work *Les commensaux et les parasites dans le règne animal* (Commensals and parasites in the animal kingdom), which was translated into English and German. In 1859 he had published a medical zoology (*Zoologie médicale*) written in collaboration with his friend Paul Gervais.

Another direction in which Van Beneden's activities found a vent was connected with the vertebrate division of the animal kingdom. During the excavations rendered necessary by the fortifying of Antwerp a number of bones of fossil whales were exposed to view. These attracted Van Beneden's attention and led him to undertake a detailed study of the group, whose characteristics were at that time very imperfectly known. On the subject of the Cetacea, living and extinct, he published a number of papers and several large works. The most important of these is his *Ostéographie des cétacés vivants et fossiles* (The osteology of cetaceans, living and fossil), which was written in collaboration with Paul Gervais and published between 1868 and 1880. His papers on the extinct species found near Antwerp were published in the *Annales du musée royal d'histoire naturelle de Bruxelles*, and with them was incorporated a description of the fossil seals which were discovered in the same neighbourhood. But it must be said that as a whole the excavations of Antwerp were rather deceiving and Van Beneden stated later on that he had the feeling of having lost some twenty years of his life in this undertaking.

Van Beneden was a foreign member of the Royal Society and also of the Linnaean, Geological, and Zoological Societies of London. He was a corresponding member of the French Academy of Sciences in 1866 and an associated foreign member in 1892. He was president of the Royal Belgian Academy in 1881, and was created Grand Officer of the Order of Leopold on the occasion of his professorial jubilee.

'He was always a devout and convinced adherent of the Catholic Church — wrote the Catholic Encyclopaedia *— though, as the writer of his obituary for the Royal Society particularly states, always exhibiting "the widest toleration for the views of others"'*. So it was Van Beneden who in 1859 pointed out to the Belgian Academy Darwin's work on the origin of species. Still a convinced fixist at that time, he later on changed gradually to a more transformist speech in his works on cetaceans.

Van Beneden was married to Rose Valcke, daughter of an Ostend citizen enriched by trading with the Far East. They had six children, five daughters and one son, Edouard, the eldest of the children. Edouard Van Beneden (1846-1910), professor at the University of Liège, was a distinguished biologist and zoologist. His greatest achievement was the discovery of the meiosis and the centrosome.[53]

6. Henri Milne-Edwards, a Frenchman born in Belgium

In 1843 Dumortier and Van Beneden wrote that their 'learned compatriot' Milne-Edwards (they spell it 'Milne Edwars')[54] had written several beautiful works on marine 'polyps'.[55] We are moreover told that in 1839 he wrote a report on a memoir of Gervais, and that he pretended in this report that the 'eggs' of *Plumatella* are equipped with hooks just as those of *Cristatella*.[56] A blatant error, of course!

Henri Milne-Edwards[57] is traditionally considered as a French naturalist although he was born in Bruges (Belgium) on 23rd October 1800. (From 1795 to 1814 Belgium was a part of the French Republic, which had annexed the country). He was the 27th of the 28 (certain sources say 29) children of the Englishman William I. Edwards, who had been a planter and colonel in the militia in Jamaica. His mother was Elizabeth Vaux, Williams second wife, and Henri was her second child (so she was just responsible for a few of the many children). Henri's parents returned from Jamaica to England, but moved soon to Bruges, Belgium, where Henri was born, passed his childhood and started his studies.

Then he left the country and was brought up in Paris by his older brother William Edwards,[58] a well-reputed physiologist, because his father had been imprisoned for several years, having helped some Englishmen to get out of Belgium during the war. After the fall of Napoleon, the father was released and then the rest of the family also moved to Paris. It seems that Henri Milne-Edwards was naturalized in 1831.[59]

Henri took a M.D. in 1823. He was, however, more interested in zoology and became a disciple of Georges Cuvier (1769-1832). In 1823, the same year his father died, he married Laura Trézel,[60] who bore him nine children and also assisted him by illustrating several of his works. Henri was appointed Professor of hygiene and natural history in 1832 (the year after he became a French citizen)[61] at the 'Ecole Centrale des Arts et Manufactures'. He moved to the Chair of Entomology at the Paris Muséum in 1841, after the death of his predecessor, friend and joint author Audouin.[62] In 1862 he was transferred to the mammological chair, which had long been vacant after the decease of Geoffroy Saint-Hilaire (1772-1844), and he also became professor of the 'Faculté des Sciences'.

Henri Milne-Edwards published numerous publications on zoological subjects. During more than fifty years he was the editor of *Annales des sciences naturelles*. He died in Paris in on 29th July 1885.

Henri was the father of Alphonse Milne-Edwards (1835-1900), who was director of the Paris Natural History Museum during the last decade of the 19th century. Alphonse studied medicine (M.D. in 1860) and biology (Dr Sci. in 1861) and was appointed assisting naturalist at the Paris Muséum in 1862. In 1864 he was appointed Professor of Zoology of the School of Pharmacy (Ecole supérieure de Pharmacie) in Paris, but succeeded his father at the division of Mammalia and Aves at the Muséum in 1876; member of the Academy of Sciences in 1879 and of the Academy of Medicine in 1885, he eventually became director of the Muséum in 1892.[63]

The name Milne was originally one of Henri's first names, but Henri put it to his family name in order to be discriminated from all his relatives. He usually did not use a hyphen between Milne and Edwards, but his son Alphonse always did and in literature references (including taxon authorship) the name of both father and son is most often hyphened.

7. Remarks concerning Gérard-Daniel Westendorp

In 1843 Gérard-Daniel Westendorp, a military physician born in The Hague on 8th March 1813 and who died in Termonde on 31st January 1868, published an article on his researches on the marine bryozoan fauna of Belgium with special reference to the coast of Ostend: *Recherches sur les polypiers flexibles de la Belgique et particulièrement des environs d'Ostende*, Bruges, F. De Pachtere, 1843, 48 p, 1 plate. This was reprinted from *Annales de la Société médico-chirurgicale de Bruges*, t. IV.[64]

In 1853 he published on the same subject a booklet with eight pages and 32 plates with mounted specimens: *Polypiers flexibles de la Belgique. Collection des bryozoaires, sertulaires, flustres et spongiaires qu'on rencontre en Belgique, et particulièrement aux environs d'Ostende*, published in Courtrai by Beyard-Feys.[65]

8. Karel Loppens and his research on bryozoans

At the beginning of the 20th century Karel Loppens began publishing a series of minor notes on marine, brackish and freshwater bryozoans of the Belgian coast. His first paper published in 1903 dealt with a variety of *Membranipora membranacea* and some other animals living in brackish water. In 1905 he first mentioned a phylactolaemate, *Plumatella repens*, that he had found in the slightly brackish water of the old canal of Veurne. In his *Contributions à l'étude des Bryozoaires de Belgique* (1906) he gives a list of the Belgian freshwater bryozoans: *P. repens*, *P. repens* var. *fungosa*, *P. lucifuga*,[66] *Fredericella sultana*, *Lophopus crystallinus*, *Cristatella mucedo* and the gymnolaemate *Paludicella ehrenbergi* = *Paludicella articulata*; the same paper also gives the list of the marine ecto- and entoprocts.

In the same year he published the paper *Bryozoaires marins et fluviatiles de la Belgique* (1906). It was followed in 1908 by *Les Bryozoaires d'eau douce*, in which he made a general analysis of the systematic problems linked to the freshwater bryozoans hitherto described; moreover he gave a complete survey of the bryozoans worldwide noted in freshwater, including ctenostome species such as *Paludicella*, *Victorella*, *Potsiella*, *Arachnidium*, and cheilostomes such as *Hislopia*, *Norodonia*, *Membranipora*. In 1909 Loppens published a catalogue of the freshwater bryozoans with a note on *Victorella pavida*, essentially a summary of the preceding paper. And so was his next publication, *Les Bryozoaires d'eau douce d'Europe* (1910), which however gave some additional hints on the preparation and conservation of bryozoans, as well as a dichotomic key with extensive synonymy of the species.

K. LOPPENS

Bryozoaires marins et fluviatiles de la Belgique

Extrait du *Bulletin des Naturalistes Belges* — Bruxelles
1948

Figure 6. K. Loppens's last publication on bryozoans (1948).

Figure 7. Karel Loppens (1875-1962) in his 40s. (from P. Van Oye, Karel Loppens, onbekend pionier van het brakwateronderzoek. *1963).*

After an interruption of several decades, Loppens published in 1948 a second and revised edition of his *Bryozoaires marins et fluviatiles de la Belgique* (Figure 6).

9. Biography of Karel Loppens

Charles (Karel) Pierre Jean Loppens (Figure 7)[67] was born on 1st January 1875 in Diksmuide (Western Flanders) as the son of a postmaster, who died when the child was two years old. Karel's widowed mother, Marie Louise Gommers, took him to Nieuwpoort, where he would spent most of his life. After an apprenticeship with one of his uncles, Karel Loppens established himself as a watchmaker in his mother's house in Nieuwport. But more attracted by natural sciences, he gave up watch-making and devoted himself to geology, physical geography and zoology, later on he turned more and more to local archaeology and history. In both fields Loppens was a self-made man.

His mother died in 1898. In 1912 he married Elisa Vandenbroecke from Ramskapelle who quickly adapted herself to his way of life and helped him in his works. The couple made numerous trips through Belgium, the Netherlands, Switzerland and France; during World War I they left for Exmouth (Devonshire), in south-western England.

From 1919 to 1920 Loppens lived with his wife in Ostend; in March 1921 he established himself in Coxyde where he remained until the end of his life. He died on 26th May 1962 in Veurne, aged 87 years old.

Before World War I, Loppens had worked for some time in the Museum for Natural History in Brussels, in the laboratory of Gustave Gilson (1859-1944),[68] the director of the museum, and in the laboratory of Overmeire directed by Ernest Rousseau (1872-1920).[69]

In the zoological field Loppens started as a malacologist, and his first communication in the annals of the *Société zoologique et malacologique de Belgique* was a note on *Petricola pholadiformis*, the American piddock, whose occurrence on the Belgian coast he was the first to report in 1902. There were some other papers, all of just regional importance. He soon concentrated on bryozoans, where his contribution to the knowledge of the marine bryozoans (fauna of the Belgian coast) clearly towered above his achievements in freshwater bryozoology.

Paul Van Oye[70] insisted on the value of Loppens's planktological work, ranging him with the great hydrobiologists of Belgium. His study *Contribution à l'étude du micro-plankton des eaux saumâtres de la Belgique* was the first of that kind in Belgium. It did not only contribute to a better understanding of the ecology of the brackish water, but revealed quite a number of species not yet recorded in Belgium. Whereas, the *Faune de Belgique* (1895) had reported 28 species of Rotatoria for the whole country, Loppens had found 34 different species alone in the brackish water of Nieuwpoort.

Loppens wrote also some vulgarising articles on biological subjects. At last he published quite a number of papers concerning the history and the archaeology of his home region.[71]

10. Paul Brien's work on bryozoans

In 1932 Paul Brien (1894-1975) (Figure 8)[72] published a two-page note on the asexual reproduction of the Phylactolaemata (in collaboration with M. Aen den Boom).[73] The next step were two reports published in 1934, the first on his studies of the statoblast development of *Fredericella sultana* and the second on the artificial regeneration in *F.*

Figure 8. Paul Brien (1894-1975) (from M. Poll and H. Herlant-Meewis, Paul Brien, 1894-1975. *in Florilège des sciences en Belgique, II,1980).*

sultana after cutting off the first cystids and polypids derived from a statoblast. In 1936, it was a more important contribution to the study of the asexual reproduction of the phylactolaemates. Then there was a long break, and Brien resumed his work on the bryozoans only at the beginning of the 1950s with as a result a paper on the affixing to the substratum and the metamorphosis of phylactolaemate larvae (*Plumatella fungosa*, *F. sultana*) in 1952. In 1954 Brien published his *Etude sur les Phylactolémates*, an important comprehensive study on bryozoans where he included the results of his own long research on topics such as: zooecial evolution and colony growth by polypidal budding, budding of the statoblast, embryogenesis and astogeny, and multiple ontogenesis (development of a new individual as a result of polypidal budding, or statoblast budding, or embryogenesis). The studied genera were: *Fredericella*, *Plumatella*, *Lophopus* and *Cristatella*.

The next papers were consecrated to general considerations on phylactolaemates (1954) or more specific subjects such as the relationship between sexual and asexual reproduction (1955), the formation and the metamorphosis of the larvae (1956). Then followed some pages on the bryozoan cystid and the colony growth (1958).

In 1960, Brien was one of the collaborators in Grassé's[74] monumental *Traité de Zoologie* where he was in charge of the chapter 'class of the Bryozoa', which he divided in the two sub-classes Phylactolaemata and Gymnolaemata. In 1959 he had already written the chapter concerning the Entoprocta or Kamptozoa. Prior to his contribution to Grassé's treatise Brien had discussed the budding of the Entoprocta and their phylogeny (1956), comparing entoprocts and ectoprocts in this respect in 1960. In 1970 he concluded with phylogenetic considerations concerning the lophophorates.

Brien was one of those who insisted on separating the entoprocts from the ectoprocts, putting them into two different phyla. In his opinion entoprocts are neotenic forms of annelids that have become aberrant as a consequence of their sessile life, whereas ectoprocts are far more advanced in their evolution and should be placed near the Phoronida.[75]

Brien was less interested in marine bryozoans. Except for his contribution to Grassé's treatise (1960) Brien had published just one article on marine bryozoans, in 1937, a study on the growth and the budding of the stolon of the Stolonifera (*Bowerbankia*), with Graziella Huysmans as joint author.[76]

11. Biography of Paul Brien

Paul Brien was born in the village Hannut (province of Liège) on 24th May 1894; he died at Watermael-Boitsfort, on the outskirts of Brussels, on 19th February 1975.[77]

The son of a primary school teacher, Paul Brien went to the teacher training school ('Ecole normale') of Nivelles, and eventually became teacher in his native village, from 1915 to 1918. Once the war was over, he decided to go to the 'Université Libre de Bruxelles' (ULB) in order to study natural sciences. He was a student of the renowned zoologist Auguste Lameere (1864-1942),[78] an excellent professor and an avant-gardist of the evolutionist ideas. Brien got his Ph.D. in 1922 with a thesis on the embryology of *Salpa maxima* (Tunicata) and he became assistant in the zoological service of Lameere. In 1926 he was charged with a biology course at the Faculty of Social Sciences, in 1928 he taught animal physiology at the science faculty, in 1930 he got a nomination as ordinary professor. When Lameere retired in 1934, Brien was put in charge of lower level courses (first and second year): elements of zoology, compared anatomy, animal biology. From 1937 on, he taught animal biology and systematic zoology (including compared anatomy and histology) to students preparing a 'licence en zoologie' (third and fourth year). Brien stayed with the ULB until his retirement in 1964.

In 1926 Brien had married Emilie Gavage, doctor of natural sciences of the ULB, and professor at the 'Athénée royal' (secondary school) of Ixelles. Their son Jean-Pierre was born in 1928.

From 1922 to 1975, Brien produced 342 scientific publications, the first one on *Salpa maxima*.[79]

Brien continued his research on the Tunicata studying biological phenomena such as regeneration and budding. He was particularly keen on the question of germplasm and somatoplasm. Noting that during budding a whole organism could be reconstituted by somatic cells alone, without intervention of the germplasm, Brien concluded that the duality of germplasm and somatoplasm, such as postulated in 1893 by August Weismann (1834-1914) in his theory of the continuity of the germplasm, does not exist in metazoans.

In 1929 Brien began studying sponges, their morphogenesis and their embryogenesis, and also, of course, their extraordinary regeneration power. So he entirely dissociated sponge cells by filtrating fragments of *Ephydatia fluviatilis* through a fine-meshed cloth, and these cells were able to fall into small spherical groups or aggregates developing in entire new sponges (somatic embryogenesis). Brien studied also the sexual reproduction of sponges.

He was then attracted by the Bryozoa and afterwards by the Hydrozoa. There his idea that there is no immortal germinal line was confirmed.

Principally an invertebrate zoologist, Brien had nevertheless the occasion to make an important contribution to the knowledge of the vertebrates, especially dipnoans. During an expedition to Katanga in 1937 he discovered a hitherto unknown dipnoan, *Protopterus brieni*, and he studied the estivation of these lungfishes. In 1957 he participated in a mission to Leopoldville, and there he had the luck to find a *Protopterus* still in its nest, so that he could describc the ethology of the reproduction of the African *Protopterus dolloi*.

Brien was the author of quite a number of successful books, among these his *Eléments de zoologie et d'anatomie comparée* with a first edition published in 1938, and a second edition in 1951 (Volume 1) and 1961 (Volume 2). Many students have appreciated his *Guide des travaux pratiques* (first edition 1938). He wrote numerous biographical or necrological notes, and in 1968 he outlined the history of zoology and animal biology in

Belgium during the 19th century and the beginning of the 20th (in *Florilège des sciences en Belgique*, edited by the Académie royale de Belgique, Classe des Sciences, Brussels) A complete list of Brien's publication is given by Poll and Herlant-Meewis (1979).

Brien was a free thinker highly interested in the problems of evolution. Fighting for the right of free expression for all, he was known in Belgium as a defender of French culture and French language.

12. A.W. Lacourt and the Belgian bryozoans

Paul Brien was not immediately involved in faunistic research on the Belgian bryozoans, Loppens's publications were outdated, so good luck would have it that the Dutch specialist Adrianus Willem Lacourt looked into the freshwater bryozoans of the neighbouring country. In 1949 he published a new survey on the phylactolaemates of Belgium: *Les Bryozoaires d'eau douce (Phylactolaemata) de la Belgique* essentially based on the examination of the collection of the Royal Institute of Natural Sciences of Belgium and some literature research. Lacourt noted the occurrence of the following eight species in Belgium: *Fredericella sultana*, *Plumatella fruticosa*, *P. emarginata*, *P. repens*, *P. fungosa*, *P.* (*Hyalinella*) *punctata*, *Lophopus crystallinus*, and *Cristatella mucedo*. Concerning the rare *P. fruticosa* (which he had not seen himself) he remarks that the occurrence of this species remains uncertain. This situation was due to Loppens who, for what in his opinion was the same species, used either the name *P. emarginata* or *P. lucifuga*, which he considered both as synonyms of *P. fruticosa* Allman. So one or the other of his *P. emarginata* or *P. lucifuga* specimens could possibly correspond to *P. fruticosa*.

It may be remarked that in 1948 Lacourt had published a note 'On two freshwater Bryozoa (Phylactolaemata) from Belgian Congo'.[80]

Adrianus Willem Lacourt was born on 6th June 1910 in Leiden where he would remain for all his life.[81] As a young man he worked for one and a half year as a volunteer in the Museum of Natural History of Leiden (Rijksmuseum van Natuurlijke Historie), where he was busy with birds, especially their skulls and skeletons.

Figure 9. Signature of Adrianus Willem Lacourt (1910-1987).

Then he got a job in the Pathological Laboratory of the University of Leiden, where he was working at the beginning in the annexed museum. Some ten years later he was gradually involved in the administration of the laboratory. He finished his career as a high-ranking administration official after 53 years of service. Unfortunately his health left much to be desired, as he suffered from angina pectoris.

Lacourt was a very active member of the Dutch Malacological Society (Nederlandse Malacologische Vereniging, N.M.V.) Indeed, since he was sixteen he had been interested by shells, then he gave himself up to the study of brachiopods, bryozoans, crustaceans, cephalopods and egg-capsules of rays and skates (mermaid's purses). He published on all these fields, in periodicals such as: Zoölogische Mededelingen (Leiden), Basteria,[82] Het Zeepaard,[83] Correspondentieblad van de Nederlandse Malacologische Vereniging).[84] After his retirement he was nominated honorary scientific collaborator of the Rijksmuseum.

As far as the molluscs were concerned, Lacourt confined himself above all to the taxonomy of marine species that he collected along the coast of the Netherlands, Great Britain, Ireland and Western France (especially Arcachon and its surroundings). Lacourt is most famous for his monograph of freshwater bryozoans he published in 1968.

Lacourt died on 2nd February 1987. In the obituary notice published by the N.M.V., we are told that Lacourt was rather lonesome, very formal, reserved and taciturn, hiding his emotions, pugnacious, not very keen on making friends, on the whole a somewhat odd character. Once as he was due to be operated for an inguinal hernia on his left side, he decided that there was too great a risk of being operated on the wrong side. So he saw to it that prior to his entering the operating theatre he had a clear inscription on his belly indicating where the operation should be done.

Of course there have been bryozoological studies and publications in Belgium after Brien and Lacourt. Our approach being historical we do not intend to make the complete list of what has been done in the last decades. In the 1970s Pierre Job initiated studies about the role of bryozoans in pollution and self-purification;[85] B. Goethals, V. Henry, C. Brene and others followed him in the 1980s.[86] But there were also studies about topics such as the chemical composition of the ectocyst and the capsule of statoblasts.[87]

13. Bryozoan studies in the Grand-Duchy of Luxembourg

The first record of a freshwater bryozoan in Luxembourg was made in 1901. At the monthly meeting of the Société des Naturalistes luxembourgeois that took place on 14th July 1901 the physician Dr Eugène Bricher[88] exhibited a bryozoan he had found in the Alzette in Pulvermühl, a suburb of the city of Luxembourg, encrusting a rusty spade in a two and a half meter's depth. He identified it as *Plumatella fungosa* Pall. (Bryozoa); astonishingly it is called a freshwater sponge (!) in the report (written by Bricher himself!).[89]

In 1956 Professor Jos Hoffmann (Figure 10)[90] encountered a colony of *Cristatella mucedo* in the Wark River near Warken, but did not publish it. From 1958 to 1959 Erik Mauch,[91] a German hydrobiologist, studied the benthos of the Moselle River, from Coblenz (Germany) to Schengen (Luxembourg). In the Luxembourg part he reported three species: *Plumatella fungosa*, *P. emarginata* and *Fredericella sultana*. His results were recorded in his doctoral thesis submitted in 1961 (University of Frankfurt), and which was published in 1962 and in 1963.[92] In 1978 and 1979 Mauch made a control examination of the Moselle between Palzem and Grevenmacher; he noted a strong regression of *P. fungosa*.[93]

A complete faunistic survey of the freshwater Bryozoa of the Grand Duchy was given in the scientific dissertation *Les Bryozoaires (Lophophoriens, Ectoproctes) du Grand-Duché de Luxembourg*,[94] which Gaby Geimer (Figure 10)[95] submitted in 1975. This was part of the examination giving access to the secondary school teacher career in Luxembourg at the end of one's practice period. The fieldwork had been done from 1974 to 1975 and had been supervised by Professor Jos Hoffmann, a former student of Paul Brien. The list of the Luxembourg freshwater Bryozoa was enlarged and included now the following

Figure 10. Looking for bryozoans in Luxembourg (1974): Jos Hoffmann (left), Gaby Geimer (centre).

species: *Fredericella sultana*, *Plumatella emarginata*, *P. fruticosa* (new record), *P. repens*, *P. fungosa*, *Hyalinella punctata* (new record), *Cristatella mucedo*, and *Paludicella articulata* (new record). The results were presented during the meeting of the section of sciences of the grand-ducal Institute of 18th February 1976.

The dissertation was not published, but was used as the basis of a new survey of the bryozoans of Luxembourg and the close Belgian, French and German borderland: *Les Bryozoaires du Grand-Duché de Luxembourg et des régions limitrophes*[96] published in 1986 by Gaby Geimer and Jos Massard.[97] A total of 172 sites had been investigated from 1983 to 1985. *Plumatella casmiana* could be added to the list of Luxembourg bryozoans. The description of the different species was completed by a SEM study of the statoblasts and hibernacula. Criteria for separating *P. repens* and *P. fungosa* by means of SEM study of the floatoblasts were given.[98] The possible confusion between hyaline forms of *P. repens* and *Hyalinella punctata* was discussed. The distribution of the species was illustrated by grid-maps, and a key to the European freshwater bryozoans was added. In the following years the same authors completed the survey and understanding of the Luxembourg bryozoan fauna.[99] And they also studied these astonishing animals in some other parts of the world.[100]

14. Acknowledgments

The authors would like to thank the International loan service of the National Library of Luxembourg, as well as the National Museum of Natural History of Luxembourg and the University Centre of Luxembourg. Thanks are extended to Sepp Simon for reading the manuscript.

Appendix 1. List of the publications on Bryozoa by Karel Loppens

(1) Publications concerning only marine bryozoans:

'Sur une variété de *Membranipora membranacea* L. et sur quelques animaux vivant dans l'eau saumâtre', *Annales de la Société royale zoologique et malacologique de Belgique,* 38 (1903), CXLII-CXLIII.

'Bryozoaires et Cnidozoaires nouveaux pour la faune belge trouvés pour la plupart pendant l'année 1903', *Annales de la Société royale zoologique et malacologique de Belgique*, 39 (1904), XLV-XLVII.

'Animaux marins vivant dans l'eau saumâtre', *Annales de la Société royale zoologique et malacologique de Belgique*, 40 (1905), VII-VIII [mentions: *Farella repens*].

'Bryozoaire nouveau pour la faune belge', *Annales de la Société royale zoologique et malacologique de Belgique*, 40 (1905), XXII [concerns: *Schizoporella hyalina*].

'Rapide multiplication de quelques Bryozoaires et Hydroïdes', *Annales de la Société royale zoologique et malacologique de Belgique*, 40 (1905), bull. séances, XXII-

XXIII [concerns: *Flustra foliacea* and other bryozoans].
'Bryozoaire nouveau pour la faune belge', *Annales de la Société royale zoologique et malacologique de Belgique*, 40 (1905), XLIX [concerns: *Alcyonidium mamillatum*].
'Sur quelques variétés de *Membranipora membranacea* vivant dans l'eau saumâtre', *Annales de biologie lacustre*, 1 (1906), 3 p. [separatum].
'Sur les caractères distinctifs entre *Alcyonidium gelatinosum* et *Alc. hirsutum'*, *Annales de la Société royale zoologique et malacologique de Belgique,* 42 (1907), 169-174.
'Influence du milieu sur la composition chimique des zoécies des Bryozoaires marins', *Annales de la Société royale zoologique et malacologique de Belgique,* 51 (1920), 91-110.

(2) Publications concerning freshwater and brackish water or marine Bryozoa:

'*Plumatella repens* L. vivant dans l'eau saumâtre', *Annales de la Société royale zoologique et malacologique de Belgique*, 40 (1905), bull. séances, XLIX-L.
'Contributions à l'étude des Bryozoaires de Belgique', *Annales de la Société royale zoologique et malacologique de Belgique,* 41 (1906), 130-136 [list of Entoprocts, Ectoprocts including Phylactolaemates and *Paludicella* (Gymnolaemates)].
'Bryozoaires marins et fluviatiles de la Belgique', *Annales de la Société royale zoologique et malacologique de Belgique*, 41 (1906), 286-321, figs 1-50.
'Contribution à l'étude du micro-plankton des eaux saumâtres de la Belgique', Annales de Biologie lacustre, 3(1908-1909), 16-53 [Bryozoa: p. 17 (*P. repens* f. *fungosa)*, p. 25ss (statoblasts of *P. repens)*].
'Les Bryozoaires d'eau douce', *Annales de biologie lacustre*, 3(1908-1909), 141-183.
'Bryozoaires', *Annales de biologie lacustre*, 3 (1909), 428-430 [review of articles on freshwater bryozoans written by Annandale, Braem, Kraepelin, Oka, Rousselet, Sollas, Weltner, Wesenberg-Lund].
'Catalogue des Bryozoaires d'eau douce avec une note sur *Victorella pavida'*, *Annales de la Société royale zoologique et malacologique de Belgique*, 44 (1909), 97-110.
'Fauna aquatica europaea. Les Bryozoaires d'eau douce d'Europe', *Annales de biologie lacustre*, 4 (1910), 139-161.
'Bryozoaires marins et fluviatiles de la Belgique', *Bulletin des Naturalistes belges*, 29 (1948), 54-63, 121-128, 138-144, 3 pl. [id. *Extrait du Bulletin des Naturalistes belges* 1948, 27 p.].

Appendix 2. List of Paul Brien's publications on Bryozoa and Entoprocta

'Reproduction asexuée chez les Phylactolémates', *Association française pour l'avancement des sciences, congrès de Bruxelles, comptes rendus de la 56e session* (1932), 278-279 (joint author: M. Aendenboom).
'Développement des statoblastes des Phylactolémates (*Fredericella sultana*). Loi

d'apparition des bourgeons d'accroissement', *Comptes rendus des séances de la Société de Biologie, Société belge de Biologie, séance du 14 juillet,* CXVII (1934), 214-217.

'Régénération artificielle, après section antérieure, des premiers cystides et polypides de *Fredericella sultana*, issus d'un statoblaste', *Comptes rendus des séances de la Société de Biologie, Société belge de Biologie, séance du 14 juillet,* 117 (1934), 217-219.

'Régénération et régulation', *11e Congrès national des Sciences, Bruxelles, 19-23 juin 1935* (1935),1086-1091.

'Contribution à l'étude de la reproduction asexuée des Phylactolémates (Origine des bourgeons polypidiaux et des cystigènes statoblastiques)', *Mémoires du Musée royal d'histoire naturelle de Belgique*, 2e série, fasc. 3 (1936), 569-625.

'La croissance et le bourgeonnement du stolon chez les Stolonifera (*Bowerbankia* Fabre). Evolution du stolon et de la zoécie chez les Bryozoaires', *Annales de la Société royale zoologique de Belgique*, 68 (1937), 13-40 (joint author: P.G. Huysmans).

'Fixation et métamorphose des larves de Phylactolémates (*Plumatella fungosa, Fredericella sultana*), *Comptes rendus des séances de l'Académie des Sciences de Paris*, 235 (1952), 1435-1437.

'Etude sur les Phylactolémates', *Annales de la Société royale zoologique de Belgique*, 84 (1953), 301-440 [published in 1954].

'A propos des Bryozoaires Phylactolémates', *Bulletin de la Société zoologique de France*, 79 (1954), 203-239.

'Les Endoproctes et la classe des Bryozoaires', *Annales de la Société royale zoologique de Belgique*, 85 (1954), 59-87.

'Relations entre les reproductions sexuée et asexuée. A propos des Phylactolémates', *Annales de la Société royale zoologique de Belgique*, 86, fasc. 2 (1955), 169-188 (joint author: C. Mordant).

'Le bourgeonnement des Endoproctes et leur phylogenèse', *Annales de la Société royale zoologique de Belgique*, 87, fasc. 1 (1956), 27-43.

'Formation et métamorphose des larves de Phylactolémates', *Proceedings of the XIVth International Congress of Zoology, Copenhagen* (1956), 519-520.

'La signification du cystide des Bryozoaires et la croissance de la colonie', *Académie royale de Belgique, Bulletin de la Classe des Sciences*, 44, 5e série (1958), 750-766.

'La reproduction sexuée', *Année biologique*, 34, fasc. 5-6 (1958), 242-297 [II. Bryozoaires].

'Classe des Endoproctes ou Kamptozoaires', in P.P. Grassé (éd,), *Traité de Zoologie*, vol. 5, fasc. 1 (1959), 927-1007 (joint author: L. Papijn).

'Classe des Bryozoaires', in P.P. Grassé (éd.), *Traité de Zoologie*, vol. 5, fasc. 2 (1960), 1054-1392

'Le bourgeonnement et la phylogenèse des Endoproctes et des Ectoproctes. Réflexions sur les processus de l'Evolution animale', *Académie royale de Belgique, Bulletin de la Classe des Sciences*, 46, 5e série (1960), 748-766.

'Kamptozoaires. Lophophoriens', *Encyclopaedia Universalis* (1969), 309-312.

'Considérations phylogénétiques à propos des Lophophoriens', *Académie royale de*

Belgique, Classe des Sciences, séance du 6 juin, 46 (1970), 565-579.

Appendix 3. List of some of A.W. Lacourt's publications on Bryozoa

'Bryozoa van Nederland', *Handelingen van de Hydrobiologische Club, Amsterdam*, 6 (1943), 22-23.

'On two Freshwater-Bryozoa (Phylactolaemata) from Belgian Congo', *Rev. Zool. Bot. Afr*, 40 (4) (1948), 229-234.

'Les Bryozoaires d'eau douce (Phylactolaemata) de la Belgique', *Bull. Inst. Roy. Sci. Nat. Belg*, 25(19) (1949), 1-9.

'Bryozoa of the Netherlands', *Archives néerlandaises de Zoo*logie, 8(3) (1949), 289-321.

'Bryozoaires', *Rés. Sci. Explor. Hydrobiol. Lac Tanganyika (1946-1947)*, 3 (1951), 21-25.

'Some remarks on *Plumatella philippinensis* Kraepelin', *Hydrobiologia*, 7(3) (1955), 261-263.

'Freshwater Bryozoa (Phylactolaemata) from Curaçao, Aruba and Bonaire', *Studies on the Fauna of Curaçao and other Caribbean Islands*, 27 (1955), 86-88.

'*Lophopodella pectinatelliformis* nov. spec. (Bryozoa, Phylactolaemata)', *Zool. Meded*, 36 (17) (1959), 273-274.

'Histochemical investigation of *Pectinatella magnifica* (Leidy, 1851)', *Proceedings of the Koninklijke Nederlandsche Academie van Wetenschappen*, serie C, 69 (1966): 22-23 (joint author: A.G.D. Willighagen).

'A monograph of the freshwater Bryozoa. Phylactolaemata', *Zoölogische Verhandelingen, Leiden*, 93 (1968), 1-159.

'De nederlandse mariene mosdiertjes, Bryozoa (Ectoprocta, Gymnolaemata)', *Wetenschappelijke mededelingen van de Koninklijke Nederlandse Natuurhistorische Vereniging,* 129 (1978), 21 p.

'Handleiding voor het projekt Bryozoa van binnenwateren', *Instrukties voor medewerkers EIS-Nederland*, Leiden, 7 (1982), 11 p.

Notes

1 B.C. Dumortier, *Bulletin de l'Académie royale des Sciences et Belles-Lettres de Bruxelles*, 2 (1835), 421-56, pl. V-VI.

2 George James Allman (1812-1898), Irish botanist and zoologist, childhood in Cork, educated at Belfast, studied at the Trinity College in Dublin, 1844 professor of botany in Dublin, then professor of natural history in Edinburgh from 1855-1870; curator at the department of invertebrate zoology at Harvard University, U.S.A. (in 1873); president of the Linnean Society of London (1874-1883). His most important work was his investigation into the morphology of the Coelenterata and the Polyzoa. See: *Proceedings of the Royal Society of London*, 75 (1905), 25-7 (obituary notice signed S J H). — The dictionary of national biography, *'The concise dictionary, part 1, from the beginnings to 1900'* (Oxford, 1953, reprint), 19. — Ilse Jahn, *Geschichte der Biologie* (Jena 1998), 763. — H.G. Hansson, *Biographies of marine*

taxonomists and their friends, http://www. tmbl.gu.se/libdb/taxon/personetymol.1.html. Allman established the fundamental separation of the bryozoans into Gymnolaemates and Phylactolaemates. See: J.L. d'Hondt, 'Histoire et chronologie des recherches sur les bryozoaires', *Bulletin d'Histoire et d'Épistémologie des Sciences de la Vie*, 2(1) (1995), 103.

3 G.J. Allman, *'A monograph of the fresh-water Polyzoa, including all the known species, both British and foreign'*, Ray Society London, 17 (1856), 66. Trembley discovered this animal in April 1741 in freshwater near The Hague, a lobed jelly-like mass, from which protruded numerous polypoid bodies, each characterized by the possession of an elegant crown of tentacula borne on the margin of a crescent-shaped disc. Because of this striking tentacular plume, Trembley naturally supposed this new kind of animal to be intimately related with the polyps; so he called it ' Polype à Panache ' in his *' Mémoires pour servir à l'histoire d'un genre de Polypes d'eau douce, à bras en forme de cornes '* (Leide, 1744). — Abraham Trembley (1740-1784), born at Geneva (Switzerland), best known for his studies on the *Hydra*; he made his chief discoveries between 1739 and 1747 while tutor to two boys at The Hague (*Encyclopaedia Britannica* 1969, vol 22: 213).

4 B.C. Dumortier and P.J. Van Beneden, 'Histoire naturelle des Polypes composés d'eau douce', *Nouveaux Mémoires de l'Académie Royale des Sciences et Belles-Lettres de Bruxelles*, 16 (1843): 8.

5 P.J. Van Beneden, 'Recherches sur les Bryozoaires fluviatiles de Belgique', *Nouveaux Mémoires de l'Académie Royale des Sciences et Belles-Lettres de Bruxelles*, 21 (1848), 22; B.C. Dumortier and P.J. Van Beneden, 'Histoire naturelle des Polypes composés d'eau douce, IIe partie', *Nouveaux Mémoires de l'Académie royale des Sciences et Belles-Lettres de Bruxelles*, 16 (complément) (1850), 65.

6 According to Allman, note 3, p. 92, the genus *Alcyonella* is characterized by the 'coalescence of the tubes into a common mass'.

7 Dumortier and Van Beneden, note 4, p. 27.

8 Dumortier, note 1, p. 6.

9 In his monograph on the German freshwater bryozoans K. Kraepelin (1887) rallied this point of view [K. Kraepelin, 'Die Deutschen Süßwasserbryozoen. I. Anatomisch-systematischer Teil', *Abhhandlungen aus dem Gebiete der Naturwissenschaften, Naturwissenschaftlicher Verein Hamburg,* 10 (1887), 1-168 (page 10)]. — Karl Kraepelin (1848-1915), German zoologist and botanist, studied natural sciences at the universities of Göttingen and Leipzig (Dr. phil. in 1873), secondary school teacher in Leipzig (1873), than at the 'Johanneum' in Hamburg (1878-1889), where he was appointed professor in 1887; director of the Natural History Museum of Hamburg (1889-1914); member of the council of professors of the Colonial Institute in Hamburg (1908). Kraepelin was a wellknown pedagogue who had written some greatly appreciated botanical and zoological schoolbooks. As a zoologist he is most known as a specialist of Tarantulidae, Uropygi, Scorpions, Scolopendrida, Palpigradi, Solifugids, etc. Karl was the brother of Emil Kraepelin (1856-1926), a wellknown psychiatrist. Ref. I. Jahn, note 2, p. 877; H. Weidner, 'Karl Kraepelin', in *Neue Deutsche Biographie*, 12 (1980), 640-641.

10 *Bulletin zoologique*, 2e section, 1835, p. 123 (cf. Dumortier and Van Beneden, note 4, p. 28).

11 Paul Gervais (1816-1879), French palaeontologist and zoologist, obtained doctorates in science and medicine, studied palaeontology as assistant to H.M.D. de Blainville (1777-1850), Cuvier's successor as professor of comparative anatomy in the Muséum National d'Histoire Naturelle (Paris). Gervais served as professor of zoology and comparative anatomy at

Montpellier (1845-1865), and in 1865 as professor of anatomy, comparative physiology and geology at the Sorbonne, in 1868 as professor of comparative anatomy in the Muséum (*Enyclopaedia Britannica*, 1969, vol. 10, 368).

12 B.C. Dumortier, *'Mémoire sur l'anatomie et la physiologie des polypes composés d'eau douce nommés Lophopodes, 2e édition'* (Tournay,1836), 84 p., 2 pl (p. 5-7).

13 cf. Dumortier and Van Beneden, note 4, p. 28, and Allman, note 3, p. 66.

14 Presently known as *Fredericella sultana*, a name created by Gervais in 1838 (with the orthograph *Fredericilla*) as an homage to the deceased Frédéric Cuvier (1773-1838), the director of the menagerie of the Muséum in Paris, member of the Academy of Sciences, the elder brother of Georges Cuvier [cf. P. Tort (dir.) *'Dictionnaire du darwinisme et de l'évolution'* (Paris, 1996), vol. 1 (A-E), 747-8]. In his *'Monographie des bryozoaires d'eau douce'* (Bulletin de la Société Zoologique de France, 10, Paris, 1885), p. 1, the French physician and zoologist Jules Jullien (1842-1897) insists that concerning his *Plumatella sultana* Dumortier was right or at least getting close to the truth. In fact, Jullien is persuaded that this form is nothing else but a monstrosity of what he calls *P. lucifuga* (which essentially corresponds to *P. fruticosa*) (cf. Jullien 1885, p. 121ss, G. Geimer and J.A. Massard, 'Les Bryozoaires du Grand-Duché de Luxembourg et des régions limitrophes', *Travaux Scientifiques du Musée d'Histoire Naturelle de Luxembourg*, 7 (1986), 1-187, p. 30). Biographical details on Jullien are given by Jean-Loup d' Hondt, 'La destinée singulière d'un naturaliste original, Jules Jullien. 112e Congrès National des Sociétés savantes, Lyon, 1987', *Histoire des Sciences et des Techniques*, 1 (1987), 181-190. Jullien's role in French marine bryozoology is dealt with by J.L. d'Hondt, 'Contributions des chercheurs français à la connaissance des Bryozoaires actuels', *Histoire et Nature*, 26-27 (1985), 3-26; and J.L. d'Hondt, 'Histoire et chronologie des recherches sur les bryozoaires', *Bulletin d'Histoire et d'Épistémologie des Sciences de la Vie*, 2(1) (1995), 98-111.

15 Dumortier, note 12, p. 20-4.

16 See hereafter.

17 Dumortier and Van Beneden, note 5, p. 78.

18 Turpin, 'Etude microscopique de la Cristatelle', *Annales des Sciences naturelles*, 7, 2e série (1836). — Pierre Jean François Turpin (1775-1840), French botanist, one of the foremost botanical painters of his day; collaborated with von Humboldt; after about 1820 he devoted his energies to the study of microscopic plants and plant organs. See: *Biographie universelle (Michaud) ancienne et moderne*, nouvelle édition, t. 42, 296, See also: M. Rix, *'The art of botanical illustration'* (New York 1990), p. 141.

19 Allman, note 3, p. 71; Jullien, note 14, p. 149-50.

20 P.J. Van Beneden, 'Quelques observations sur les polypes d'eau douce', *Bulletin de l'Académie Royale des Sciences de Bruxelles*, 6 (9) (1839), 276-9. The same text was published in: *Annales des Sciences Naturelles (Paris),* 2e série, 14 (1840), 222-4; it was also reprinted in the report on the October 5, 1839 meeting of the Belgian Academy by *L'Institut, journal général des sociétés et travaux scientifiques de la France et de l'étranger (Paris),* 1ère section, t. 8 (1840), 153-4 (Zoologie: Polypes d'eau douce).

21 Allman, note 3, p. 68.

22 'Histoire naturelle: M. Van Beneden, correspondant de l'académie, présente trois mémoires de sa composition', *Bulletin de l'Académie royale des sciences et Belles-Lettres de Bruxelles*, 7 (2) (1840): 163-4 (concerns: freshwater bryozoans, development of cephalopods, development of slugs).

23 P.J. Van Beneden, 'Polypes d'eau douce', *Journal de l'Institut* 1840, 153-4.

24 P.J. Van Beneden, 'Recherches sur la structure de l'oeuf dans un nouveau genre de Polype (genre Hydractinie)', *Bulletin de l'Académie royale des sciences et Belles-Lettres de Bruxelles,* 6 (1) (1841), 89-93, 1 pl. In this article on the structure of the egg of *Hydractinia* (Hydrozoa), Van Beneden included a remark concerning *Alcyonella* and its so-called mobile eggs (pp. 90 and 93, fig B). Normally the term 'egg' was used for a 'statoblast'. Van Beneden's so-called 'mobile eggs' correspond to the ciliated larvae; this is clearly shown by the figure and the commentaries he inserted in this note in order to react to an objection made by Dumortier in December 1840.

25 P.J. Van Beneden, 'Mémoire sur le développement des Tubulaires, etc.', *Mémoires de l'Académie des Sciences de Bruxelles,* 17 (1844).

26 P.J. Van Beneden, 'Sur la classification des Bryozoaires', *Journal de l'Institut*, 18 (1850).

27 P.J. Van Beneden, 'Recherches sur les Bryozoaires fluviatiles de Belgique', *Nouveaux Mémoires de l'Académie royale de Belgique*, 21 (1848), 1-33, pl. VI-VII.

28 Allman, note 3, p. 70.

29 Kraepelin, note 9, p. 12.

30 N. Annandale, Freshwater sponges, hydroids and polyzoa, in *The Fauna of British India, London*: I-VIII, 177.

31 '*Recherches sur l'organisation des* Laguncula *et l'histoire naturelle des différents Polypes Bryozoaires qui habitent la côte d'Ostende'* (Bruxelles, Hayez impr., 1845), 29 p., 3 pl. (Extr. des Mémoires de l'Académie des Sciences, t. 18) [*Laguncula = Laguncula repens = Farella repens*, a Ctenostome Bryozoan].

'Recherches sur l'anatomie, la physiologie et le développement des Bryozoaires qui habitent la côte d'Ostende' (Bruxelles, 1845), 8 p.

'Recherches sur l'anatomie, la physiologie et le développement des Bryozoaires qui habitent la côte d'Ostende' (Bruxelles, 1845), [3]-8,89,[13] p., X pl. (a reprint of three articles from Nouveaux Mémoires de l'Académie royale des Sciences et Belles-Lettres, t. 18 and 19, with an introduction) (reference: The National Union Catalog Pre-1956 Imprints, vol. 46, 158).

'Recherches sur l'anatomie, la physiologie et le développement des Bryozoaires qui habitent la côte d'Ostende' (Bruxelles, Hayez impr., 1845) (Extr. des Mémoires de l'Académie des Sciences, t. 19).

'Recherches sur les Polypes Bryozoaires de la Mer du Nord' (Bruxelles, 1848), 16 p., 1 pl. (Extr. des Bulletins de l'Académie des Sciences de Bruxelles, t. XV, 67-82).

'Recherches sur les Polypes Bryozoaires de la Mer du Nord (suite) et projet d'une classification des animaux de ce groupe' (Bruxelles, 1849), 15 p., 2 pl. (Extr. des Bulletins de l'Académie des Sciences de Bruxelles, t. XVI).

32 'Recherches sur l'anatomie, la physiologie et le développement des Bryozoaires qui habitent la côte d'Ostende: Histoire naturelle du genre *Pedicellina'*, *Nouveaux Mémoires de l'Académie royale des Sciences de Bruxelles*, 19 (1845), 31 p., pl. I and II. Cf. A. Kemna, *'P.-J. Van Beneden. La vie et l'oeuvre d'un zoologiste'* (Anvers, 1897), p. 105; P.J. Van Beneden and Hesse, 'Recherches sur les Bdellodes (Hirudinées) et les Trématodes marins', *Mémoires de l'Académie royale des Sciences et Belles-Lettres de Bruxelles*, 34 (1) (1864),1-150 [discovery of the entoproct *Cyclatella annelidicola = Loxosoma annelidicola*, first considered as a trematode (p. 83-4, pl. XII); cf. M. Prenant and G. Bobin, 'Bryozoaires. Première partie. Entoproctes, Phylactolèmes, Cténostomes', *Faune de France*, 60 (1956), p. 34-6; and K. Nielsen, 'Entoprocts', *Synopses of the British Fauna (New Series),* 41 (1989), 78].

33 B.C. Dumortier and P.J. Beneden, 'Histoire naturelle des Polypes composés d'eau douce', *Nouveaux Mémoires de l'Académie Royale des Sciences et Belles-Lettres de Bruxelles*, 16 (1843), 1-33.

34 Allman, note 3, p. 69.

35 B.C. Dumortier and P.J. Beneden, *Histoire naturelle des Polypes composés d'eau douce, IIe partie. Descriptions. (Séance du 9 mai 1848). Mémoire servant de complément au tome XVI des Mémoires de l'Académie royale des sciences et belles-lettres de Bruxelles* (Bruxelles, 1850), 96 p. — Both parts of this memoir were published in one volume entitled: *Histoire naturelle des Polypes composés d'eau douce, ou des Bryozoaires fluviatiles* (Bruxelles, Impr. Hayez, 1850), 130 p. (National Union Catalogue 151, p. 490).

36 We have seen that years later Jullien was still strongly annoyed by this gross error (see: note 19).

37 Allman, note 3, p. 71.

38 Kraepelin, note 9, p. 12.

39 G.J. Allman, 'On the present state of our knowledge of the freshwater Polyzoa', *Reports of the British Association for the Advancement of Science*, 19 (1850), 306. — When writing this statement, Allman seemingly did not yet have the second part of Dumortier and Van Beneden's memoir which is not cited in the references.

40 F. Crépin, Notice biographique sur Barthélemy-Charles-Joseph Du Mortier, président de la Société Royale de Botanique de Belgique, *Bulletin de la Société royale de Botanique de Belgique*, 18 (1879), 7-49. — E. de Wildeman and L. Hauman, 'Barthélemy-Charles-Joseph Du Mortier', in *Biographie Nationale de Belgique*, 30, supplém., t. 2 (1959), col. 611-28. — G. Lebrun (dir.), *Grandes figures de la Belgique indépendante*, 3e éd. (Bruxelles, 1936), 309-10. — A. Lawalrée, 'Barthélemy Dumortier, 1797-1878', in *Florilège des sciences en Belgique (II)*, Académie royale de Belgique, classe des sciences (Bruxelles 1980), 541-58. — T. Denoël, *Le nouveau dictionnaire des Belges* (Bruxelles, 1992), 269.

41 F.L. Lefort, 'Contribution à l'histoire botanique du Luxembourg', *Bulletin de la Société des Naturalistes luxembourgeois*, 54 (1949), 116-8. — A. Lawalrée (1980), 545 (see note 40). — J.A. Massard, 'Historisch-naturwissenschaftlicher Streifzug durch den Kanton Echternach', *Nos Cahiers*, 19 (2-3) (1998), 367-70. — J.A. Massard, 'Aspects historiques de l'histoire naturelle du Mullerthal et de ses environs', *Annuaire de la Ville d'Echternach*, 2001, p. 41-3.

42 cf. Jahn, note 2, p. 312 (where Dumortier is cited in the first place in the chronology of the observation of cell division, then Morren and finally von Mohl). Hugo von Mohl (1805-1872) was a German botanist, professor for physiology at the university of Bern in 1832; professor of botany at Tübingen 1835-1872 (I. Jahn et al., [1985], 312 and 905. — *Encyclopaedia Britannica*, 15 [1969], 647). Charles Morren (1807-1858) was born in Ghent; he was professor of botany and agrononomy at the university of Liège; this compatriot of Dumortier had seen the cell divison in 1832 and he published his observation in 1836 in his *'Mémoires sur les Clostéries'* (I. Jahn et al. [1985], 312 and 906). See: J. Lebrun, 'Esquisse d'une histoire de la botanique et des botanistes belges pendant le XIXe siècle et le début du XXe, in: *Florilège des Sciences en Belgique* (Bruxelles 1968), 595-634 (p. 600).

43 de Wildeman and Hauman, note 40, p. 614.

44 B.C. Dumortier, 'Mémoire sur le delphinorhynque microptère échoué à Ostende', *Académie royale des Sciences, des Lettres et des Beaux-Arts de Belgique*, nouveaux mémoires, 12 (1839), 13 p. *Delphinorhynchus micropterus* (F. Cuvier, 1836) is nowadays called *Mesoplodon bidens* (Sowerby, 1804). Dumortier's article is cited by V.V. de Buffrénil, '*Mesoplodon bidens*

(Sowerby, 1804) - Sowerby-Zweizahnwal', in J. Niethammer and F. Krapp: *Handbuch der Säugetiere Europas*, Bd. 6, Teil 1B (Wiesbaden, 1995), 546, 559.

45 References for the biblio-biographical data: WHF, 'Obituary notice: P.-J. Van Beneden', *Proceedings of the Royal Society*, 57 (1894-95), XX-XXI. — Ad. Kemna, *'P.-J. Van Beneden. La vie et l'oeuvre d'un zoologiste'* (Anvers, 1897), 137 p. — Dictionnaire des Ecrivains belges, 'Pierre-Joseph Van Beneden', *Bibliographie Nationale: Dictionnaire des Ecrivains belges et Catalogue de leurs publications, 1830-1880*, vol. 4 (1910), 20-30. — G. Lebrun (dir.), *'Grandes figures de la Belgique indépendante'*, 3e éd. (Bruxelles, A. 1936), 319-320. — A. Lameere, 'Pierre-Joseph Van Beneden', *Biographie nationale publiée par l'Académie royale des sciences, des lettres et des beaux-arts de Belgique*, 26 (1936-1938), 184-91. — P. Brien, 'Pierre-Joseph Van Beneden', in *Florilège des sciences en Belgique*, Académie royale de Belgique, classe des sciences (Bruxelles 1968), 824-51. — T. Denoël, *'Le nouveau dictionnaire des Belges'* (Bruxelles,1992), 701. — B.C.A. Windle (transcribed by M.T. Barrett), 'Pierre-Joseph Van Beneden', *The Catholic Encyclopaedia, Online edition* (1999), http://www.newadvent.org.cathen/15266a.htm (The Catholic Encyclopaedia, vol. XV, R. Appleton Company 1912). — P. Tort, Patrick (dir.), *'Dictionnaire du darwinisme et de l'évolution'* (Paris, 1996), vol. 3 (O-Z), 4402-3.

46 cf. S.P. Parker (ed.), *'Synopsis and classification of living organisms'*, vol. 2 (New York, 1982), 726.

47 R. Siewing, (ed.), *'Lehrbuch der Zoologie*, Bd. 2: Systematik, 3. Aufl. (Stuttgart, 1985), 761.

48 cf. C. Nielsen, *'Animal evolution. Interrelationships of the living phyla'*, (Oxford, 1995), 164.

49 cf. Nielsen, note 48.

50 cf. Kemna, note 32, p. 84-5.

51 Friedrich Küchenmeister (1821-1890), German physician in Zittau (Saxony) since 1846, then from 1859 on in Dresden; he published on cestodes (1853) and on human parasites (1855) (Jahn, note 2, p. 879).

52 Kemna, note 32, p. 82-3. Cf. G. Theves, 'Helminthologie vétérinaire', *Annales de Médecine vétérinaire,* 139 (1995), 241-8.

53 See G. Hamoir, 'La découverte de la méiose et du centrosome par Edouard Van Beneden', *Académie royale de Belgique, Mémoire de la Classe des Sciences, Collection in-octo*, 3e série, tome 8 (1994), 1-128. For biographical data on Ed. Van Beneden see: R. de Winiwarter, 'Edouard-Joseph-Louis-Marie Van Beneden', *Biographie nationale publiée par l'Académie royale des sciences, des lettres et des beaux-arts de Belgique*, 26 (1936-1938), 174-184.

54 Dumortier and Van Beneden, note 4, p. 30.

55 Published in the *Annales des Sciences naturelles*: Mémoire sur les Alcyonides (1835); Recherches anatomiques, physiologiques et zoologiques sur les Escharres (1836); Sur les Polypes du genre Tubulipore (1837); Sur la nature et le mode de croissance des Polypes (1838).

56 H. Milne-Edwards, Rapport sur un mémoire de M. Gervais, intitulé: Observations pour servir à l'histoire naturelle des Polypes d'eau douce, fait à l'Académie des Sciences, le 8 avril 1839, *Annales des sciences naturelles*, 2e série, 11 (1839), 179-185 (p. 182). — P. Gervais, Observations sur les Polypes d'eau douce, *Ann. franç. et étrang. Anat. Physiol.*, 3 (1839), 129-176.

57 *Nouvelle biographie générale depuis les temps les plus reculés jusqu'à 1850-60 publiée par MM. Firmin Didot frères,* XXXV-XXXVI (Copenhague, 1968, reprint), 551-2. Patrick Tort (dir.), *Dictionnaire du darwinisme et de l'évolution*, vol. 2 (Paris, 1996), 2957-60. H.G. Hansson, *Biographical etymology of marine organism names*, http://www.tmbl.gu.se/libdb/

taxon/personetymol./petymol.a.html

58 William-Frédéric Edwards (Jamaica 1777 - Paris 1842), after a number of years in Bruges, he had left for France during the Revolution; doctor of medicine in Paris (1815), important research work on anatomy, pathological physiology and compared anatomy; member of the Institut de France (1832); founding member of the 'Société ethnologique' (1839, father of French ethnology. Cf. *Nouvelle biographie générale depuis les temps les plus reculés jusqu'à 1850-60 publiée par MM. Firmin Didot frères,* XXXV-XXXVI (Copenhague, 1968, reprint), 552-3.

59 Hannsson, note 57.

60 Daughter of general Camille Alphonse Trézel (1780-1860), who had participated in Napoleon's campaigns in Spain and Russia. Under Louis-Philippe he successfully fought in Algeria (victory of Macta in 1835), and was later on Minister of War and Governor of Paris (*Dictionnaire Encyclopédique Quillet*, volume 6, 1958, p. 5512).

61 cf. Hansson, note 57.

62 Jean-Victor Audouin (1797-1841), French naturalist, born in Paris, where he studied medicine, natural history and pharmaceutics; disciple of Cuvier, assistant at the Muséum National d'Histoire Naturelle in Paris (1825), professor of entomology at the Muséum in 1833.

63 Year indicated by P. Tort and by d'Hondt (1990); other authors indicate 1890 or 1891.

64 Bibliographie Nationale, *'Dictionnaire des Ecrivains belges et Catalogue de leurs publications, 1830-1880'* (Bruxelles, 1910), 4: 339-40.

65 The National Union Catalog Pre-1956 Imprints, vol. 657, 479.

66 = *P. emarginata* and/or *P. fruticosa.* See chapter about Lacourt.

67 H.R. Berquin, 'Op bezoek bij Karel Loppens', *Bachten de Kupe, Heemkundig tijdschrift*, 1 (1959), 3-6, Nieuwport (includes an autobiographical note by Loppens: 'Korte levensbeschrijving'). — H.R. Berquin, 'In Memoriam Karel Loppens', *Bachten de Kupe, Heemkundig tijdschrift*, 4 (1962), 62-67. — P. van Oye, 'Karel Loppens, onbekend pionier van het brakwateronderzoek', *Mededelingen van de Koninklijke Vlaamse Academie voor Wetenschappen, Letteren en Schone Kunsten van België. Klasse der Wetenschappen,* 25 (7) (1963): 1-34, Brussel.

68 See about Gilson: P. Brien, Esquisse d'une histoire de la zoologie et de la biologie animale en Belgique durant le XIXe siècle et le début du XXe, in: *Florilège des sciences en Belgique, Académie royale de Belgique, classe des sciences, Bruxelles*, p. 757-8.

69 Ernest Rousseau (1872-1920), physician, entomologist, hydrobiologist, curator of the Museum of natural history, founder of *'Les Annales de Biologie lacustre'*, founded the limnological station of Overmeire (station de biologie lacustre) with the financial support of Charles Torley (†1926), an industrialist interested in natural sciences. See Brien, note 68.

70 Paul van Oye (Ostende, 1886-1969), professor at the university of Ghent; he was a member of honour of the 'Société des Naturalistes luxembourgeois' (*Bulletin de la Société des Naturalistes luxembourgeois*, 71-75 (1966-1970), 157.

71 A list of these publications is given by van Oye, note 67, p. 28.

72 See: M. Poll and H. Herlant-Meewis, 'Notice sur Paul Brien, membre de l'Académie', *Annuaire de l'Académie royale de Belgique*, vol. 145 (1979) (notices biographiques), 39-141. — M. Poll, 'Paul Brien', *Nouvelle Biographie nationale, Acad. r. Belg.*, 2 (1990), 55-58. — Mme Henriette Herlant-Meewis: doctor of natural sciences, 'chef de travaux' at the University of Brussels, associated member of the Belgian zoological society in 1932 [*Annales Soc zool Belg.*, 84 (1953), 489]. Max Pol: see hereafter.

73 In 1933 M. Aen den Boom published some thirty pages on this subject: 'Contribution à l'étude de la reproduction asexuelle chez les Bryozoaires d'eau douce', *Recueil de l'Institut zoologique Torley-Rousseau*, 4 (1933), 165-195. (The Zoological Institute Torley-Rousseau was a foundation directed by Lameere; his successor was Brien. Cf. Poll and Herlant-Meewis, note 72, p. 44.). — In 1933 Marcelle Aen den Boom, then a zoology student, got married to Max Poll; later on she became biology teacher at the 'Ecole Normale Berckendael' at Brussels. Max Pol (1908-1991), was an assistant of Lameere and Brien; he entered the 'Musée du Congo' at Tervuren (now Royal Museum of Central Africa), where he became head of the zoology department in 1948; he retreated in 1973; he was known as the best specialist of freshwater fishes of Black Africa; also professor at the ULB (P. Basilewsky, *'Max Poll'*, http://angelicus.netliberte.org/articles/celebrites/poll.htm, © 1999 Angelicus, 7 p.).

74 Pierre-Paul Grassé (Périgueux 1895 - Carlux 1985), French zoologist, professor at the Sorbonne, worked on protozoans and termites, as well as on evolution (including man); internationally known for his *'Traité de Zoologie'* (first volume published in 1948).

75 Poll and Herlant-Meewis, note 72, 64-5. The systematic and phylogenetic position of the Entoprocts is still a controversial topic. For more details see: J.L. d'Hondt, 'Progrès récents dans la connaissance des "Lophophoriens" (sensu lato)', *Bulletin de la Société Zoologique de France*, 107 (1982), 179-183. P. Emscherman, 'Les Kamptozoaires: Etat actuel de nos connaissances sur leur anatomie, leur développement, leur biologie et leur position phylogénétique', *Bulletin de la Société Zoologique de France*, 107 (1982), 317-344. P. Emschermann, 'Kamptozoa', in *Süßwasserfauna von Mitteleuropa*, 1 (2+3) (1995), 119. C. Nielsen, *'Animal evolution. Interrelationships of the living phyla'* (Oxford, 1995), 182-191. C.C. Emig, 'Les Lophophorates constituent-ils un embranchement?', *Bulletin de la Société Zoologique de France* , 122 (1997), 279-288. C. Nielsen, 'The phylogenetic position of Entoprocts and Ectoprocts', Proceedings of the 11th International Bryozoology Association Conference, Panama (2000), 66-73 (in this paper Nielsen concludes that the phylum Bryozoa has a phylogenetic position close to the base of the spiralians, and that it is possible that the two phyla, i.e. ectoprocts and entoprocts, are sister groups, such that the old classification, with Bryozoa consisting of Entoprocta and Ectoprocta, could be reinstated).

76 cf. *Annales de la Société zoologique de Belgique*, 84 (1953), 486: Graziella Brasseur-Huysmans, licenciée en sciences zoologiques, La Louvière, associated member in 1936.

77 Poll and Herlant-Meewis, note 72, 39-141. — M. Poll, 'Paul Brien', Nouvelle Biographie nationale, Académie royale de Belgique, 2 (1980), 55-58. — M. Poll and H. Herlant-Meewis, 'Paul Brien, 1894-1975', in *Florilège des sciences en Belgique (II),* Académie royale de Belgique, classe des sciences (Bruxelles 1980), 641-52.

78 Auguste Lameere (Ixelles, 1864 - Brussels, 1942); Belgian zoologist, author of *'Manuel de la Faune de Belgique'* (3 vol., 1895-1907) and *'Précis de Zoologie'* (7 vol., 1927-1942). See: P. Brien, 'Auguste Lameere', in: *Biographie nationale de Belgique*, 31, suppl. t. III, fasc. 1 (1961), 529-37.- P. Brien, 'Auguste Lameere', in *Florilège des sciences en Belgique*, Académie royale de Belgique, classe des sciences (Bruxelles 1968), 879-906.

79 'Recherches sur l'embryogénie de *Salpa maxima'*, *Bulletin de la classe des sciences de l'Académie royale de Belgique, séance du 4 novembre*, pp. 616-27.

80 *Rev. zool. bot. Afr.*, 40 (1948), 229-234.

81 A. Verduin, 'In memoriam: A.W. Lacourt (6.6.1910 - 19.2.1987)', *Correspondentieblad van de Nederlandse Malacologische Vereniging*, 236 (1987), 263-265. We would like to thank Dr Maya Borel Best (Nationaal Natuurhistorisch Museum, Leiden) for sending us a copy of this

obituary, and extend these thanks to Drs W. van Weiten, general director of the museum.

82 Scientific journal of the Dutch Malacological Society; issued twice a year.

83 Organ of the coast working group (Strandwerkgemeenschap) of the Royal Dutch Natural History Society (*Koninklijke Nederlandse Natuurhistorische Vereniging, K.N.N.V.)*

84 A bimonthly journal for and made by members of the Dutch Malacological Society.

85 P. Job, 'Intervention des populations de *Plumatella fungosa* (Pallas) (Bryozoaire Phylactolème) dans l'autoépuration des eaux d'un étang et d'un ruisseau', Hydrobiologia, 48 (3) (1976), 257-261.

86 B. Goethals, 'Les *Plumatella* (Bryozoaires d'eau douce): quelques aspects de leur morphologie, de leur composition chimique et leur incidence écologique en rivières polluées', Univ. Liège, Mém. Lic. Sci. zool. (1982), 84 p.; V. Henry, *Bryozoaires de la Meuse liégeoise: biomasse et réactions vis-à-vis de certains polluants.* Univ. Liège, Mém. Lic. Sci. zool. (1984), 55 p; C. Brene, *Contribution à l'écologie des Bryozoaires Phylactolèmes et application au problème de l'entretien des aéroréfrigérants d'une centrale électrique.* Mémoire de licence en sciences zoologiques, Univ. de Liège (1988), 1-45.

87 B. Goethals, M.F. Voss-Foucart and G. Goffinet, 'Composition chimique et ultrastructurale de l'ectocyste et de la coque des statoblastes de *Plumatella repens* et *Plumatella fungosa* (Bryozoaires Phylactolèmes)', *Ann. Sic. nat. Zool.*, Paris, 13, 6 (1984), 197-206.

88 Eugène Bricher (1875-1937): born in Weimerskirch in the outskirts of Luxembourg-City on February 6, 1875; qualified as a doctor in Luxembourg in November 1898; practice in Luxembourg-City; died in Kehlen on 17th December 1937. He was secretary of the Société des Naturalistes luxembourgeois from 1901 to 1905 Cf. H. Kugener, *'Die zivilen und militärischen Ärzte und Chirurgen in Luxemburg* (Luxemburg, 1995), 97-9. — J.A. Massard, 'La Société des Naturalistes Luxembourgeois du point de vue historique', *Bulletin de la Société des Naturalistes luxembourgeois*, 91 (1990), 158.

89 Eugène Bricher, 'Sitzung vom 14. Juli 1901', *Bulletin de la Société des Naturalistes luxembourgeois*, 11 (1901), 285.

90 Joseph Hoffmann, born on 12th December 1911 in Reckange (Mersch), professor of zoology and biology at the 'Cours universitaires' in Luxembourg-City, author of numerous publications on the Invertebrates of Luxembourg, one of the leading 20th century zoologists in Luxembourg, died in Luxembourg on December 16, 2000, aged eighty-nine. J.A. Massard and G. Geimer, 'Les professeurs de sciences et leur oeuvre', in: *Festschrëft 150 Joër Iechternacher Kolléisch (1841-1991),* Luxembourg, 1992, 475-6.

91 Erik Mauch, born in 1934 in Mannheim (Germany); biological studies at the University of Frankfurt/Main, Ph.D. in 1961; hydrobiologist in the water management administration ('*Wasserwirtschaftsverwaltung*') in Baden-Württemberg, then in Bavaria, where he was active in river quality mapping, with special devotion to saprobic classification and fundamental aspects of biological analysis of waters; retreat in 1999; author of numerous articles; since 1989 editor of the limnological journal '*Lauterbornia*'.

92 E. Mauch, *'Untersuchungen über das Benthos der deutschen Mosel unter besonderer Berücksichtigung der Wassergüte',* Bundesanstalt für Gewässerkunde (Koblenz, 1962), 335 p. Anl. 1-6 [hectogr.]. [Bryozoa: 231-235]. [Diss. Frankfurt/M. (1961)]; E. Mauch, 'Untersuchungen über das Benthos der deutschen Mosel unter besonderer Berücksichtigung der Wassergüte', *Mitt. Zool. Mus. Berlin*, 39 (1963): 3-172.

93 E. Mauch, 'Der Einfluß des Aufstaus und des Ausbaus der deutschen Mosel auf das biologische Bild und den Gütezustand,' *DVWK-Schriften*, 45 (1981), 39-137 [Bryozoa: 75-6, tab. 3, p. 53-

54].

94 G. Geimer, *'La faune des Bryozoaires (Lophophoriens ectoproctes) du Grand-Duché de Luxembourg',* (Luxembourg, 1975), 104 p., 29 pl. (Mémoire scientifique).

95 Gaby (Gabrielle) Geimer, born in Luxembourg on 29th June 1950; secondary school teacher (biology) in Echternach. See: International Biographical Centre, *'International Who's Who of Intellectuals'*, 10th Edition 1993/94 (Cambridge, 1993), 158. — International Biographical Centre, '*Dictionary of International Biography*, 24th edition 1996 (Cambridge, 1995), 111; and 25th edition 1997 (Cambridge, 1997), 131.

96 Geimer and Massard, note 14.

97 Jos. A. Massard, born at Tétange (Luxembourg) on 3rd December 1944, secondary school teacher in Echternach (biology) and part-time lecturer at the University Centre of Luxembourg (history and philosophy of sciences, history of medicine). See: International Biographical Centre, *'International Who's Who of Intellectuals'*, 10th Edition 1993/94 (Cambridge, 1993), 314. — *International Biographical Centre, 'Men of Achievement'*, 16th edition (Cambridge, 1995), 293; 17th edition (Cambridge, 1997), 331.- International Biographical Centre, *'Dictionary of International Biography'*, 24th edition 1996 (Cambridge, 1995), 221; and 25th edition 1997 (Cambridge, 1997), 261.

98 See also: G. Geimer and J.A. Massard, 'Note sur les caractères distinctifs de *Plumatella repens* (Linné, 1758) et de *Plumatella fungosa* (Pallas, 1768) (Bryozoa, Phylactolaemata)', *Archives de l' Institut Grand-Ducal de Luxembourg, section des sciences*, N.S., 40 (1987), 41-46.

99 Other publications of J.A. Massard and G. Geimer concerning the freshwater bryozoans of Luxembourg: 'Notice sur les Bryozoaires du lac d'Echternach et du lac de la Haute-Sûre ainsi que sur la présence de l'Hydrozoaire *Cordylophora caspia* (Pallas, 1771) dans la centrale nucléaire de Cattenom', *Bull. Soc. Nat. luxemb.*, 90 (1990), 163-168. — 'Note sur les Bryozoaires d'eau douce trouvés en diverses stations luxembourgeoises et belges, avec des considérations sur la bryozoofaune de la Sûre et des remarques concernant les statoblastes de *Hyalinella punctata* (Hancock, 1850)', *Bull. Soc. Nat. luxemb.*, 92 (1991), 131-148. — 'Signalement du Bryozoaire d'eau douce *Plumatella casmiana* Oka, 1907 (Phylactolaemata) en Lorraine', *Bull. Soc. Nat. luxemb.*, 98 (1997), 237-238. — 'Découverte de Plumatella fungosa (Pallas, 1768) (Bryozoaires, Phylactolémates) dans le <lac> d'Echternach (Luxembourg)', *Archs Inst. g.-d. Luxemb. Sect. Sci. nat. phy. math.*, NS 44 (2002), 101-105.

100 Publications of J.A. Massard and G. Geimer concerning freshwater bryozoans of other countries: 'Note on the freshwater Bryozoa of Israel (Phylactolaemata)', in: F. Bigey (ed.) and J.L. d'Hondt (collab.): Bryozoaires actuels et fossiles : Bryozoa living and fossil. — *Bull. Soc. Sci. nat. Ouest Fr.*, Mém. HS 1 (1991), 243-253. — 'Note on the freshwater Bryozoa (Ectoprocta, Phylactolaemata) of Tenerife', *Vieraea*, 19 (1990), 327-338. — 'Additional note on the freshwater Bryozoa of Tenerife (Phylactolaemata)', *Bull. Soc. Nat. luxemb.*, 92 (1991): 149 -157. — Additional note on the fresh and brackish water Bryozoa of Israel (Phylactolaemata, Gymnolaemata)', *Bull. Soc. Nat. luxemb*, 93 (1992): 199-214 (joint authors: H.J. Bromley and C. Dimentman). — 'Distribution of freshwater and brackish-water Bryozoa (Phylactolaemata, Gymnolaemata) in Israel', in: P.J. Hayward, J.S. Ryland and P.D. Taylor: *Biology and Palaeobiology of Bryozoans* (Fredensborg, 1994), 117. — 'Das Moostierchen *Fredericella indica* kommt in Bayern vor (Bryozoa, Phylactolaemata)', *Lauterbornia*, 20 (1995): 99-101. — 'On the occurrence of *Fredericella*

indica Annandale, 1909 (Phylactolaemata) in Europe', in: D.P. Gordon, A.M. Smith and J.A. Grant-Mackie: *Bryozoans in space and time* (Wellington, 1996), 187-192. — 'Freshwater bryozoans of New Zealand: a preliminary survey', *New Zealand Journal of Marine and Freshwater Research*, 32 (1998), 639-648 (authors T.S. Wood, L.J. Wood, G. Geimer and J. Massard). — 'Occurrence of *Pectinatella magnifica* (Leidy, 1851) (Bryozoa, Phylactolaemata) in the German-Luxembourg border region near Bech-Kleinmacher (Luxembourg) and Nennig (Germany)', *Archs Inst. g.-d. Luxemb. Sect. Sci. nat. phys. math.*, NS 44 (2002), 107-120. — 'Phylactolaemates of St Stephen's Green in Dublin (Ireland)', *Archs Inst. g.-d. Luxemb. Sect. Sci. nat. phys. math.*, NS 44 (2002), 121-126. — 'Note on the occurrence of the freshwater bryozoan *Plumatella casmiana* Oka, 1907 (Bryozoa, Phylactolaemata) in Norway', *Archs Inst. g.-d. Luxemb. Sect. Sci. nat. phys. math.*, NS 44 (2002), 127-143 (joint authors: K.A. Økland & J. Økland).

Towards a general history of the south-eastern Pacific Bryozoa

Hugo I. Moyano G.
Departamento de Zoología, Universidad de Concepción Casilla 160-C, Concepción, Chile

1. Introduction: the place

The south-eastern Pacific represents a large oceanic area containing only a few islands like the Galápagos, Easter, Sala y Gómez, Desventuradas and Juan Fernández. Its eastern limit is the relatively straight, western south American coast that extends from the Gulf of Guayaquil to Chiloé Island, where it changes abruptly becoming an intricate collection of thousands of islands, channels and fjords whose coastal line increases greatly. The significance of this is that there is a large area of submarine space available for species to live and to evolve, and this has resulted in a high diversity. This is corroborated by comparing the known diversity of disparate zoological groups, such as Annelida, Polychaeta, Mollusca, Foraminifera and Bryozoa, north of Chiloé Island where the coast line is straight (42°S-6°S) with the diversity of those groups south of Chiloé Island (42°S-56°S) where the coastal line and continental shelf is larger. In this part of the world, the biological diversity seems to decrease to the tropics and to increase to the poles.[1]

Along the almost 50° latitudinal coastal line, from the Gulf of Guayaquil to the area of Cape Horn, most of the zoogeographers that have explored and studied the western south American coast have revealed only two huge faunal provinces: the Peruvian-Chilean (6°S-42°S) and the Magellan (42°S-56°S).[2] This paucity of centers of endemism is due to the uniforming effect of the West Wind Drift cold waters that affects the western coast of South America at 40°S. It gives rise to two cold currents: the austral Cape Horn current and the northern cold Humboldt current. The translation of this into zoological terms is shown by the presence and range of three of four species in the penguin genus *Spheniscus*: *S. magellanicus* and the Cape Horn current characterize the subantarctic Magellanic province; *S. humboldti* littorally follows the Humbolt current from central Chile to most of the Peruvian coast, and *S. mendiculus* is found on the austral set of the Galápagos islands, that is in the northermost part still influenced by relatively cold waters derived from the Humbold current. Opposing this poor zoogeographical provinciality each island or island group over the Nazca plate might represent a faunal province *per se*. Faunal and zooegeographic studies of them have demonstrated so far, that there are at least four insular provinces: Galápagos, Easter and Sala y Gómez, Juan Fernández and the Desventuradas, and the bathial Nazcaplatensis.[3] This number could be increased to five provinces if the present studies on the marine diversity of the Desventuradas demonstrates a species endemism higher than ten per cent.[4]

The immense marine area of the triangle situated between Easter Island, the Gulf of Guayaquil and the southernmost tip of South America contains part of the eastern Pacific rise which is the western-most natural limit of the South Eastern Pacific. This is a spreading geological submarine zone that generates the Nazca plate eastward and the huge Pacific plate westward. From this submarine rise, emerge Easter and Sala y Gómez Islands near its spreading centre. These islands represent the visible and emerged western part of a large submarine guyot line that encompasses the submersed Sala y Gómez and Nazca ridges.[5] Very close to the eastern end of the Sala y Gómez ridge emerge the Desventuradas Islands and six degrees latitude further south, facing central Chile, lies the Juan Fernández archipelago. These islands: Galápagos, Easter, Sala y Gómez, Desventuradas (San Félix and San Ambrosio) and Juan Fernández archipelago (Róbinson Crusoe, Santa Clara and Alejandro Selkirk), the almost straight western south American coast from the equatorial line to Chiloé Island (42°S), and the huge archipelago area south of 42°S in southern Chilean coast, constitutes the frame, base and core of this paper. From here several bryozoan faunas have been described and studied since the beginning of the nineteenth century.

2. Pre-Hispanic epoch

Littoral conch deposits along the Peruvian and Chilean coasts contain large mytilid and taidiid conchs that are frequently colonized by several encrusting and boring bryozoan species. The earliest collectors of these shells were probably unaware of the existence or significance of the bryozoan encrustations. Further south, nevertheless, there is a greater probability that the indigenous peoples, who lived amid the channels and fjords, and who obtained their meals by diving in the cold subantarctic waters of the Magellan Strait area, had discovered and named large bryozoan colonies. Even today it is possible to collect large pieces of the giant *Alcyonidium australe* d'Hondt & Moyano that measure more than

Figure 1. Chilean postal stamp commemorating Juan Ignacio Molina (1740-1829)

half a metre long, on beaches of Magellanic South America.[6] It has been also possible to buy large specimens of *Aspidostoma giganteum*, that measured up to 30 cm diameter, from a small shop on Chiloé Island.

3. The Hispanic period

From 1536 onwards, the year in which the Spanish discovered Chile, several naturalists or colonial writers have described Chilean natural history from the geographical, ethnic, botanical and zoological points of view.[7] They principally described the most obvious and large terrestrial animals and plants. Marine animals were also surveyed, including mammals, birds, fishes, large molluscs and crustaceans. By the end of this period the Jesuite priest, Juan Ignacio Molina (1740-1829) (Figure 1), who can be regarded as the first "modern" naturalist who adopted the Linnean binomial system in the area, was active. He described for the first time the most common plants and animals from central Chile, including both vertebrate and invertebrate marine faunal elements. His principal contribution, *Saggio sulla storia naturale del Cile* which was written in Bolonia, Italy and published in 1782[8] was based firstly on some notes taken with him when he and all the Jesuites were ejected from the Spanish possessions, but also on his memories of Chilean nature explored in his juvenile years. Although many common invertebrates such as *Loxechinus albus* (Molina) (regular echinoid), *Aulacomya ater* (Molina) and *Choromytilus chorus* (Molina) (large coastal mytilids), and *Pyura chilensis* Molina (edible ascidian) were described, no bryozoans or colonial animals were included in Molina's work. The species *Escharoides molinai* Moyano was erected in his memory in 1983;[9] this represents the only hitherto known member of this bryozoan genus in the south-eastern Pacific Ocean.

4. First half of the nineteenth century

This period saw the development of three large European scientific enterprises that pioneered and greatly promoted the development of marine and terrestrial zoology in western and southern South America.

On 27th December 1831 the H.M.S. *Beagle* commanded by Robert FitzRoy set sail from Devonport, England, on its long five year voyage around the world. On board was the young Charles Darwin. One year later the *Beagle* and its crew spent two months around Tierra del Fuego from 16th December 1832 to 26th February 1833, and one year later again she spent another month in the same place from 29th January to 7th March 1834. Finally, from middle June 1834 to the end of October 1835, the *Beagle* spent more than a year sailing from Tierra del Fuego to Galapagos Islands along the eastern limit of the south western Pacific. During that year Darwin visited and explored coasts, valleys and the Andes of southern and central Chile, and collected plants and animals.[10] Some of the Magellan bryozoan species described by George Busk in three catalogues[11] on the bryozoan collections of the British Museum (Natural History) were collected by Darwin. Almost at the same time the Voyage dans l'Amérique Meridionale (1826-1833) took place. It explored the republics of Brazil, Uruguay, Argentine, Bolivia, Peru and Chile. The zoophytes collected during this French expedition were described by Alcide d'Orbigny and included several common bryozoan species from along the Peruvian and Chilean coasts, such as *Membranipora isabelleana* (d'Orbigny), *Jellyella tuberculata* (Bosc), *Celleporella bougaivillei* (d'Orbigny), *Fenestrulina cornuta* (d'Orbigny), *Cellaria ornata* (d'Orbigny), *Hornera americana* (d'Orbigny), *Galeopsis pentagonus* (d'Orbigny), *Umbonula alvareziana* (d'Orbigny) and *Terebripora gramosa* (d'Orbigny).[12] d'Orbigny's importance issues from having described for the first time littoral bryozoan species affected by the Humboldt cold current from Chiloé Island to the south and central Peruvian coasts. It is also worth mentioning the description of *Terebripora ramosa*, the first known boring bryozoan on Calyptraeidae shells collected in Arica (18°S).

By the end of the first third of 19th century, the Chilean Goverment contracted the French naturalist, Claude Gay, to explore, collect and describe the Chilean nature and to edit the results in Spanish. He arrived in Chile in December 1828. From 1844 to 1865, 30 volumes of the *Historia Física y Política de Chile* were edited and published in Paris; the product of minuciose work of Claude Gay (Gay,1854),[13] and many other French naturalists. This encyclopedic work describes all kind of invertebrates but unfortunately no bryozoans.

5. Second half of the nineteenth century

The final years of this period were marked by a rapid increase in knowledge of the bryozoans from the southern tip of South America, both from Valpara'so on the Pacific side and from Montevideo and Buenos Aires on the Atlantic side to the Cape Horn Area. This was due to the arrival of the first worldwide oceanographic expedition accomplished on board of H.M.S. *Challenger* (1872-1876); it explored for organisms all around the world at depths below 300 fathoms. Results of the extensive and intensive dredging and trawling demonstrated the presence of incredible biological diversity at all depths and latitudes, and refuted Edward Forbes' assumption, that life was wanting under 300

Table I. Challenger *stations in the south eastern Pacific that yielded bryozoans.*

Station number	Latitude; Longitude	Bryozoan species
299	33°31'S; 74°43'W	*Menipea pateriformis*
		Kinetoskias pocillum
		Bugula reticulata
		Flustra biseriata
303	45°31'S; 78°09'W	*Menipea benemunita*
		Carbasea ovoidea
		Menipea aculeata
		Cribrilina monoceros
		Cellepora eatonensis
		Bugula reticulata
304	46°53'S; 75°12'W	*Aetea anguina*
		Salicornaria variabilis
		Salicornaria clavata
		Salicornaria malvinensis
		Cellepora signata
308	50°8'30"S; 74°41'W	*Menipea benemunita*
		Menipea aculeata
		Bugulata reticulata
		Carbasea ovoidea
		Cribrilina monoceros
		Cellepora eatonensis
		Cellepora armata erecta
312	53°37'S; 70°56'W	*Carbasea ovoidea*

fathoms.[14] The *Challenger* expedition sailed to the eastern Pacific via Tahiti, and the Juan Fernández Islands and arrived in Valparaíso in November 1875.

From this point the ship navigated south to enter the Gulf of Penas, a latitude at which glaciers touch sea level, sailed along Messier Channel to pass through the Magellan Strait to the south Atlantic.

Table I shows the *Challenger* collecting stations that yielded bryozoans on the Nazca Plate from near Easter Island to the eastern part of the Magellan Strait. These were described by George Busk in 1884.[15] There are only sixteen species in comparison with more than 200 species known so far today. Curiously, the survey of some *Challenger* stations (*viz.* 320, 314, 315 and 135) in the south western Atlantic gives more information on the actual bryozoan diversity of the southern-most part of the south eastern Pacific. The reason for this apparent bryozoan paucity is probably related to the constant bad weather conditions in the Chilean fjordland, from 42°S to 56°S, that have constrained the collecting activities of scientific vessels from the *Challenger* times until today. When rare good weather conditions permit dredging and trawling in that area, however, the gathered species indicate a basic common bryozoan fauna in both the south western Atlantic and south eastern Pacific.[16]

In the meantime, the German naturalist, Rudolph Amandus Philippi, who had arrived in Chile on 18th October 1851 at the age of 43 years, had undertaken many investigations on the fauna, flora and fossils which he had collected in several trips in central and northern Chile. In depicting fossils obtained from Tertiary outcrops in the Atacama desert, he described and illustrated a *Lunulites*, the first Chilean fossil bryozoan, in 1887.[17]

6. The end of the nineteenth century and the Antarctic connection

The end of the 19th century was marked by an increase of European interests in southern seas including the Antarctic Ocean and continent. Many scientific ships passed through the Magellanic region or the Cape Horn area collecting or stopping there to explore, draw maps, or collect all kind of scientific data including benthic zoological collections from littoral to the deep sea. As a consequence large and increasing collections of Bryozoa from the south western Atlantic, the subantarctic Magellanic area, the Scotia Arch archipelagos and the Antarctic peninsula begun to accumulate in European natural history museums. Among the scientific expeditions that explored the Magellanic area after the *Challenger* Expedition, the Mission Scientifique du Cap Horn and the Hamburger Magalhaesischen Samelreise stand out. Under the auspice of the "Ministères de la Marine et de l'Instruction Publique" the French carried out the "Mission Scientifique du Cap Horn 1882-1883". This expedition settled for two years in Orange Bay in Hoste Island situated south of Beagle Channel. There, many investigations in zoology, botany and ethnology were made. The bryozoans collected were studied by Jules Jullien and published in 1888.[18] This author, after criticizing the use of the colonial form for stablishing generic entities made by Busk and other authors, decided to fundamentally use the structures of isolated zooecia. Jullien described as new two *Pedicellina* species, five ctenostomes, 22 cheilostomes and one cyclostome totalling 30 new species out of 56 in the whole collection. In 1904 Louis Calvet published in German descriptions of 61 marine and two freshwater bryozoans presented to him by Dr W. Michaelsen of the Komitee der Hamburger Magalhaensischen Sammelreise, that had been obtained from Chiloé (42°S) to Tierra del Fuego (56°S) between 1892 and 1893.[19] Calvet described ten new species: eight cheilostomes and two *Barentsia* (entoprocts) among the 61 marine bryozoans. He recorded ten species of cyclostomes, two ctenostomes and for the first time two phylactolaemates (freshwater bryozoans) in Tierra del Fuego. These reports described up *circa* 100 species - much more than those of the *Challenger* - and this work began to reveal the real nature and diversity of the Magellan Pacific bryozoans.

7. First half of the twentieth century

The increasing European interest in exploring the subantarctic and Antarctic seas, materialized in scientific expeditions such as those of the Swedish Antarctic Expedition (1901-1903) and those of the Discovery Committee, that resulted in monumental and fundamental bryozoological publications. In 1944 Folke Borg published on the stenolaematous Bryozoa as part of the zoological results of the Swedish Antarctic Expedition 1901-1903 directed by Dr Otto Nordenskjöld.[20] However part of these results came from a previous work: his 1926 academical dissertation published in Uppsala.[21] In both works many species described before by d'Orbigny and Busk plus the new ones depicted by Borg have been discovered later in the Magellanic south Pacific. Borg's work[22] plus those of d'Orbigny,[23] Busk,[24] Jullien[25] and Calvet[26] constitute the framework and basis of the present bryozoan knowledge in both the south western Atlantic and south eastern Pacific. Nevertheless this fundamental base was lacking the survey of the enormous collections amassed by the ships of the *Discovery* Committee in many years of exploring and collecting in the southern seas. A first attempt to study those collections and others obtained by several British ships was carried out by Anna Hastings who in 1943 published an important monograph which included 120 species of 6 families of cellularine bryozoans.[27] Although most collecting stations are in the south western Atlantic, many cellularine species are also present in the south eastern Pacific. On the other hand, in the first quarter of the twentieth century scientific exploration of the sporadic islands and archipelagos of the Nazca plate in the south eastern Pacific began. The first two archipelagos to be surveyed were Galápagos and Juan Fernández. The Galápagos Islands that represent the northernmost island set on the Nazca Plate and also the northernmost outcrop of tropical bryozoans of the eastern south Pacific were investigated by the *Albatross* of the U.S. Fisheries Commission in 1888 and 1891 that collected 50 species studied by Canu and Bassler.[28] Hastings in 1930 added other 37 taxa[29] and Osburn[30] and Soule in Osburn[31] in the 1950s aggregated several new records issued from the collecting activities of the *Velero* III expeditions of the Allan Hancock Foundation. Finally, Banta and Redden[32] in 1990 listed 184 bryozoan species whose zoogeographical affinities are with the Panamanian faunas more than with the Peruvian or Magellanic ones. The first survey of bryozoans obtained from the Juan Fernández area by several German collectors was published by Marcus in 1921. The nineteen species studied were collected during the Swedish Juan Fernández Expedition 1916-1917 commanded by Carl Skottsberg. Although only two new species were described, subsequent studies[33] have demonstrated that the so-called *Haswellia auriculata* Busk and *Fasciculipora ramosa* d'Orbigny were new species of *Galeopsis* and *Frondipora* respectively.

8. Second half of the twentieth century

Similar to the second half of the foregoing century, this period was marked by an explosive development and diversification of the bryozoan studies in the south eastern Pacific.

Nevertheless, there are significative differences: researchers have not been only European but also Chilean and the scientific vessels - that have collected along the Chilean coasts and the oceanic eastern Pacific islands - have been Chilean, European and American.

Chilean Expeditions: Since 1960 onwards the Chilean Navy has devoted time and efforts to carry out a series of oceanographic expeditions from the border with Peru (18°S) to the Drake Passage (55°S-60°S). The first two named as Mar-Chile I (1960) and Mar-Chile II (1962) were very productive in terms of plankton and benthos collections gathered from Arica to Chiloé area. In 1964 , the Instituto de Fomento Pesquero, a new Goverment Institution dedicated to study the fishery potential of the south American coasts of the south eastern Pacific, organized the IFOP-01 expedition. The collecting stations of this expedition started at Coquimbo (30°S) continued along the Chilean coast to Chiloé island and also some of them were carried out at Juan Fernández archipelago, allowing the gathering of fine and abundant bryozoan samples. By the end of the century the Chilean Navy Oceanographaphic Institute has organized the CIMAR expeditions in which many scientists of the Chilean Universities have participated obtaining bottom samples from Chiloé to the Magellan Strait and also from Easter, Sala y Gómez, Desventuradas and Juan Fernández.

Hero Expeditions: As part of activities of the U.S. National Science Foundation research vessel *Hero* in Antarctic and subantarctic areas, two legs of a general survey of the southern part of the Chilean Fiordland from Golfo de Penas (46°S-47°S) to the Magellan strait were carried out in October 1972. During the second leg (between the Pacific entrance of the Magellan strait and the Golfo de Penas) amid many benthic collecting stations, the original *Challenger* expedition stations were revisited, and rarely seen animals collected previously by the *Challenger*, were examined. These included living specimens of *Cephalodiscus dodecalophus* (Hemichordata, Pterobranchia) and the weird sea urchin *Dermechinus horridus* among many others. The bryozoans were represented by species hitherto known from the western south Atlantic such as *Cellarinella dubia* and new genera such as *Parafigularia* and *Flustrapora*, and many new species.[34]

Italian Magellano I Expedition: This expedition, as an extention to the subantarctic area of the Italian Antarctic Program, was carried out on board of the *Cariboo* between 20th February and 3rd March 1991 along the whole length of the Magellan Strait. The abundant and diverse bryozoan samples yielded 76 species which again demonstrated the similarity and unity of the bryozoan fauna to both sides of southernmost tip of South America.[35]

German Magellanic expeditions: In October 1994 the Joint Chilean-German-Italian Magellan "Victor Hensen" Campaign took place, that made possible the acquisition of a bryozoan collection similar to that collected by the *Cariboo* (see above). Nevertheless the most important source of new and unexpected taxa have been the sporadic collecting stations realized by the Alfred Wegener Institut research vessel *Polarstern* in the shelf and slope of the Cape Horn area as part of its 1996 subantarctic and Antarctic Program. Part of the bryozoan material collected has been already studied[36] and other material will be published later on. The processed material has revealed both a new familiy, a new genus and several new species, and the presence of families, genera and species already described from the western south Atlantic and also from distant places such as New Zealand and Australia, in the southernmost eastern part of the south Eastern Pacific. This work has revealed an apparent relict bryozoan fauna of tropical affinities due to the presence of genera such as *Sinupetraliella*, *Chiastosella* and *Adeonella*.

Among other collecting trips made along the Chilean coast those of Carlos Viviani deserve special mention.[37] He described 67 species of ectoprocts and seventeen of endoprocts collected by him between Arica (18°S) and Chiloé (42-44°S). These data are found in a doctoral dissertation, unfortunately unpublished so far, with the exception of species in family Hippothoidae.[38] In 1973, Moyano[39] uncovered part of the hitherto unknown bryozoan fauna from Easter Island and in 1982 he defined bryozoogeographically the Magellanic region, and made a comparison of it with the bryozoans of other south American and Antarctic faunas.[40] Later on in 1983, 1985, and 1987 Moyano added new information on bryozoans from Juan Fernández[41] (previously studied by Ernst Marcus in 1921)[42] and from Easter Island, and compared the south eastern Pacific bryozoan fauna with the whole Pacific basin bryozoan fauna. Finally in 1991, 1995 and 1996 he extended this approach to define five bryozoogeographic provinces: Magellanic, Perú-Chilean, Bathyal-Center-Chilean, Juan Fernández and Easter, and compared them with the bryozoan faunas from the Antarctic peninsula and from the New Zealand and Australian realms.[43] Along the last two decades of the twentieth century additional work on the subantarctic Magellanic area and hence of the south western Pacific Ocean was carried out by Peter Hayward and collaborators.[44] They continue the previous and pioneering work of Anna Hastings in describing the immense collections collected by the ships of the Discovery Committee during the first half of the twentieth century.

The collecting activities just reviewed, comprising the last forty years, have raised the known bryozoan diversity from more than 80 species principally from the Magellanic area, the continental slope in front of central Chile and the oceanic islands of the Nazca plate.[45] The total number of known bryozoan species within the Nazca plate boundaries excepting the Galápagos Islands is approximately 400, a figure that would increase to nearer 600, if those of the Galápagos recorded by Banta and Reddin in 1990 are included.[46] The newest approach to bryozoan studies in the south eastern Pacific is an ecological one that has been developed by J.M. Cancino and his collaborators who study the eco-physiology of littoral bryozoans.[47]

9. The future

Four issues are expected to be developed in the near future: (a) to continue the bryozoan diversity inventory especially along the Peruvian coasts, the western south American continental slope and the oceanic islands on the Nazca plate plus their associate guyot systems; (b) to increase the eco-physiology studies of larvae, colonies and communities of littoral and neritic species; (c) to make regional and Panpacific molecular biological studies of species having ample distribution patterns; and (d) to fortify the until now timid scientific cooperation between bryozoan workers of both western and eastern Pacific Ocean coasts.

Notes

1 H.I. Moyano G., 'Magellan Bryozoa: a review of the diversity and of the Subantarctic and Antarctic zoogeographical links', in *Magellan-Antarctic: ecosystems that drifted apart*, edited by W.E. Arntz and K.C. Rios. *Sci. Mar.*, 63 (supl.1) (1999), 219-226; H.I. Moyano G., 'Bryozoa from the Magellanic continental slope near Cape Horn: an unexpected collection', in *Proceedings of the 11th International Bryozoology Conference*, edited by A. Herrera Cubilla and J.B.C. Jackson (Smithsonian Tropical Research Institute, Republic of Panama, 2000), 298-304.

2 J.C. Briggs, *Marine Zoogeography* (McGraw-Hill Book Co., New York, 1974); H. Brattström and A. Johansen, 'Ecological and regional zoogeography of the marine benthic fauna of Chile', Report No. 49 of the Lund University Chile Expedition 1948-1949. *Sarsia*, 68 (1983), 289-339.

3 N.V. Parin, A.N. Mironov and K.N. Nesis, 'Biology of Nazca and Sala y Gómez submarine ridges, an outpost of the Indo-west Pacific fauna in the Eastern Pacific Ocean: composition and distribution of the fauna, its communities and history', *Advances in Marine Biology*, 32 (1997), 145-242

4 Briggs, note 2.

5 Parin, Mironov and Nesis, note 3.

6 J.-L. Hondt and H.I. Moyano G., 'Notes sur trois espèces chiliennes du genre *Alcyonidium* (Bryozoa-Ctenostomida)', *Cahiers de Biologie marine*, 20 (1979), 361-369.

7 A. Ovalle, *Histórica relación del Reyno de Chile* (Francisco Cavallo, Rome, 1646) [Reprint: Editorial Universitaria, Santiago 1969, 503 pp].

8 J.I. Molina, *Saggio sulla Storia Naturale del Cile* (Bologna, 1782). This went into several foreign editions: German 1786, and French 1788. After his exile he became Professor of Natural History in Bologna.

9 H.I. Moyano G., 'Southern Pacific Bryozoa: A general view with emphasis on Chilean species, *Gayana Zoología*, 46 (1983), 1-45. Molina is also honoured in the name *Molinia*, a genus of grass, and he also had a town in Chile named for him.

10 A. Moorehead, *Darwin and the* Beagle (George Rainbird Ltd., London, 1969).

11 Busk, G. *Catalogue of the marine Polyzoa in the collection of the British Museum, I. Cheilostomata (part).* (Trustees of the British Museum (Natural History), London, 1852), 54 pp.; Busk, G. *Catalogue of the marine Polyzoa in the collection of the British Museum, II. Cheilostomata (part).* (Trustees of the British Museum (Natural History), London, 1854). 55-120; Busk, G. *Catalogue of the marine Polyzoa in the collection of the British Museum, III. Cyclostomata. (Trustees of the British Museum (Natural History), London, 1875), pp. 1-41.*

12 A. d'Orbigny, 'Zoophytes', *Voyage dans l'Amérique éridionale* 5(4): 7-28 (1847); Atlas 9, pl. 1,3,5 (1841), pl. 2,4,6-13 (1842).

13 C. Gay, *Atlas de la Historia Física y Política de Chile*, Volume 2 (Thunot et Cie, Paris, 1859).
14 E. Linklater, *The voyage of the* Challenger (George Rainbird Ltd. London, 1972).
15 G. Busk, 'Report on the Polyzoa - the Cheilostomata', *Scientific Results of the* Challenger *Expedition, Zoology*10 (30) (1884), 1-216; G. Busk, 'Report on the Polyzoa - Cyclostomata, Ctenostomata and Pedicellinea', *Scientific Results of the* Challenger *Expedition, Zoology* 17 (50) (1886), 1-47.
16 H.I. Moyano G., 'Bryozoa Marinos Chilenos VIII: una síntesis zoogeográfica con consideraciones sistemáticas y la descripción de diez especies y dos géneros nuevos', *Gayana Zoología*, 55 (4) (1991), 305-389.
17 R.A. Philippi, *Die Tertiären und quartären Versteinerungen Chiles* (Leipzig, 1887); P. Fürstenberg, 'Dr. Rudolph Amandus Philippi, seine Leben und seine Werke', *Sonderabdruck aus den Verhandlungen des Deutschen Wissenschaftlichen Vereins in Santiago*, 5, 1-39.
18 J. Jullien, 'Bryozoaires', *Mission scientifique du Cap Horn 1882-1883*, 6 (3) (1888), 3-92.
19 L. Calvet, 'Bryozoen', *Ergeb. Hamb. Magalh Sammelreise 1892-1893*, 3 (1904),1-45.
20 F. Borg, 'The Stenolematous Bryozoa', *Further Zoological Results of the Swedish Antarctic Expedition, 1901-1903*, 3 (5) (1944),1-276.
21 F. Borg, 'Studies on Recent Cyclostomatous Bryozoa', *Zoologiska Bidrag från Uppsala*, 10 (1926), 181-507.
22 Borg, note 19 and 20.
23 d'Orbigny, note 11.
24 Busk, note 14.
25 Jullien, note 17.
26 Calvet, note 18.
27 A.B. Hastings, 'Polyzoa (Bryozoa) I. Scrupocellariidae, Epistomiidae, Farciminariidae, Bicellariellidae, Aeteidae, Scrupariidae', *Discovery Reports*, 22 (1943), 301-510.
28 F. Canu and R.S. Bassler, 'The bryozoan fauna of the Galápagos Islands', *Proceedings of the U.S. National Museum*, 76 (13) (1930), 1-78.
29 A.B. Hastings, 'Cheilostomatous Bryozoa from the vicinity of the Panamá Canal collected by Dr. C. Crossland on the cruise of the S.Y. "St. George"', *Proceedings of the Zoological Society of London*, 1929 (4) (1930), 1-17.
30 R.C. Osburn, *Bryozoa of the Pacific Coast of America Part 1, Cheilostomata - Anasca.* Allan Hancock Pacific Expeditions, 14(1) (University of Southern California, Los Angeles 1950), 1-270; R.C. Osburn, *Bryozoa of the Pacific Coast of America. Part 2, Cheilostomata – Ascophora.* Allan Hancock Pacific Expeditions, 14(2). (University of Southern California, Los Angeles 1952), 271-611; R.C. Osburn, *Bryozoa of the Pacific Coast of America. Part 3, Cyclostomata, Ctenostomata, Entoprocta and Addenda.* Allan Hancock Pacific Expeditions, 14(3). (University of Southern California, Los Angeles 1953), 613-841.
31 J.D. Soule, 'Suborder Ctenostomata', in R.C. Osburn, *Bryozoa of the Pacific Coast of America.* Pt. 3. Allan Hancock Pacific Expeditions, 14(3) (University of Southern California, Los Angeles, 1953), 726-784.
32 W.C. Banta and J.C. Redden, 'A checklist of the Bryozoa of the Galapagos', *Proceedings of the Biological Society of Washington*, 103 (4) (1990), 789-802.
33 Moyano, note 8.
34 Moyano, note 15.
35 H.I. Moyano, G., 'Bryozoa de la Expedición Italiana al Estrecho de Magallanes, Febrero Marzo de 1991: evaluación preliminar', in *Oceanografia in Antartide*, edited by V.A. Gallardo, O.

Ferretti and H.I. Moyano (ENEA - Progetto Antartide - Italia; Centro EULA - `Universidad de Concepción, Concepción, Chile, 1992), 509-516.

36 Moyano, note 1 (2000).

37 C.A. Viviani, *Die Bryozoen (Ento- und Ectoprocta) des chilenischen Litorals.* (Inaugural-Dissertation zur Erlangung des Doktorgrades der naturwissentschaftlichen Fakultät der JustusLiebig-Universität Giessen, 1969).

38 C.A. Viviani, 'Briozoos del Litoral Chileno. Las especies del género *Hippothoa* (Ascophora)', *Medio Ambiente*, 2 (2) (1977), 38-52.

39 H.I. Moyano, G., 'Briozoos Marinos Chilenos I. Briozoos de la Isla de Pascua I', *Gayana Zoología*, 26 (1973), 1-22.

40 H.I. Moyano, G., 'Magellanic Bryozoa: some ecological and zoogeographical aspects', *Marine Biology*, 67 (1982), 81-96.

41 Moyano, note 8; H.I. Moyano, G., 'Bryozoa Lekythoporidae: discusión general y nuevas especies de los géneros *Catadysis* y *Orthoporidra* de Chile Austral y de la Antártica', *Gayana Zoología*, 49 (1985), 103-149; H.I. Moyano, G., 'Bryozoa Marinos Chilenos VI. Cheilostomata Hippthoidae: South Eastern Pacific species', *Boletin de la Sociedad de Biología de Concepción*, 57 (1987), 89-135.

42 E. Marcus, 'Bryozoa von den Juan Fernández-Inseln', in *The natural history of Juan Fernández and Easter Island*, Volume 3, edited by K. Skottsberg (Uppsala, 1921), 93-124.

43 Moyano, note 15; H.I. Moyano G., 'Bryozoa', in *Diversidad Biológica de Chile*, edited by J.A. Simonetti (Santiago de Chile, Conicyt, 1995), 163-173; H.I. Moyano G., 'Holocene bryozoan links between Australia, New Zealand, southern South America, and Antarctica', in *Bryozoans in Space and Time*, edited by D.P. Gordon, A.M. Smith and J.A. Grant-Mackie (National Institute of Water and Atmospheric Research, Wellington, 1996), 207-219.

44 P.J. Hayward, 'Cheilostomata (Bryozoa) from the South Atlantic', *Journal of Natural History*, 14 (1980), 701-722; P.J. Hayward, 'The recent speçies of *Adeonella* (Bryozoa: Cheilostomata) including descriptions of fifteen new species', *Zoological Journal of the Linnean Society*, 94 (1988) 111-191; P.J. Hayward, and J.S. Ryland, 'Some Antarctic and Subantarctic species of Microporellidae (Bryozoa: Cheilostomata)', *Journal of Natural History*, 24 (1990), 1263-1287; P.J. Hayward and J.P. Thorpe, 'Species of *Arachnopusia* collected by *Discovery* investigations', *Journal of Natural History*, 22 (1988), 773-799; P.J. Hayward and J.P. Thorpe, 'Membraniporoidea, Microporoidea and Cellarioidea (Bryozoa, Cheilostomata) collected by *Discovery* investigations', *Journal of Natural History*, 23 (1989), 913-959; P.J. Hayward and J.P. Thorpe, 'Some Antarctic and sub-Antarctic species of Smittinidae (Bryozoa: Cheilostomata)', *Journal of Zoology*, 222 (1990), 137-175.

45 Moyano, notes 15 and 42 (1995).

46 Banta and Reddin, note 31.

47 M. Muñoz and H.I Moyano G., 'Distribución espacial de epibiontes coloniales sobre *Macrocystis pyrifera* en tres localidades de la VIII Región, Chile', *Boletín de la Sociedad de Biología de Concepción*, 59 (1988), 115-132; M. Muñoz and J.M. Cancino, 'Consecuencias del tamaño colonial en la tasa metabólica de *Caulorhamphus spiniferum.*', *Revista Chilena de Historia Natural*, 62 (1989), 205-216; J.M. Cancino, B. Catañeda and M.C. Orellana, 'Reproductive strategies in bryozoans: experimental test of effects of conspecific neighbours', in *Bryozoaires Actuels et Fossiles: Bryozoa Living and Fossil*, edited by F.P. Bigey and J.-L. d'Hondt, Bulletin de la Société des Sciences Naturelles de l'Ouest de la France, Nantes. Mem. HIS (1991), 81-88; C.C. Ramírez and J.M. Cancino, 'Respuestas a la luz y conducta de asentamiento de larvas

de *Celleporella hyalina* (L.) (Bryozoa)', *Revista Chilena de Historia Natural* 64(1) (1991), 29-35; J.M. Cancino, R.N. Hughes and M.C. Orellana, 'A comparative study of larval release in bryozoans', in *Biology and Palaeobiology of Bryozoans*, edited by P.J. Hayward, J.S. Ryland and P.D. Taylor (Olsen and Olsen, Fredensborg, 1994), 41-45; M.C. Orellana, J.M. Cancino and R.N. Hughes, 'Is settlement of in lecithotrophic bryozoan larvae constrained by energy reserves?', in *Bryozoans in Space and Time*, edited by D.P. Gordon, A.M. Smith and J.A. Grant-Mackie (National Institute of Water and Atmospheric Research, Wellington, 1996), 221-226; J.M. Cancino and J.A. Gallardo, 'Effects of light on bryozoan larval settlement', in *Proceedings of the 11th International Bryozoology Conference*, edited by A. Herrera Cubilla and J.B.C. Jackson (Smithsonian Tropical Research Institute, Republic of Panama, 2000), 168-175.

Colonial behaviour and group zooidal reactions in Bryozoa: history of research

Andrew N. Ostrovsky* and Natalia N. Shunatova
Department of Invertebrate Zoology, Faculty of Biology and Soil Science, St. Petersburg State University, Universitetskaja nab. 7/9, St. Petersburg, 199034, Russia
**Present address: Institut für Geowissenschaften, Christian-Albrechts-Universität zu Kiel, Olshausenstr. 40, 24118 Kiel, Germany*

1. Introduction

Investigations of modular systems encompass many colonial organisms with a high level of morphological and physiological integration. Bryozoan colonies are a very promising model in this respect, and studies of colonial behaviour are one of the approaches in evaluating the degree of colonial integration. Bryozoans demonstrate a variety of group behavioural reactions which are connected with different vital functions. Co-ordinated activities of polypides are regulated neurophysiologically, physiologically, structurally and, possibly, hormonally. Different manifestations of the regulatory mechanisms have been studied previously, and we here present a review that discusses the history of these investigations.

2. Early studies

Henry Baker was probably the first who noted in passing colonial behavioural reactions in fresh-water bryozoans (possibly, *Plumatella*).[1] Barthelemy Dumortier described retraction of all polypides in a colony of the phylactolaemate *Lophopus crystallinus* (Pallas) after a stimulus was applied to a 'coenoecium'.[2] Contact with a single lophophore resulted only in its retraction. The first description of synchronized polypide protrusion and retraction in colonies of marine bryozoans was published by George Johnston in 1838.

Similarly to Dumortier, he also mentioned that 'if, when many [polypides] are expanded [in a colony], one is...touched with a sharp instrument, it alone feels an injury and retires, without any others being conscious of the danger'.[3] But in spite of the intensive studies on different taxa of these colonial animals using live specimens (Hassall,[4] Dalyell,[5] Gosse,[6] Busk,[7] Allman,[8] Smitt,[9,10] Hyatt,[11] Nitsche,[12-15] Vigelius,[16] Kraepelin,[17] Pergens,[18] Oka,[19] Proucho,[20] Calvet,[21] Gerwerzhagen,[22] Borg,[23] etc.)[24] there are only occasional remarks concerning group autozooidal behaviour in the literature published. For instance, Thomas Hincks[25] and Fritz Müller[26] mentioned 'common and simultaneous movements' of the autozooids in the ctenostome *Mimosella gracilis* Hincks.[27,28,29]

3. Twentieth century studies

Folke Borg described removal of released eggs by their consecutive transference from lophophore to lophophore in a colony of the cheilostome *Electra pilosa* (Linnaeus).[30] Ernst Marcus made some observations on bryozoan behaviour during his experiments on several species of both marine and fresh-water bryozoans, including the cheilostome *E. pilosa,* the ctenostome *Farrella repens* (Farre), and the phylactolaemates *Cristatella mucedo* Cuvier and *Fredericella sultana* (Blumenbach).[31,32] One of his main conclusions was that there is no colonial co-ordination in gymnolaemate Bryozoa, and that, as in the case of Johnston,[33] no autozooid responded to a stimulus applied to any other.[34] In contrast, the transmissions of stimuli from one member of a colony to another was demonstrated by Marcus for the phylactolaemates mentioned.[35] Results of Marcus's studies were regarded as being of the greatest importance by several authors[36,37,38] in spite of the opinion of Georges Bronstein who claimed to have observed colonial responses in the ctenostome *Bowerbankia* sp.[39] Also the data of Stanislaw Hiller, who described the interzooidal nervous connections in *E. pilosa* and *E. crustulenta* (Pallas) (both as *Membranipora*) and group polypide retractions after the stimulation of the frontal cystid wall,[40] were considered as insufficient to cast doubt upon the conclusions of Marcus.[41] Collective movements of entire zooids and whole zoarial branches resulting in a change of their orientation were investigated in detail by Lars Silén[42] in ctenostomes of the genera *Mimosella* and *Farrella* and the cheilostome *Kinetoskias* correspondingly.[43] However, Silén supported Marcus's[44] view, stating that mechanical stimulation of a zooid does not trigger any response from adjacent zooids.[45,46] Thus, the opinion about the absence of colonial co-ordination of zooidal activity survived for about 50 years in spite of the facts that there are momentary simultaneous polypide retractions after stimulation[47-52] as well as movements of cheilostome bryozoan vibracula that can beat in unison.[53-59] For instance, Silén believed that synchronous action of the vibracula in the cheilostome *Caberea* is a kind of metachronal chain reaction during which very long vibracular setae inevitably touch neighbouring ones that thereupon respond.[60,61,62]

Further investigations led researchers to suppose that the behaviour of Bryozoa is a very flexible and sensitive system of reactions that combines diverse individual and group zooidal activities depending on the situation. At the present time, there are a lot of both morphological and behavioural evidences in favour of existence of the presumed colonial co-ordination in all main Recent bryozoan taxa. In addition to the anatomical data of earlier authors [63-71] many new facts that are indicative of nervous connections and transmission of nervous signals between zooids were obtained by Geneviève Lutaud,[72-80] who supported the hypothesis of interzooidal information spreading through a colony and leading to colonial responses.[81]

In the phylactolaemate *Plumatella fungosa* (Pallas) George Mackie recorded the phenomenon in which 'the lophophores over wide regions become orientated by muscular action so that they point in the same direction'.[82] He also mentioned co-ordinated withdrawal responses in *Cristatella mucedo*. Colonial behaviour and nervous responses were studied in some species of free-living cheilostome.[83-87] Colonies under investigation showed a high degree of co-ordination of avicularian setae movement owing to which they were able to stop themselves being buried in sand, to return back to a normal position after being placed upside down, to travel across the aquarium preferably towards the light and to climb one over another.[88] A high level of co-ordination was also recorded between autozooids, and between autozooids and avicularia, when feeding periods ending in simultaneous polypide retraction are alternated with simultaneous setal cleaning movement after which the colonial feeding is resumed. Co-ordinated avicularian activity in Recent lunulitiform bryozoans (Cupuladriidae) was also noted by Ernst and Eveline Marcus,[89] Patricia Cook[90] and Ronald Greeley.[91] Judith Winston and Patricia Cook recorded synchronous movements of avicularian setae in different species of Cheilostomatida.[92-95]

Collective activity of another type of bryozoan polymorphs was described by Lars Silén and Jean-Georges Harmelin who observed rocking and circular movements of the single tentacle of nanozooids in cyclostomes of the genera *Plagioecia* and *Diplosolen*.[96] These movements were not synchronized but were simultaneous, and cleaned a colony surface after induced obstruction by silt. It was noted that the tentacles of nanozooids were the first to appear in the colony after strong disturbance, whereas autozooids expanded their lophophores much later. Similar 'sentry' behaviour performed solely by autozooids was also recorded in several cheilostomates.[97]

Some neurophysiological aspects of intra- and intercolonial co-ordination of behavioural responces were investigated in *Membranipora membranacea* (Linnaeus), *Electra pilosa*, *Selenaria maculata* (Busk) and *Flustrellidra hispida* (Fabricius).[98-101] The long-term controversy (see above) was rejected at last. It was found in experiments that 'mechanical [as well as electrical] stimulation of an extended lophophore...resulted solely in the retraction of that polypide'.[102] In contrast, stimulation of the frontal membrane triggers 'the immediate rapid withdrawal of all the extended lophophores within some distance of the zooid stimulated'.[103, 104, 105] This is in good agreement with observations of Winston who described group retractions during removal of large particles from a colony.[106] Conduction of nerve pulses through the colony accompanied by quick consecutive retractions of polypides was proved both within the same colony[107-110] and within complexes of adjoining colonies.[111] Moreover, such conduction was also shown in a representative of carnosan ctenostomes (*F. hispida*),[112] although interzooidal nervous connections were not discovered by Lutaud who investigated *Alcyonidium polyoum* (Hassall).[113] Michael Berry and Peter Hayward suggested that 'the nervous activity responsible for co-ordinating locomotion would spread throughout the entire colony' after their experiments with free-living *S. maculata* where strong illumination caused the colonies to start moving.[114] Noticeably, in contrast with earlier observations[115, 116, 117] John Thorpe found no obvious evidence for co-ordinated lophophore retractions in phylactolaemates.[118]

However, it has been found that co-ordinated autozooidal behaviour in Bryozoa is not limited to simultaneous protrusion-retraction excursions. Bryozoans are colonial organisms and, therefore, the effects of large numbers of individual lophophores may be readily seen. Collective feeding leads to several problems, e.g. how to prevent a repeated water filtration, and how to remove filtered water, sediment and wastes from a colony where an area with incurrent flow is many times larger than the excurrent area. This inequality potentially must result in an appearance of zones with high water pressure where normal zooidal functioning might be hampered. In such conditions it is conceivable that autozooids will somehow react in groups to alleviate the problems, and this might be considered as a consequence of colonial regulation. Colonial effects on feeding and behaviour of marine bryozoans have been intensively studied since the work of William Banta, Ken McKinney and Russell Zimmer who observed colonies of the cheilostome *Membranipora* sp. and noted that most of the colony was covered by extended polypides, but that there were numerous blank spaces ('chimneys').[119] In these areas lophophores were not extended, while those around the edge of the chimney leaned away from its centre having the obliquely-truncated tentacle crowns with the longest tentacles bordering a chimney. The authors concluded that the incurrent water passed through the tentacles and down between the zooids exiting via the colony edge or a chimney. They also suggested that a formation of excurrent water outlets was connected with a colony size. In their opinion, in large colonies 'lophophores…interfere with each other unless excurrent channels such as chimneys are present'.[120] These researchers proposed that the monticules (maculae) known in some fossil stenolaemates might be such chimneys reflected in a skeleton, and this idea was supported and exploited by many researchers.[121-132]

Removal of filtered water from the colonies and co-ordinated autozooidal activity was further studied in detail by Patricia Cook and Jim Chimonides,[133, 134] who examined 17 different species of the Recent marine bryozoans. For multiserial encrusting colonies they noted three patterns of excurrent flow: (1) active centripetal and centrifugal colony-wide flows, (2) "active" local excurrent flows from chimneys ringed by functioning autozooids with obliquely-truncated lophophores and (3) 'passive' chimneys or local exhalant flows over areas devoid of actively feeding zooids. Their observations on *M. membranacea* led them to the opinion that both the position of chimneys in a colony and the position of obliquely-truncated lophophores in a chimney are astogenetically determined. In general, these researchers supposed that both astogenetic and ontogenetic (connected with a polypide recycling) changes are involved in production of various forms of filtered water outflows, and agreed with previous authors[135, 136] that some reproductive processes might take part in chimney formation.[137] Cook was the first to describe chimneys bordered by obliquely-truncated polypides that were tilted towards the chimney centre.[138] Scott Lidgard, who conducted his experiments on *Membranipora villosa* Hincks, concluded that chimneys are an adaptation which allows colonies to minimise the recirculation of previously filtered water, making the colony more efficient.[139] Matthew Dick proposed a hypothetical mechanism for chimney formation, suggesting that this process may involve both hydrodynamic and astogenetic control as well as their combination.[140] He also showed the ways of transformation from obliquely-truncated lophophore to equitentacled one, and vice versa, in his study on *Holoporella brunnea* (Hincks) and *M. serrilamella* Osburn, and connected the reason for this transformation with lophophore position respectively to excurrent flow. Some problems of the spatial arrangement of feeding zooids and their interference in encrusing colonies were also investigated and discussed by John Thorpe and John Ryland[141] and later by Daniel Grünbaum.[142]

Judith Winston investigated 79 species of Recent marine bryozoans and recognized six patterns of bryozoan colonial behaviour in relation to the morphology of polypides, their grouping, colony form and structure, and water currents produced.[143, 144, 145] She distinguished species where colonies were characterized by (1) predominantly individualized polypide behaviour,[146] (2) separated polypides whose orientation is controlled by the colony skeleton, (3) polypides forming temporary clusters, (4) polypides forming fixed clusters unreflected in the colony skeleton, (5) polypides forming fixed clusters enchanced by irregular skeletal patterning, (6) polypides forming fixed clusters enhanced by regular patterning of the colony skeleton. She also observed several earlier unknown manifestations of group autozooidal behaviour, and supported the hypothesis of Banta, McKinney and Zimmer[147] recording an excurrent flow of filtered water above raised areas (knobs or monticules) in the skeleton of some cheilostomes.[148, 149] This and subsequent work was reviewed by Frank McKinney[150, 151, 152] who classified the feeding of 91 species of Bryozoa into four main categories. The first category includes those bryozoans that behave essentially as individuals when they feed. This often occurs when lophophores are sufficiently separated and do not act together with neighbours. Typically such lophophores are supported on long introverts (or elevated by tubular and erect distal ends of the cystids), radially symmetrical and equitentacled. The second category includes species where lophophores form temporary feeding clusters. In such cases only a fraction of the colony feeds at any time and clusters of lophophores form around an initially emerging lophophore. Alternatively, in the third category, lophophores are in fixed groups across a continuous colony surface, and in such colonies there is usually a high degree of integrated behaviour. Invariably in such designs there are areas of outflow from the colony and in this category of Bryozoa there are often asymmetric lophophores found in these regions. The last category in which bryozoan colonies may be considered is that of skeletally separated lophophores, where a branching structure facilitates unidirectional flow towards the colony and between the branches. The lophophores may cover the area between the branches and lean over the space. There may also be a high degree of asymmetry in these lophophores and all of them typically have short introverts.

In her book, published in 1985, Patricia Cook mentioned the colonial behaviour in several cheilostomates. She was the first to describe the 'colony-wide episodes of tentacle 'reversal'', when 'the tentacle crowns were widely extended and the tips directed towards the colony surface'.[153] Cook suggested that this behaviour pattern might be a colonial rejection, 'correlated with reversal of ciliary movement, and may be associated with cleaning of detritus,[154] sperm release or other, at present unknown functions'.[155]

Natalia Shunatova and Andrew Ostrovsky recorded group autozooidal reactions and colonial behaviour in 13 species and subspecies of marine bryozoans.[156] Several collective reactions (synchronized scanning behaviour, repeated particle transference by circular water currents, feeding and cleaning of the colony surface by 'chains' of inclined lophophores) were described for the first time. In large colonies of *Tegella armifera* (Hincks) a new type of chimney was discovered. Chimneys are formed by temporary retraction of 10-12 neighbouring polypides, and the blank space is surrounded by equitentacled lophophores standing vertically. Chimneys associated with elevated areas on the colony surface (monticules) were found in *Porella smitti* (Kluge) and *Schizomavella lineata* (Nordgaard). But, in contrast to all previously published speculations and observations, the monticules were often places of incurrent rather than excurrent flow, and water outlets were formed in depressions between monticules. In *S. lineata,* monticules change their function from incurrent to excurrent after polypide degeneration. Shunatova and Ostrovsky suggested that elevated areas formed by frontal budding can probably change their function at least four times: (1) they can probably work as excurrent zones during the initial stages of formation when young monticules consist of only incipient zooids with non-functional polypides, (2) later they start to function as incurrent zones, (3) after polypide degeneration, they become excurrent zones, and (4) after polypide regeneration they function as incurrent zones again. The depressions between monticules would probably follow exactly the opposite succession. Thus, the hypothesis of Robert Anstey,[157] who speculated that in some Palaeozoic stenolaemates the hydrodynamical regime around the monticules could change, was made more plausible.

Colony-wide water currents in relation to ambient flow, colony growth form and polypide morphology, distribution and behaviour were also intensively studied by Frank McKinney with co-authors in some Recent marine Bryozoa.[158-163] The results fit well with previously proposed theoretical speculations concerning the interrelationship of the ambient and cilia-generated water currents and growth forms in different fossil bryozoans.[164-170] It was accepted that feeding was a 'major sculptor of bryozoan form'.[171, 172] Colony-wide water currents have been also considered to be very important in intra and interspecific competition.[173, 174, 175]

Beth Okamura and Julian Partridge supported Dick´s hypothesis[176] of pressure build-up under the lophophore sheet in experiments with *Membranipora membranacea*. They showed "a trend towards [lophophore] miniaturization with increased flow: the zooids were less elongate, the lophophores were smaller in diameter and had fewer tentacles, and the distances between excurrent jets were shorter".[177]

Colonial effects on feeding of some phylactolaemates have been studied by George Mackie,[178] John Bishop and Leonard Bahr,[179] Beth Okamura and Lita Doolan[180] and Irina Antipenko.[181, 182]

4. Conclusions

In spite of many investigations, the only attempt to quantitatively estimate the degree of integration in bryozoan colonies was made by Frank McKinney.[183] He used the extent of the colonial feeding current interactions for this purpose. Judith Winston proposed that the data on group behaviour of autozooids could be used for such estimation, and worked up a "matrix" where "levels of morphological and behavioural integration (with respect to feeding and current producing activities)" are considered.[184, 185] Thus, previous research showed that bryozoans are a convenient model for investigations of different aspects of integrative processes. The data accumulated at the moment can be considered as a strong background for future research, and further observations on bryozoan behaviour as well as mathematical formalization should help to make progress in this direction.

5. Acknowledgements

We are greatly indebted to Professor Dr Claus Nielsen and Dr Mary E. Petersen, Zoological Museum, University of Copenhagen, and Mary Spencer Jones, The Natural History Museum, London, for kind help with literature. Our thanks to Drs William G. Sanderson, Michael A. Best and John P. Thorpe, Department of Environmental and Evolutionary Biology, University of Liverpool, Port Erin Marine Laboratory, for unpublished review on bryozoan feeding. We are grateful to Dr Ken McKinney, Appalachian State University and Dr Paul D. Taylor, The Natural History Museum, for reading and criticising earlier drafts of the manuscript, useful comments and much encouragement. Andrew N. Ostrovsky thanks the Danish Ministry of Education for a post-doctoral stipend which made work in the University of Copenhagen Library possible. The Otto Kinne Foundation is acknowledged for financial support.

Notes

1 H. Baker, *Employment for the microscope* (London, 1753), xiv, 442.
2 B.-C. Dumortier, 'Recherches sur l'anatomie et la physiologie des polypiers composés d'eau douce', *Bulletins de l'Académie Royale des Sciences, Bruxelles*, 2 (1835), 421-454 (422).
3 G. Johnston, '*A history of the British zoophytes*' (Edinburgh, London, Dublin, 1838), 341 (36).
4 See: A.H. Hassall, 'Supplement to a catalogue of Irish zoophytes', *The Annals and Magazine of Natural History,* 7(44, 45) (1841), 276-287, 363-373 (Plate VI, figure 1).
5 J.G. Dalyell, *Rare and remarkable animals of Scotland, represented from living subjects: with practical observations on their nature,* 2 vols (Van Voorst, London, 1847-1848), vol 1, xii, 268, vol 2, iv, 322.
6 P.H. Gosse, 'Notes on some new or little-known marine animals', *The Annals and Magazine of Natural History,* 2 Serie, 16 (1855), 27-36.
7 G. Busk, 'Zoophytology', *Quarterly Journal of Microscopical Science*, 4 (1856), 308-312.
8 G. Allman, *A monograph of the fresh-water Polyzoa, including all the known species, both british and foreign* (Ray Society, London, 1856), 120.
9 F.A. Smitt, 'Om Hafs-Bryozoernas utveckling och fettkroppar', *Öfversigt af Kongliga Vetenskapsakademien Förhandling,* 1 (1865), 5-50.
10 F.A. Smitt, 'Kritisk fürteckning öfver Skandinaviens Hafs-Bryozoer', *Öfversigt af Kongliga Vetenskapsakademien Förhandling,* 2 (1865), 115-533.
11 A. Hyatt, 'Observations on Polyzoa. Suborder Phylactolaemata', *Proceedings of the Essex Institute*, vols 4 and 5 (Salem, 1866-1868), iv, 103.
12 H. Nitsche, *Beiträge zur Anatomie und Entwickelungsgeschichte der phylactolaemen Süsswasserbryozoen, insbesondere von* Alcyonella fungosa *Pall.*, Akademisch Dissertation (Grimm and Maass, Berlin, 1868), 57.
13 H. Nitsche, 'Beiträge zur Kenntnis der Bryozoen. I. Beobachtung über die Entwickelungsgeschichte einiger cheilostomen Bryozoen', *Zeitschrift für wissenschaftliche Zoologie*, 20(1) (1869), 1-13.
14 H. Nitsche, 'Beiträge zur Kenntnis der Bryozoen. IV. Über die Morphologie der Bryozoen', *Zeitschrift für wissenschaftliche Zoologie*, 21 (1871), 92-119.
15 H. Nitsche, 'Beiträge zur Kenntnis der Bryozoen. V. Über die Knospung der Bryozoen', *Zeitschrift für wissenschaftliche Zoologie*, 25(supplement) (1876) 121-180.
16 W.J. Vigelius, 'Morphologische Untersuchungen über *Flustra membranaceo-truncata* Smitt', *Biologisches Centrablatt*, 23 (1884), 705-721.
17 K. Kraepelin, 'Die Deutschen Süsswasser-Bryozoen. Anatomish-systematischer Teil', *Abhandlungen aus dem Gebiete der Naturwissenschaft herausgeben von dem Naturwissenschaftlicher Verein in Hamburg,* 10 (1887), 1-168.
18 E. Pergens, 'Untersuchungen an Seebryozoen', *Zoologischer Anzeiger* 12 (1889), 504-510, 526-533.
19 A. Oka, 'Observations on fresh-water Polyzoa. (*Pectinatella gelatinosa*, nov. sp.)', *Journal of the College of Science, Imperial University of Japan, Tokyo,* 4 (1891), 89-150.
20 H. Proucho, 'Contribution à l'histoire des Bryozoaires', *Archives de Zoologie expérimentale et générale,* 10 (1892), 557-656.
21 L. Calvet, 'Contribution à l'histoire naturelle des Bryozoaires Ectoproctes marins', *Travailes de l'Institut de Zoologie de l'Université de Montpellier,* Nouvelle Série, 8 (1900), 1-488.
22 A. Gerwerzhagen, 'Untersuchungen an Bryozoen', *Sitzungberichte der Heidelberger Akademie*

der Wissenschaften, Matematische-naturwissenschaftliche Klasse B, 9 (1913), 1-16.
23 F. Borg, 'On the structure of cyclostomatous Bryozoa', *Arkiv für Zoologie* 15 (1923), 1-17.
24 For review see also: A.N. Ostrovsky, N.N. Shunatova, I.I. Antipenko, this volume.
25 T. Hincks, 'Notes on British Zoophytes', *Annals and Magazine of Natural History*, 2 Serie, 8 (1851), 353-362.
26 F. Müller, 'On the common nervous system (Kolonialnervensystem) of the Bryozoa (Polyzoa), exemplified in *Serialaria Coutinhii*, n. sp.', *Quarterly Journal of Microscopical Science,* New Series, 1 (1861), 300-305 (300).
27 See also H. Nitsche, 'Beiträge zur Kenntnis der Bryozoen. III. Über die Anatomie und Entwickelungsgeschichte von *Flustra membranacea*', *Zeitschrift für wissenschaftliche Zoologie*, 21 (1871), 37-91.
28 T. Hincks, 'Contributions to history of Polyzoa.', *Quarterly Journal of Microscopical Science*, (N. S.) 13 (1873), 13-37.
29 T. Hincks, *A history of the British marine Polyzoa*, 2 volumes (Van Voorst, London, 1880), 601.
30 F. Borg, 'Studies on recent cyclostomatous Bryozoa', *Zoologiska Bidrag från Uppsala*, 10 (1926), 181-150.
31 E. Marcus, 'Beobachtungen und Versuche an lebenden Meersbryozoen', *Zoologische Jahrbücher Abteilung für Systematik, Ökologie und Geographie der Tiere*, 52 (1926), 1-102.
32 E. Marcus, 'Beobachtungen und Versuche an lebeden Sübwasserbryozoen', *Zoologische Jahrbücher Abteilung für Systematik, Ökologie und Geographie der Tiere*, 52 (1926), 279-350.
33 Johnston, note 3.
34 Marcus, note 31.
35 Marcus, note 32.
36 L.H. Hyman, *The invertebrates: smaller coelomate groups*, 5 (McGraw-Hill Book Company, New York, Toronto, London, 1959), 275-515.
37 P. Brien, 'Classe des Bryozoaires', in *Traité de Zoologie*, edited by P.-P. Grasse, 5(2) (Masson et Sie, Paris, 1960), 1053-1335.
38 J.S. Ryland, *Bryozoans* (Hutchinson University Library, London, 1970), 175.
39 G. Bronstein, 'Étude du système nerveux de quelques bryozoaires gymnolémides', *Travaux de la Station Biologique de Roscoff*, 15 (1937), 155-174.
40 S. Hiller, 'The so-called 'colonial nervous system' in Bryozoa', *Nature*, 143 (3634) (1939), 1069-1070.
41 Summarized in J.P. Thorpe, 'Bryozoa', in *Electrical conduction and behaviour in 'simple' invertebrates*, edited by G.A.B. Shelton (Clarendon Press, Oxford, 1982), 393-439.
42 L. Silén, 'On the motility of entire zooids in Bryozoa', *Acta Zoologica*, 31 (1950), 349-386.
43 Reviewed by Hyman, note 36.
44 Marcus, note 31.
45 Silén, note 42.
46 L. Silén, 'Polymorphism', in *Biology of Bryozoans*, edited by R.M. Woollacott and R.L. Zimmer (Academic Press, London, New York, 1977), 183-231.
47 See Dumortier, note 2.
48 Johnston, note 3.
49 G. Johnston, '*A history of the British zoophytes*', 2nd edition (Van Voorst, London, 1847), xiv, 488.
50 Bronstein, note 39.
51 Hiller, note 40.

52 Summarized in T.H. Bullock and G.A. Horridge, 'Bryozoa', in *Structure and function in the nervous systems of invertebrates*, 1 (W. H. Freeman and Company, San Francisco, London, 1965), 632-647.
53 C.R. Darwin, *Journal of researches into the geology and natural history of the various countries visited by H.M.S.* Beagle *under the command of Capt. Fitzroy, R.N., from 1832 to 1836* (Henry Colborn, London, 1839), 379.
54 C.R. Darwin, *The origin of species by means of natural selection or the preservation of favoured races in the struggle for life* (John Murray, London, 1872), 539.
55 G. Busk, 'Remarks on the structure and function of the avicularian and vibracular organs of the Polyzoa; and on their value as diagnostic characters in the classification of those creatures', *Transactions of the Microscopical Society*, New Series, 2 (1854), 26-34.
56 T. Hincks, 'Note on the movements of the vibracula in *Caberea Boryi*, and on the supposed common nervous system in the Polyzoa', *Quarterly Journal of Microscopical Science,* New Series, 18 (1878), 7-9.
57 Hincks, note 29.
58 A. Forbes, 'Conditions affecting the response of the avicularia of *Bugula*', *Biological Bulletin*, 65 (1933), 469-479.
59 Summarized in Hyman, note 36.
60 Silén, note 42.
61 Silén, note 46.
62 See also P.L. Cook, 'Bryozoa from Ghana', *Zoologische Wetenschappen Musee Royal l'Afrique centrale Tervuren, Belgique*, 238 (1985), 1-315.
63 A. Gerwerzhagen, 'Beiträge zur Kenntnis der Bryozooen. I. Das Nervensystem von *Cristatella mucedo* Cuv.', *Zeitschrift für Wissenschaftliche Zoologie*, 107 (1913), 309-45.
64 Marcus, note 31.
65 Marcus, note 32.
66 E. Marcus, 'Über *Lophopus crystallinus* (Pall.)', *Zoologische Jahrbücher Abteilung für Anatomie und Ontogenie der Tiere*, 58 (1934), 501-606.
67 Bronstein, note 39.
68 Hiller, note 40.
69 C.I. Cori, 'Ordnung der Tentaculata: Bryozoa', in *Handbuch der Zoologie* III, 2(5) (Walter de Gruyter and Company, Berlin, 1941), 263-374, 375-503.
70 See also C.W.F. Krukenberg, *Vergleichend-physiologische Studien, Experimentelle Untersuchungen von C.F.W. Krukenberg* (Winter, Heidelberg, 1887), 517.
71 Brien, note 37.
72 G. Lutaud, 'Le "plexus" parietal de Hiller et la coloration du système nerveux par le bleu de méthylène chez quelques Bryozoaires Chilostomes', *Zeitschrift für Zellforschung und mikroskopische Anatomie Abteilung*, 99 (1969), 302-314.
73 G. Lutaud, 'Le plexus parietal des Ctenostomes chez *Bowerbankia gracilis* Leydi (Vesicularines)', *Cahiers de Biologie Marine*, 15 (1974), 403-408.
74 G. Lutaud, 'L'innervation des parois de la loge chez *Flustra papyracea* (Ellis and Solander), (Bryozoaire Chilostome)', *Cahiers de Biologie Marine*, 17 (1976), 337-346.
75 G. Lutaud, 'The bryozoan nervous system', in *Biology of Bryozoans*, edited by R.M. Woollacott and R.L. Zimmer (Academic Press, London, New York, 1977), 377-410.
76 G. Lutaud, 'The probability of a plexus in the calcified wall of *Crisidia cornuta* (Linné)', in *Advances in Bryozoology*, edited by G.P. Larwood and M.B. Abbot (Academic Press, London,

1979), 33-46.
77 G. Lutaud, 'Etude ultrastructurale du 'plexus colonial' et recherche de connexions nerveuses interzoidiales chez le Bryozoaire Chilostome *Electra pilosa* (Linné)', *Cahiers de Biologie Marine*, 20 (1979), 315-324.
78 Summarized in Thorpe, note 41.
79 R.S. Boardman, A.H. Cheetham, D.B. Blake, J. Utgaard, O.L. Karklins, P.L. Cook, P.A. Sandberg, G. Lutaud, and T.S. Wood, 'Bryozoa' (Part G, revised), in *Treatise on Invertebrate Paleontology*, edited by R.A. Robison, 1 (Geological Society of America, Boulder, Colorado and University of Kansas, Lawrence, 1983), 1-625.
80 H. Mukai, K. Terakado and C.G. Reed, 'Bryozoa', in *Microscopic Anatomy of Invertebrates,* edited by F.W. Harrison, 13 (Wiley-Liss, New York, 1997), 45-206.
81 Lutaud, note 75.
82 G.O. Mackie, 'Siphonophores, bud colonies and superorganisms', in *The Lower Metazoa Comparative Biology and Phylogeny*, edited by E.C. Dougherty (University of California Press, Berkeley, 1963), 329-336 (332).
83 P.L. Cook and P.J. Chimonides, 'Observations on living colonies of *Selenaria* (Bryozoa, Cheilostomata). I', *Cahiers de Biologie Marine*, 19 (1978), 147-158.
84 J.E. Winston, 'Polypide morphology and feeding behaviour in marine ectoprocts', *Bulletin of Marine Science*, 28(1) (1978), 1-31.
85 P.J. Chimonides and P.L. Cook, 'Observations on living colonies of *Selenaria* (Bryozoa, Cheilostomata). II', *Cahiers de Biologie Marine*, 22 (1981), 207-219.
86 M.S. Berry and P.J. Hayward, 'Nervous and behavioural responses to light in colonies of the free-living bryozoan, *Selenaria maculata* (Busk)', *Experientia*, 40 (1984), 108-110.
87 For discussion see also: P.L. Cook, 'Some problems in interpretation of heteromorphy and colony integration in Bryozoa', in *Biology and Systematics of Colonial Organisms,* edited by G.P. Larwood and B.R. Rosen, Systematics Association Special vol 11 (Academic Press, London, New York, 1979), 193-210.
88 Summarized in Thorpe, note 41.
89 E. Marcus and Ev. Marcus, 'On some lunulitiform Bryozoa', *Boletim da Faculdade de Filosofia, Ciências e Letras, Universidade de São Paulo, Zoologia*, 24 (1962), 281-312.
90 P.L. Cook, 'Observations on live lunulitiform zoaria of Polyzoa', *Cahiers de Biologie Marine*, 4 (1963), 407-413.
91 R. Greeley, 'Natural orientation of lunulitiform bryozoans', *Bulletin of the Geological Society of America*, 78 (1967), 1179-1182.
92 J.E. Winston, 'Why bryozoans have avicularia – a review of the evidence', *Novitates of the American Museum of Natural History*, 2789 (1984), 1-26.
93 J.E. Winston, 'Victims of avicularia', *Marine Ecology*, 7 (1986), 193-199.
94 J.E. Winston, 'Avicularian behaviour – a progress report', in *Bryozoaires Actuels et Fossiles: Bryozoa Living and Fossil*, edited by F.P. Bigey, *Bulletin de la Société des Sciences Naturelles de l'Ouest de la France,* Mémoire Hors Série, 1 (1991), 531-540.
95 Cook, note 62.
96 L. Silén and J.-G. Harmelin, 'Observations on living Diastoporidae (Bryozoa Cyclostomata), with special regard to polymorphism', *Acta Zoologica*, 55 (1974), 81-96.
97 N.N. Shunatova and A.N. Ostrovsky, 'Group autozooidal behaviour and chimneys in marine bryozoans', *Marine Biology*, 140 (2002), 503-518.
98 Reviewed in Thorpe, note 41.

99 Mukai, Terakado and Reed, note 80.
100 For discussion see also: J.S. Ryland, 'Physiology and ecology of marine bryozoans', *Advances in Marine Biology*, 14 (1976), 285-443.
101 J.S. Ryland, 'Structural and physiological aspects of coloniality in Bryozoa', in *Biology and Systematics of Colonial Organisms*, edited by G.P. Larwood and B.R. Rosen, Systematics Association Special vol 11 (Academic Press, London, New York, 1979), 211-242.
102 J.P. Thorpe, G.A.B. Shelton and M.S. Laverack, 'Electrophysiology and co-ordinated behavioural responses in the colonial bryozoan *Membranipora membranacea* (L.)', *Journal of Experimental Biology*, 62 (1975), 389-404 (393).
103 J.P. Thorpe, 'Behaviour and colonial activity in *Membranipora membranacea* (L.)', in *Bryozoa 1974*, edited by S. Pouyet, *Documents des Laboratoires de Géologie de la Faculte des Sciences de Lyon*, 3 (1975), 115-121.
104 Thorpe, Shelton and Laverack, note 102, p. 393.
105 J.P. Thorpe, G.A.B. Shelton, and M.S. Laverack, 'Colonial nervous control of lophophore retraction in cheilostome Bryozoa', *Science*, 189 (1975), 80-1.
106 Winston, note 84.
107 Thorpe, note 103.
108 Thorpe, Shelton and Laverack, note 102.
109 Thorpe, Shelton and Laverack, note 105.
110 D.F. Shapiro, 'Intercolony coordination of zooid behaviour and a new class of pore plates in a marine bryozoans', *Biological Bulletin*, 182 (1992), 221-30.
111 Shapiro, note 110.
112 Thorpe, note 41.
113 G. Lutaud, 'The innervation of the external wall in the carnosan ctenostome *Alcyonidium polyoum* (Hassall)', in *Recent and Fossil Bryozoa*, edited by G.P. Larwood and C. Nielsen (Olsen and Olsen, Fredensborg, 1981), 144-150.
114 Berry and Hayward, note 86, p. 109.
115 Dumortier, note 2.
116 Marcus, note 32.
117 Mackie, note 82.
118 Thorpe, note 41.
119 W.C. Banta, F.K. McKinney and R.L. Zimmer, 'Bryozoan monticules: excurrent water outlets?', *Science*, 185 (1974), 783-784.
120 Banta, McKinney and Zimmer, note 119, p. 783.
121 P.D. Taylor, 'Monticules in a Jurassic cyclostomatous bryozoan', *Geological Magazine*, 112 (1975), 601-6.
122 P.D. Taylor, 'The inference of extrazooidal feeding currents in fossil bryozoan colonies', *Lethaia*, 12 (1979), 47-56.
123 P.D. Taylor, 'Functional significance of contrasting colony form in two Mesozoic encrusting bryozoans', *Palaeogeography Palaeoclimatology Palaeoecology*, 26 (1979), 151-158.
124 F.K. McKinney, 'Functional interpretation of lyre-shaped Bryozoa', *Paleobiology*, 3 (1977), 90-97.
125 F.K. McKinney, 'Evolution of erect marine bryozoan faunas: repeated success of unilaminate species', *American Naturalist* 128 (1986), 795-809.
126 F.K. McKinney, 'Historical record of erect bryozoan growth forms', *Proceedings of the Royal Society, London*, 228 (1986), 133-148.

127 R.L. Anstey, 'Zooid orientation structures and water flow patterns in Paleozoic bryozoan colonies', *Lethaia*, 14 (1981), 287-302.
128 R.L. Anstey, 'Colony patterning and functional morphology of water flow in Paleozoic stenolaemate bryozoans', in *Bryozoa: Present and Past*, edited by J.R.P. Ross (Western Washington University, Bellingham, WA, 1987), 1-8.
129 R.L. Anstey, 'Astogeny and phylogeny: evolutionary heterochrony in Paleozoic bryozoans', *Paleobiology*, 13 (1987), 20-43.
130 R.L. Anstey, J.F. Pachut, and D.R. Prezbindowski, 'Morphogenetic gradients in Paleozoic bryozoan colonies', *Paleobiology*, 2 (1976), 131-146.
131 M.E. Patzkowsky, 'Inferred water flow patterns in the fossil *Fistulipora* M'Coy (Cystoporata, Bryozoa)', in *Bryozoa: Present and Past*, edited by J.R.P. Ross (Western Washington University, Bellingham, WA, 1987), 213-219.
132 F.K. McKinney and J.B.C. Jackson, *Bryozoan evolution* (University of Chicago Press, Chicago, London, 1989), 238.
133 P.L. Cook, 'Colony-wide water currents in living Bryozoa', *Cahiers de Biologie Marine*, 18 (1977), 31-47.
134 P. L. Cook and P.J. Chimonides, 'Further observations on water current patterns in living Bryozoa', *Cahiers de Biologie Marine*, 21 (1980), 393-402.
135 Banta, McKinney and Zimmer, note 119.
136 Taylor, note 121.
137 For review and discussion see Cook, note 87.
138 Cook, note 133.
139 S. Lidgard, 'Water flow, feeding and colony form in an encrusting cheilostome', in *Recent and Fossil Bryozoa*, edited by G.P. Larwood and C. Nielsen (Olsen and Olsen, Fredensborg, 1981), 175-182.
140 M.H. Dick, 'A proposed mechanism for chimney formation in encrusting bryozoan colonies', in *Bryozoa: Present and Past*, edited by J.R.P. Ross (Western Washington University, Bellingham, WA, 1987), 73-80.
141 J.P. Thorpe and J.S. Ryland, 'Some theoretical limitations on the arrangement of zooids in encrusting Bryozoa', in *Bryozoa: Present and Past*, edited by J.R.P. Ross (Western Washington University, Bellingham, WA, 1987), 276-283.
142 D. Grünbaum, 'A model of feeding currents in encrusting bryozoans shows interference between zooids within a colony', *Journal of Theoretical Biology* 174 (1995), 409-425.
143 Winston, note 84.
144 J.E. Winston, 'Current-related morphology and behaviour in some Pacific coast bryozoans', in *Advances in Bryozoology*, edited by G.P. Larwood and M.B. Abbott, Systematics Association Special vol 13 (Academic Press, London, New York, 1979), 247-268.
145 J.E. Winston, 'Feeding behaviour of modern bryozoans', in *Studies in Geology 5: Lophophorates. Notes for a Short Course*, edited by T.W. Broadhead (University of Tennessee, 1981), 1-21.
146 See also Cook, note 62.
147 Banta, McKinney and Zimmer, note 119.
148 Winston, note 84.
149 Winston, note 144.
150 F.K. McKinney, 'Feeding and associated colonial morphology in marine bryozoans', *Reviews in Aquatical Science*, 2 (1990), 255-280.

151 F.K. McKinney, 'Feeding currents of gymnolaemate bryozoans: better organization with higher colonial integration', *Bulletin of Marine Science*, 34 (1984), 315-319.
152 See also P.D. Taylor, 'Bryozoans', in *Functional Morphology of the Invertebrate Skeleton*, edited by E. Savazzi (John Wiley and Sons Ltd, 1999), 623-46.
153 Cook, note 62, p. 29.
154 See description of the similar individual reaction in: N.N. Shunatova and A.N. Ostrovsky, 'Individual autozooidal behaviour and feeding in marine bryozoans', *Sarsia*, 86 (2001), 113-142.
155 Cook, note 62, p. 29.
156 Shunatova and Ostrovsky, note 97.
157 Anstey, note 127.
158 F.K. McKinney, M.R.A. Listokin and C.D. Phifer, 'Flow and polypide distribution in the cheilostome *Bugula* and their inference in *Archimedes*', *Lethaia*, 19 (1986), 81-93.
159 F.K. McKinney, M.J. McKinney and M.R.A. Listokin, 'Erect bryozoans are more than baffling: enchanced sedimentation rate by a living unilaminate branched bryozoan and possible implications for fenestrate bryozoan mudmounds', *Palaios*, 2 (1987), 41-47.
160 F.K. McKinney, 'Planar branch systems in colonial suspension feeders', *Paleobiology*, 7 (1981), 344-354.
161 F.K. McKinney, 'Two patterns of colonial water flow in an erect bilaminate bryozoan, the cheilostome *Schizotheca serratimargo* (Hincks, 1886)', *Cahiers de Biologie Marine* 30 (1989), 35-48.
162 F.K. McKinney, 'Colonial feeding currents of *Exidmonea atlantica* (Cyclostomata)', in *Bryozoaires Actuels et Fossiles: Bryozoa Living and Fossil*, edited by F.P. Bigey, *Bulletin de la Société des Sciences Naturelles de l'Ouest de la France,* Mémoire Hors Série, 1 (1991), 263-270.
163 F.K. McKinney, 'How phylogeny limits function - the example of *Exidmonea*', *National Geographic Research and Exploration*, 7 (1991), 432-441.
164 McKinney, note 124.
165 McKinney, note 160.
166 McKinney, note 125.
167 McKinney, note 126.
168 McKinney, Listokin and Phifer, note 158.
169 See also R. Cowen, J. Rider, 'Functional analysis of fenestellid bryozoan colonies', *Lethaia*, 5 (1972), 145-164.
170 J.F. Stratton and A.S. Horowitz, 'Studies of the flow of water through models of *Polypora*', in *Bryozoa 1974*, edited by S. Pouyet, *Documents des Laboratoires de Géologie de la Faculte des Sciences de Lyon*, 3 (1975), 425-438.
171 McKinney, note 150.
172 McKinney and Jackson, note 132, p. 238 (Chapter 6).
173 L.W. Buss, 'Bryozoan overgrowth interactions – the interdependence of competition for space and food', *Nature*, 281 (5731), (1980), 475-477.
174 J.E. Winston and J.B.C. Jackson, 'Life histories of cryptic reef bryozoans', in *Recent and Fossil Bryozoa*, edited by G.P. Larwood and C. Nielsen (Olsen and Olsen, Fredensborg, 1981), 317-318.
175 J.B.C. Jackson and J.E. Winston, 'Ecology of criptic coral reef communities I. Distribution and abundance of major groups of encrusting organisms', *Journal of Experimental Marine*

Biology and Ecology, 57 (1982), 135-147.

176 Dick, note 140.

177 B. Okamura and J.C. Partridge, 'Suspension feeding adaptations to extreme flow environments in a marine bryozoan', *Biological Bulletin* 196 (1999), 205-215 (205).

178 Mackie, note 82.

179 J.W. Bishop and L.M. Bahr, 'Effects of colony size on feeding by *Lophopodella carteri* (Hyatt)', in *Animal Colonies: Development and Function through Time*, edited by R.S. Boardman, A.H. Cheetham and W.A. Oliver (Dowden, Hutchinson and Ross, Stroudsberg, PA, 1973), 433-437.

180 B. Okamura and L.A. Doolan, 'Patterns of suspension feeding in the freshwater bryozoan *Plumatella repens*', *Biological Bulletin*, 184 (1993), 52-56.

181 I.I. Antipenko, *Feeding and behaviour of some fresh-water bryozoan species (Bryozoa, Phylactolaemata),* Unpublished MSc Thesis, (St. Petersburg State University, 1998), 112 (in Russian).

182 For further discussion see Shunatova and Ostrovsky, note 97.

183 McKinney, note 151.

184 Winston, note 84, p. 30.

185 Winston, note 144.

Historical review on individual autozooidal behaviour and feeding mechanisms in Bryozoa

Andrew N. Ostrovsky*, Natalia N. Shunatova and Irina I. Antipenko
Department of Invertebrate Zoology, Faculty of Biology and Soil Science, St. Petersburg State University, Universitetskaja nab. 7/9, St. Petersburg, 199034, Russia
**Present address*: *Institut für Geowissenschaften, Christian-Albrechts-Universität zu Kiel, Olshausenstr. 40, 24118 Kiel, Germany*

1. Introduction
2. Eighteenth century studies
3. Nineteenth century studies
4. Twentieth century studies
5. Conclusions
6. Acknowledgements

1. Introduction

During our studies on bryozoan autozooidal behaviour[1,2] we, following Winston, realized that 'the 19th [as well as 18th] century knowledge of individual (polypide) behaviour was never codified'.[3] This negligence in respect to the work of the old authors led to two main consequences. Firstly, the names of many brilliant naturalists, whose observations and detailed descriptions were, as a rule, perfect and whose results are the basement of modern knowledge, were either undeservedly forgotten or mentioned only in passing. Secondly, 'double work' was carried out since many of previously recorded phenomena and structures were newly redescribed, which is intimately connected with the problem of scientific priority. So, this paper was written with the aim to correct the situation in question at least partially. We present a historical survey on the development of views on individual autozooidal behaviour and feeding mechanisms in Bryozoa, with the emphasis on the priority in the recording of one or other type of activities and feeding/sensory structures.

2. Eighteenth century studies

The first description of bryozoan behaviour was given by Abraham Trembley,[4] who observed the phylactolaemate bryozoan *Lophophus*[5] in 1741. A brief summary of Trembley's notes was published first by Henry Baker,[6] who later made and published his own observations of phylactolaemates.[7] They both observed polypide protrusion and retraction, water currents carrying small animals into the lophophore, various movements of the lophophore and tentacles connected with both feeding and particle rejection (for instance, Trembley recorded tentacle flicks, lophophore expansion, accompanying the particle rejection, and particle transference along the oral grooves of lophophore), and defecation. Trembley[8] and later Baker attributed movements of water within and around the lophophore only to the perpetual swinging or 'vibrating motion' of the tentacles.[9] In contrast to Trembley, who believed that the tentacles did not touch the particles and bent inside the lophophore in order to create turbulent currents, Baker described particle capture by tentacles when these particles had escaped from the feeding current. Judging from the description, he was the first who recorded this kind of feeding by tentacular flicks when the previous mechanical contact between tentacle and particle was absent.[10] August Rösel von Rosenhof also worked on phylactolaemates and correctly remarked that during production of currents the tentacles were often motionless.[11] He supposed that the currents were produced by a stream issuing from the contracting mouth.

John Ellis was apparently the first to publish descriptions and drawings of living marine bryozoans, briefly mentioning that he saw their feeding and polypide excursions.[12] Leendert Bomme encountered cilia on the tentacles in freshwater as well as marine bryozoans and connected 'a strong whirlpool, sucking all nearby objects to the animal into its stomach' with the ciliary activity 'permanent movement from down upwards, across the tip, and downwards again'.[13] It should be noticed that Bomme was the first to record the travel of the ciliary movements over the tentacle tips in phylactolaemates. He criticized the inferences of Trembley,[14] but since Bomme's brief report was apparently little known at that time, some later investigators neglected it and followed Trembley's opinion (for example, Lazzaro Spallanzani).[15] Filippo Cavolini[16] was criticized by Lucien Joliet[17] for attributing the water currents to permanent tentacle movements, mainly because he did not recognize the tentacular cilia. However, precise examination of Cavolini's text shows that though he connected the lophophore with the origination of currents and often mentioned that 'tentacoli perpetuamente oscillano', he never wrote that their movements created these currents.[18] Cavolini gave the most detailed descriptions of marine bryozoan feeding in the 18th century, recording some behavioural reactions (pharyngeal rejection, single expansion of lophophore, bending of a single or 2-3 tentacles deep inside the lophophore, waving tentacular movements, bringing straight tentacles together) for the first time.

3. Nineteenth century studies

After Rösel von Rosenhof's book mentioned above,[19] several works on freshwater bryozoans with brief descriptions of their feeding behaviour were published during the next fifty years (summarized by William Sharpey[20] and George Allman[21]), and tentacular ciliature was redescribed by Johann Eichhorn.[22] Johann Steinbuch distinguished the cilia and their motion as the cause of water movements in *Plumatella repens* (Linnaeus) (as *Tubularia*), but, observing frequent retractions of polypides caused by an addition of extraneous particles, he inferred that the currents served chiefly, if not solely, for respiration.[23] Also, Steinbuch described and illustrated what seems to be water refiltration in the form of circular currents that had been already mentioned by Baker, who compared these currents with 'the turning of a mill-wheel'.[24]

As is evident from the old published descriptions and illustrations, living marine as well as freshwater bryozoans were often observed by investigators in the XIX century (Thompson,[25] de Blainville,[26] Gervais,[27] Nordmann,[28] Hassall,[29] Van Beneden,[30] Reid,[31] Hancock,[32] Leidy[33-37] Gosse,[38] Busk,[39] Huxley,[40] Hincks,[41] Müller,[42] Smitt,[43, 44] Reichert,[45] Claparède,[46] Repiachoff,[47, 48] Ehlers,[49] Joliet,[50] Vigelius,[51] Ostroumoff,[52] Pergens,[53] Prouho).[54] Following Bomme,[55] many observers thought that effective strokes of lateral cilia were directed upwards along one side of a tentacle and downwards along the other (Grant,[56] Lister,[57] Dalyell,[58] Allman,[59] Hincks).[60] Based on this opinion, some authors concluded that there were ascending and descending water currents along consistent tentacle surfaces (for instance, John Fleming,[61, 62] and George Johnston).[63, 64] In contrast, Alpheus Hyatt was sure that cilia were 'constantly vibrating towards the centre of the crown', observing apparently their beating across the tentacle.[65] Similar statements were made by Heinrich Nitsche[66] and Karl Kraepelin[67] (see also below). Asajiro Oka apparently observed refiltering currents and thought that the cilia of the outer (abfrontal) tentacle surface generated the ascending current and those on the oral (frontal) surface produced the descending current.[68] The last four authors all worked with phylactolaemates.

Some investigators did not find the tentacular cilia, apparently due to the imperfection of microscopes. Noteworthy, Barthelemy Dumortier, who published one of the best anatomical descriptions of fresh-water bryozoans at that time, could not distinguish cilia on the tentacles in some cases (see also papers of François Raspail)[69, 70] and made supposed absence of cilia the principal character of his new genus *Lophopus*.[71]

Robert Grant summarized previous works and gave a more detailed and correct description of marine bryozoan feeding. He was apparently the first to describe single and group tentacle flicks as a result of particle impacts in gymnolaemates.[72] John Lister, who also worked with marine bryozoans, was the first to describe transport of particles along the frontal tentacle surface, their concentration in the pharynx before swallowing or rejection, and single and paired 'hairs' (setae) on the abfrontal tentacular surface. Lister's illustrations show that he also found the groups of rigid cilia on the tips of the tentacles. Moreover, he was one of the first researchers who worked with living cyclostomates.[73]

The most prominent investigation of this period on the feeding and behaviour of marine bryozoans was made by Arthur Farre.[74] He gave an excellent description of feeding in seven ctenostomate and cheilostomate species. He was one of the first to use particles of carmine in his experinents to detect feeding currents. Many structures and reactions were recorded and beautifully illustrated by him for the first time: obliquely-truncate lophophores, scanning activity, testing-position, different variants of polypide protrusion, retraction and tentacle movements (for instance, writhing, curling, flattening of the lophophore when tentacles are almost perpendicular to the introvert), cessation of ciliature beating, etc. Moreover, he was the first to use the word 'vortex' which became very popular later to describe water turbulence within and nearby the lophophore (Bomme,[75] for instance, had used the Dutch word 'voortbrengen' = whirlpool) and proposed the filtration mechanism of bryozoan feeding. After Trembley,[76] Baker[77, 78] and Bomme's[79] observations it was generally accepted that the feeding process included a directed water transference into the tentacle crown by ciliary activity as well as the activity of tentacles which 'assist the ciliary currents, by their varied movements, in obtaining food'.[80] Some observers believed that the tentacles 'catch' the food,[81] whereas others were convinced that they 'arrest [retain] the prey' to 'conduct it to the mouth'.[82] Despite the fact that Farre described various tentacle movements too, he thought they worked like 'a sieve, of a degree of fineness proportioned to the number of the arms and the consequent width of their intermediate spaces'.[83] According to his opinion, the size of filtered particles is regulated by the size of the mouth and the gaps between the tentacles. The former prohibits the swallowing of too large particles, and the latter, forming the sieve, lets small particles pass while retaining those of an intermediate size. Also Farre mentioned particle selection inside the pharynx resulting in swallowing or rejection.

John Dalyell gave very accurate and detailed descriptions of the behaviour in about 30 gymnolaemate and 4 phylactolaemate species.[84] He recorded many reactions, some of which were not described before. Following Arthur Hassall,[85] who pictured a bending of all the tentacles of the lophophore towards the mouth, Dalyell described and illustrated this reaction. In fresh-water bryozoans he recorded and described for the first time the pharyngeal rejection, concentration of particles in the pharynx before swallowing, and faster ciliature beat as a result of an increase of particle concentration. Dalyell compared feeding in bryozoans with that in sedentary polychaetes, thinking they both were capable of the 'absorption of muddy matter' and fed on 'small animals' only incidentally.[86] He was likely one of the first to perform experiments with bryozoans under light and low temperature influences.

George Allman observed feeding in about 20 species of phylactolaemates, and summarized much research on this topic in his monograph.[87] In *Fredericella sultana* (Blumenbach) he also illustrated the bending together only of tips of the tentacles for the first time.[88]

Heinrich Nitsche encountered frontal cilia in sections of the tentacles of the freshwater bryozoan *Plumatella fungosa* (Pallas) (as *Alcyonella*).[89] He made his own observations on living colonies and was able to distinguish ciliary beating along the frontal surface of the tentacles.[90] In his opinion, the entire lateral ciliature performed an ascending-descending movement (see above), but like Hyatt[91] and Kraepelin,[92] Nitsche thought that the cilia beat across the longitudinal axis of the tentacle. He first mentioned phase differences in the beating of the cilia, meaning a metachronal wave, and he found long rigid cilia both on frontal and abfrontal sides of the tentacles in the phylactolaemate bryozoan studied.[93] Samuel Silbermann summarized the existing data on presumed sensory structures in bryozoans. He described and illustrated paired and non-paired setae on the outer (abfrontal) side of tentacles in *Alcyonidium mytili* Dalyell.[94] He also encountered the fusiform cells with stiff cilia between ciliated cells on the inner side of the tentacle, but it is difficult to judge from his description and illustrations[95] whether or not these were the latero-frontal cells later described by Geneviève Lutaud[96] in cheilostomates (see below). In phylactolaemates similar fusiform 'sensory' cells with a hairy terminal process were described later by Adolf Gerwerzhagen[97] in *Cristatella mucedo* Cuvier and by Paul Brien[98] in *C. mucedo* and *Plumatella fungosa*. Gewerzhagen reported that the cells found were irregularly distributed in bunches in the epithelial layer of the tentacle, whereas Brien believed that they are located on either side of a group of frontal cells with motile cilia, or even among them.[99, 100]

Alpheus Hyatt described feeding in the fresh-water bryozoan *Fredericella sultana*.[101] He was the first to report 'cagemaking',[102] bending of tentacles to place a particle directly to the mouth, and rejection of the particles away by tentacles. Hyatt stressed that tentacles are not only feeding, but also sensory organs. Thomas Hincks summarized the existing data on bryozoan behaviour, and also made many of his own observations.[103] He described, in particular, some different lophophore shapes and tried to apply them to the current classification. Louis Calvet, using sections, found frontal cilia on the tentacles of some gymnolaemates. He believed that lateral cilia beat from outside towards the centre of the lophophore and back, across the tentacle and simultaneously from the top to its base. According to him, this beating created a turbulent water current towards the mouth and 'préhension' of particles, and the polypide then retracted, swallowing them.[104] Similar phenomenon was first described by Baker,[105] but Calvet thought that each swallowing was accompanied by retraction.[106] Calvet was also the first to mention repeated expansions and contractions of lophophore with straight tentacles, resembling pulsation.

4. Twentieth century studies

In the first half of the twentieth century, the most important investigations of marine bryozoan feeding and behaviour were made by Folke Borg. In 1923 he published a preliminary report of his research on bryozoan ciliary feeding,[107] and then gave a detailed analysis of the processes, with the description of many reactions.[108] Borg was the first to ascertain an oblique beating of the lateral cilia in respect to longitudinal tentacle axis, with effective strokes towards the outside. He compared the activity of these cilia, which differ in phase, with the movement of an undulating membrane and thought that such undulation passes through the tentacle tips, as well as from one tentacle base to another. Borg correctly described the directions of water currents inside the lophophore and assigned an important role to the change of tentacle position and cilia beat direction as a reason for hydrodynamical changes inside and around the lophophore connected with feeding and particle rejection. In his opinion, the strongest current 'arises in the medial line of the tentacular crown in a direction straight downwards towards the mouth'[109] and 'mostly only those [particles which are contained in the current] in or near the medial line ... reach the mouth'.[110] The particles are then ingested through action of the stomodaeum musculature that 'acts as a suction pump',[111, 112] assisted by the pharyngeal ciliature.[113] Borg recorded reversal, cessation, and deceleration of ciliary beating. Also he noted that the frontal cilia of tentacles played a subordinate role in the feeding process. Frontal cilia, beating from above downwards, rarely caused a slow transference of particles along the frontal surface towards the mouth. The mechanism of ciliary feeding described is enhanced by tentacular activities such as bending of individual tentacles directing particles towards the central current and lophophore contractions. Borg stressed that bryozoans perceived the nature of suspended particles and avoided ingesting inedible ones. In contrast, no food preferences were noted and particles were swallowed without apparent selection in experiments of Erich Ries.[114, 115] For the first time Borg recorded a single lophophore contraction looking like a 'clap', brief series of tentacular pulsating movements during a rejection of particles, and bringing together only tips of the tentacles (accompanying a particle rejection in some cases).[116]

Ernst Marcus made many observations on marine as well as freshwater bryozoans,[117-121] but his main aim was experimental work with changes of temperature, salinity and mechanical stimuli,[122] and he did not study their feeding behaviour specifically. Marcus believed that the frontal cilia were immovable and stiff in at least one ctenostomate, *Farrella repens* (Farre).[123] Chandler Brooks was probably the first to describe scanning behaviour in Phylactolaemata.[124]

Daphne Atkins compared the tentacle ciliary activity and the water currents created in Entoprocta and marine Ectoprocta.[125] She was the first to use the term 'metachronal wave' in describing the beating of the lateral cilia, and studied the operation of the ventral rejection tract in the ctenostomate *Flustrellidra hispida* (Fabricius) first described by Henry Proucho.[126] In *Pherusella tubulosa* (Ellis and Solander) (as *Pherusa*),[127, 128] Atkins also divided the reactions connected with feeding behaviour into three groups that serve to: (1) increase the number of food particles, (2) prevent unwanted particles from reaching the mouth, (3) reject useless particles which have already reached the region of the mouth. Following Borg,[129] she suggested an important role of the frontal cilia situated at the base of tentacles in forming and directing water currents to the mouth.

In spite of two centuries of intensive observations, Carl Cori gave credit to the frontal cilia in creating the main water current.[130] He thought that their beat was across the tentacle whereas lateral cilia were motionless or beat incidentally. This conclusion was strange because Cori saw the ascending-descending lateral ciliary movements, and described and illustrated them in some ctenostomates, comparing them with clock-wise ciliary movements in Kamptozoa and Phoronis.[131] Using his own observations on the ctenostomate *Zoobotryon verticillatum* (Delle Chiaje) (as *Z. pellucidum*) and the data of Heinz Graupner, who investigated the anatomy of gymnolaemate tentacles,[132] Cori suggested that the immobile lateral cilia formed a filter, closing narrow intertentacular gaps between the bases of all tentacles.[133] Sucking action of the frontal cilia resulted in water transference through the filter where suspended particles were retained, and, becoming associated with the mucus strands, conveyed to the pharynx. Like Lister,[134] Cori illustrated the stiff apical tentacle setae in *Z. verticillatum* without mentioning them in the text.[135] He also recorded the pushing away of particles by tips of tentacles for the first time.

Lars Silén investigated the functioning of the digestive tract of gymnolaemates in detail.[136] According to him the delivery of captured particles to the mouth proper is mainly performed by ciliary beating in the area around the mouth, stomodaeum and neighbouring part of the pharynx. Periodical contractions of the mouth may assist this transfer and take place independently from pharyngeal dilations and contractions resulting in food suction and swallowing.[137, 138] Later Silén described some zooidal behavioural reactions involving tentacles and entire polypides, and serving oviposition [139] and sperm release [140] (see also papers of Marcus,[141] Bullivant,[142] Cook,[143] Gordon,[144] Ström,[145] Chimonides and Cook,[146] Dyrynda and King,[147] summarized in the paper of Shunatova and Ostrovsky).[148]

Despite numerous studies there was no clear concept of a bryozoan feeding mechanism. Dividing suspension feeders into groups Bernhard Werner marked out 'filterers' and 'whirlers' in his paper.[149] In the latter group, cilia generate a current and water with suspended particles is transferred along the collecting surface where the particles are retained. Bryozoans were considered to belong to a group of 'non-filtering suspension feeders'. In good agreement with Cori's suggestion,[150] Carl Jørgensen speculated that mucus played an important role in particle retention.[151] In contrast, John Bullivant suggested that in all lophophorates a sharp deflection of water current passing through the lophophore causes particles to be thrown towards the mouth in a way which he compared with impingement separation used in industrial processes, and claimed that mucus plays no part in the feeding of marine bryozoans.[152, 153] He also supposed that laterofrontal cilia found by Lutaud[154] on the tentacles of the cheilostomate *Pentapora foliacea* (Ellis and Solander) (as *Hippodiplosia*)[155, 156] may either detect suspended particles in the water and initiate polypide retraction, tentacle flicking or reversal of lateral cilia, with the aim of rejection, or act as a sieve which retain the particles. Bullivant found and measured the laterofrontal cilia in the cyclostomate *Disporella separata* Osburn for the first time and used the term 'laeoplectic' to describe metachronal waves formed during lateral cilia beating.[157, 158, 159] Following Bomme[160] and Marcus,[161] he noted that metachronal waves transmitted over the tips of tentacles in phylactolaemates, but he doubted that this took place in marine bryozoans. In *Plumatella repens*, Bullivant recorded the abfrontal cilia and described particle rejection along the frontal ciliary tract on the tentacles. He also confirmed[162] observations of Günther Becker who described the rejection of particles by ciliature reversal of all of the tentacles in marine bryozoans.[163]

Dennis Gordon investigated tentacular ciliature in the gymnolaemate *Cryptosula pallasiana* (Moll) and found that laterofrontal cilia performed occasional flicking movements.[164] He also described abfrontal cilia of two types that alternate with each other - short tufts of several immobile cilia and solitary cilia, and found the 'tuft of cilia at the tentacle tip'.[165]

Bullivant's hypothesis was met with enthusiasm by some researchers, and John Ryland called bryozoans 'impingement feeders'.[166] Thomas Gilmour proposed a more detailed explanation of lophophorate feeding on the basis of the hypothesis in question and some theoretical generalizations.[167, 168] He considered the process as 'simultaneous acceptance of food particles and rejection of inedible wastes probably' depending 'on the impingement or inertial impaction of heavy, potentially inedible particles on rejectory surfaces and the filtration or direct interception of light, potentially edible particles by cilia' on the lophophore.[169] In his opinion light particles are either thrown directly to the mouth with the incoming current or, if they move with a deflecting component of the current, are caught by the oral area cilia in gymnolaemates. Heavy particles are collected on the frontal surfaces by impingement and conveyed by the reversal of their cilia beating towards the tentacle tips. Following Bullivant,[170] Gilmour described the particle rejection along the frontal surface of the tentacles in the fresh-water bryozoan *Plumatella* sp., and also found this type of rejection in *Membranipora* sp. Gilmour also described the laterofrontal cilia on the tentacles of both gymnolaemate and phylactolaemate species (see above), but he supposed they work as sensory organs only.

The impingement model of lophophorate feeding was strongly criticized by Richard Strathmann,[171, 172, 173] who argued that the velocities and the particle sizes and densities are too low for feeding by inertial impaction to operate.[174, 175, 176] Instead, Strathmann speculated that suspended particles are captured by a combination of local reversal of the lateral cilia, which produces a local reversal of the water current, and inward flicking of the tentacles. The reversed current transfers the particles to the frontal side of the tentacle and helps to retain and transport them along the frontal surface towards the mouth (so-called up-stream collecting system according to terminology of Claus Nielsen and Jørgen Rostgaard).[177, 178] It was suggested that this reversal must be induced by particles contacting either the lateral cilia themselves or the laterofrontal cilia. The latter may act as a sieve at the same time. This mechanism was first proposed for echinoderm larvae[179] and then extrapolated to lophophorates,[180, 181] and Michael Hart presented a possible physiological explanation for its existence.[182] In the ctenostomate *Flustrellidra hispida*,

Strathmann used high speed cinefilms to observe local alterations of metachrony of ciliary bands in places where the particles arrived. Though local reversal of the effective ciliary stroke could not be directly observed, these alterations were suggested to be the consequences of the reversals.[183] Strathmann did not report the transport of particles by frontal cilia in bryozoans, only that the transport along the frontal tentacle surface is aided by the reversal of the lateral cilia. He doubted that frontal cilia could transport the particles 'since only a few [of them] would be moving a particle against the action of many lateral cilia'.[184] Strathmann also was the first to measure the length of the lateral cilia in *F. hispida*,[185] to record frontal transport of particles by means of jumps and to investigate in detail flicking activity, which sometimes accompanied the proximal particle transport along the flicking tentacle.[186]

Reactions to Strathmann's hypothesis were mixed. Some authors accepted it, but gave no direct evidence in support (Ryland,[187] Winston,[188, 189] LaBarbera,[190] McKinney and Jackson,[191] Hart,[192, 193]). John Ryland tried to reach a compromise between the hypothesis of Bullivant (see above) and that of Strathmann.[194] He wrote that 'heavier particles may, as Bullivant suggested, be projected towards the mouth; but mostly it seems that particles are retained ... by local reversal of the lateral cilia'.[195] Ryland also discussed the significance of the width of the intertentacular gap and the length of lateral cilia in relation to Strathmann's opinions on particle capture.[196, 196] According to Ryland's measurements of lophophores in different marine bryozoans, filtration should be fully effective in approximately the lower half of the tentacle crown since the lateral cilia should close the intertentacular gaps.[198, 199] This idea was thereafter questioned and modified. James Markham and John Ryland[200] and William Sanderson, John Thorpe and Andrew Clarke[201] found that the lateral cilia on the tentacles of *Flustrellidra hispida* and *Himantozoum antarcticum* (Calvet) correspondingly, 'exhibited a subterminal maximum length, decreasing proximally over more than half the tentacle and distally in about the last quarter'. Thus it was concluded that the actual closure of much of the intertentacular gap by the lateral cilia occurs along less than a quarter of the tentacle length or even lesser. The mechanism of local reversal would in principle be possible here as well, but it probably occurs in addition to other particle capture methods (see below). Filtration would be less effective in the upper part of the tentacle crown, where the distance between the tentacles is larger, but flicking might supplement the food collecting mechanism.

Michael Best and John Thorpe[202] denied the major role of local reversals in particle capture and transport towards the mouth.[203, 204] Working on clearance rates and velocities of feeding currents in some gymnolaemates, they observed a frontal transport of particles downward and assumed that it might be a result of local reversals.[205] At the same time Best and Thorpe came to the conclusion that the feeding process in the ctenostomate *Flustrellidra hispida* cannot be entirely explained by the local reversal-flicking hypothesis.[206] They calculated, for instance, that the rate with which the pharynx is normally filled and the quantity of swallowed food particles would imply several flicks and more than 1000 local ciliary reversals per second, which appears to be impossible. Following Borg[207] and Atkins,[208] Best and Thorpe supposed that long frontal cilia at the bases of the tentacles take an important part in bryozoan feeding. These cilia are used to direct the main feeding current towards the mouth, and 'the bulk of particles travel rapidly downwards in the current in the central area of the lophophore cone and proceed directly to the area above the mouth where they gather and are continuously ingested'.[209] Local reversals and particle transport by frontal cilia are considered to be possible supplementary mechanisms.[210, 211] Best and Thorpe were the first to measure current velocities inside the lophophore.[212]

The most exhaustive research on individual as well as group behaviour and associated structures was made by Judith Winston, who investigated about 80 species of marine bryozoans.[213-216] She described and classified most of the feeding behavioural reactions known, recording many of them (collection of particles in a group inside the expanded lophophore near the mouth area, bending of all tentacle tips inside and outside the lophophore used for rejection at high particle concentrations, twisting activity, polypide retraction from under the particle, 'avoidance' retraction) for the first time. Winston also summarized the data on inferred sensory structures [217] and noted that the metachronal wave did not travel over the tentacle tips [218] (a fact which Bullivant doubted).[219] According to the prevailing reactions she distinguished 'filterers', 'tentacle feeders', 'scanners', 'cage-captors' and 'particle jugglers', and suggested that behavioural differences depend both on the physical and chemical characteristics of the food particles and behavioural characteristics of the various species.[220] She also investigated parameters of the lophophore and its shape, showed similar trends in its symmetry within three main taxa of marine bryozoans and discussed their siginificance in relation to zooidal activity, colony structure (zooid arrangement and growth form) and environment. Winston argued that the complexity of zooidal behaviour was correlated with the level of colonial integration. The most complex individual behaviour was mainly found in poorly integrated colonies, and the most complex forms of group behaviour were discovered in highly integrated ones. The variety of reactions also often depended on polypide size: larger polypides perform a greater range of reactions involving different movements of tentacles and the entire polypide, whereas in small ones ciliary activity was the most important component of feeding behaviour as it was shown in cyclostomates and ctenostomates. Winston stressed that during particle rejection polypides could 'utilize a combination of methods depending on the size and concentration of the particles'.[221]

Matthew Dick added to Winston's observations, recording for the first time in gymnolaemates an autocleaning of the colony surface by special activity of polypides in association with stiff abfrontal tentacular setae.[222] He described different types of rejection processes (particle flinging and partial retraction of polypide accompanying by pharyngeal rejection were described for the first time) and distinguished 'ejection' and 'rejection', and was the first to discuss the possible evolution of some behavioural reactions in bryozoans. Following Winston, who suggested that abfrontal ciliary structures 'could have some role in controlling polypide orientation relative to other polypides',[223] Dick speculated that 'long abfrontal cilia' perceive 'tactile stimulation from the tentacles of surrounding polypides' as well as 'vectors' of flow around the lophophore.[224]

Patricia Cook observed feeding behaviour of about 40 cheilostomates.[225] In some species she described different variants of a lophophore protrusion and retraction in details, noting the specific sensitivity relative to the external irritants. Cook pictured several tentacle crown shapes,[226] and described a temporary lophophore expansion with the tips of tentacles directed towards the colony surface for the first time. Since this reaction was "colony-wide", the author suggested that it might be a colonial rejection, accompanied by ciliary reversal and connected with cleaning, sperm release or other unknown functions.

Ewan Hunter and Roger Hughes described an unusual variant of feeding in the cheilostomate *Celleporella hyalina* (Linnaeus) under high particle concentrations when 'food particles were drawn from below the lophophore into the feeding crown by a current generated by the lateral cilia'.[227] According to their observations 'particles continued to move upwards and out of the lophophore, unless ciliary reversal occurred, bringing particles down' to where finally they 'were drawn into the mouth on a current generated by cilia within the pharynx'. A similar feeding method was earlier described by Borg.[228] It is obvious from the texts that all of the authors postulate simultaneously changing ciliary movement on the tentacles, but not the local reversals proposed by Strathmann.[229, 230] Hunter and Hughes also recorded a retaining of particles inside the lophophore when the tentacles cramed them into the mouth.[231]

Hans Riisgård, Patricio Manrìques[232] and Claus Nielsen[233] reviewed a wide range of literature dealing mainly with physical aspects of particle capture by different suspension-feeders.[234-245] Using theoretical considerations and non-contradictory ideas from conflicting theories of bryozoan feeding mechanism (see above) as well as their own data they proposed that the laterofrontal cilia form a mechanical filter which 'strains the water while the central current, created by the special lophophore pump-design, and the action of flicking tentacles in co-operation clean the filter and transport the particles towards the mouth'.[246] The laterofrontal cilia are considered to be multifunctional sensors that trigger tentacle flicking and local reversal of lateral cilia and serve as a mechanical sieve. By this means three main components of particle capture in bryozoans are effected: the direct transfer of the particles within the middle part of the central current to the mouth, particle impacts to a tentacle resulting in a downward transport by frontal cilia, and tentacle flicks conveying particles into the central lophophore current and triggered by the particle impacts with laterofrontal cilia. Flicking is stated to prevail in catching the particles in comparison with frontal transport,[247] and the latter being impossible in at least some cyclostomates since frontal cilia are absent.[248, 249] This is why every particle, that collides with a tentacle, is captured by a single tentacle flick directing it into the central current in *Crisia eburnea* (Linnaeus). In all species investigated the frequency of flicking depends on concentration of particles. Riisgård and Manrìques did not observe the capture of particles by local reversals and disturbances in ciliary metachrony and doubted compatibility of their proposed mechanism with Strathmann's model, but did not deny it entirely.[250] They also noted that some results obtained by Best and Thorpe[251] and Sanderson and Thorpe[252] may be disputed because they had used too high concentrations of particles, tending to disturb normal feeding behaviour.[253]

Natalia Shunatova and Andrew Ostrovsky gave comparative descriptions of the behaviour in 40 species and subspecies of marine bryozoans.[254] Some of the activities (cleaning of the colony surface using ascending water current created by reversal of the lateral cilia; cleaning of the colony surface by the inclined polypide; bending of the tentacle tips inside the lophophore during polypide protrusion; particle capture by tentacle expansion; pulsation of polypides bringing tentacles together within a lower part of the crown; particle rejection by bringing straight tentacles together; particle rejection in clumps by a series of repeated retractions and protrusions of polypide) were recorded for the first time. Behavioural reactions described were classified using a morphological approach based on the structure (tentacular ciliature, tentacles and entire polypide) performing the reaction. Authors discussed both normal and so-called spontaneous reactions as well as differences and similarities in individual autozooidal behaviour and their probable causes. They also suggested that in some species polypides can detect particles at some distance and that polypide protrusion and retraction might be the basis of the origin of some other individual activities. In marine bryozoans individual autozooidal behaviour was 'considered to be a flexible and sensitive system of reactions in which the activities can be performed in different combinations and successions, and can be switched depending on the situation'.[255]

Irina Antipenko studied in detail feeding of four fresh-water bryozoans recording many activities for the first time.[256] For instance, testing-position was recorded for Phylactolaemata for the first time (in *Fredericella sultana*). She showed that, in contrast with marine bryozoans, frontal particle transport is one of the main feeding mechanisms (together with mechanical-filtering and tentacle flicking) in phylactolaemates. Small particles slide quickly whereas middle-sized ones usually move in jumps along the tentacle frontal surface toward the mouth. Instead, large particles are rejected being lifted by a reversal of frontal ciliature beat. When such a particle reaches a tentacle tip the rejecting tentacle sometimes sharply deflects outward shaking the particle off. If a particle is of 'intermediate' size, it often starts to travel down and up before it is either swallowed or rejected. Large particles can also be rejected by a brief reversal of lateral ciliature of all tentacles, expansion of lophophore, bending either of a single or several tentacles outside, sharp shaking of all of the tentacles. In *Cristatella mucedo* rejecting particles can be rolled from tentacle to tentacle towards the end of the lophophore 'arm'. The polypide can cross 'arms' in this species, and perform a 'clap' to catch more particles. Finally, Antipenko stressed the great similarity in autozooidal behaviour of *F. sultana* (possessing ring lophophore) and that of marine bryozoans.

Quantitative characteristics of feeding in marine bryozoans have been investigated since the works of John Bullivant[257] and N. Ravindranatha Menon[258] who measured clearance rates by calculating theoretical volumes of water cleared of particles per unit time, and Judith Dudley [Winston][259] and John Ryland[260] who studied dimensional parametres of the lophophore.[261-266] Best and Thorpe,[267, 268, 269] Sanderson *et al.*,[270] Sanderson and Thorpe[271] and Riisgård and Manrìques[272] made direct measurements of feeding current velocities for various marine Bryozoa. These studies indicate that velocities vary across the diameter of the lophophore and are the strongest axially.[273, 274] Velocities of feeding current may also vary considerably depending on a number of parameters including food type, food particle size and concentration and the extent to which the colony has been starved.[275, 276, 277] Both intraspecific and interspecific variations in lophophore behaviour and morphology were documented, although generally parameters of lophophore width, tentacle length and tentacle number correlate quite strongly.[278-283] It was supposed that interspecific (and intergeneric) differences in behaviour and morphology of the lophophore probably result in corresponding differences in diet,[284-292] and 'that bryozoans [with] ... a substantial difference in the size of the lophophore and tentacle number ... may be able to coexist by means of niche partitioning'.[293, 294] Phytoplankton (mainly, diatoms) is almost certainly the most important food source for shallow-water bryozoans, but various motile protists (like ciliates and flagelliates), small planktonic arthropods, detritus, bacterial suspensions and dissolved organic matter are considered to be potential food sources as well.[295-306] Seasonal and tidal changes in food composition and concentrations have been recorded[307-310] and intraspecific behavioural variability may be partially explained by these factors.[311] Feeding is also influenced by an increased temperature.[312-316]

Current regimes have been shown to influence feeding of marine bryozoans. Relationships between ambient flow velocity, particle sizes and feeding success in the colonies of different growth forms were intensively studied by Beth Okamura.[317-320] She found that the capture of particles of different sizes may be affected in different ways under various conditions of flow and suggested a possible switch in feeding behaviour from filtering to capturing by tentacles.[321] It has been also shown that bryozoan feeding was significantly affected by the interacting influences of the flow velocity and degree of association with neighbouring suspension-feeders, for instance, other bryozoans,[322, 323] and these conclusions conform with the results of previous work by Leo Buss[324] and Michael Best and John Thorpe[325, 326] on feeding-current interactions among neighbouring colonies.[327, 328, 329] Feeding of some marine bryozoans was also investigated in laminar[330] and turbulent flow.[331] Both studies showed that feeding rates increase with increasing flow velocity. Unfortunately, phylactolaemates are almost non-studied in these respects.[332, 333]

5. Conclusions

In conclusion, we would like to stress that in spite of more than 250 years of research our knowledge of bryozoan feeding behaviour is still far from satisfactory, and implied feeding mechanisms are based more on hypotheses and suggestions than on facts. Structural uniformity of the bryozoan polypide and tentacles implies common feeding mechanisms. It is generally accepted that bryozoans probably use several mechanisms depending on polypide morphology and size within different taxa, and changes of environment.[334-340] Supposedly, the basic lophophore filter-feeding principle is added by tentacle flicking, frontal transport and, may be, local reversals. Some of these mechanisms can apparently be used simultaneously, but their importance for feeding is, probably, different. Further investigations are necessary to prove these speculations. Microscopical size and the impossibility to observe bryozoans in natural habitats are the main reasons of many existing problems and theoretical controversies. We hope that the complex approach of using sophisticated videoequipment, carefully planned and accomplished quantitative experiments and anatomical work [341, 342, 343] could improve this situation.

6. Acknowledgements

We are greatly indebted to Professor Dr Claus Nielsen and Dr Mary E. Petersen, Zoological Museum, University of Copenhagen, and Mary Spencer Jones, The Natural History Museum, London, for kind help with literature. Sincere thanks are given to Dr Margit Jensen, Zoological Museum, University of Copenhagen, Dr Harry A. ten Hove, Institute for Systematics and Population Biology, Zoological Museum, University of Amsterdam, Ekaterina V. Lobastova, Biological Scientific Research Institute, St. Petersburg State University, Natalia V. Lentsman, Dept. of Invertebrate Zoology, St. Petersburg State University, Dr Elena A. Istomina, Seacamp Association, Florida, and Tatjana V. Ostrovskaja, Leningrad Zoo, for their assistance with translating. Our thanks to Drs William G. Sanderson, Michael A. Best and John P. Thorpe, Department of Environmental and Evolutionary Biology, University of Liverpool, Port Erin Marine Laboratory, for an unpublished review on bryozoan feeding. We are grateful to Dr Judith E. Winston, Virginia Museum of Natural History, Professor Dr Roger N. Hughes, School of Biological Sciences, University of Wales, Bangor, Dr Ken McKinney, Appalachian State University and Dr Paul D. Taylor, The Natural History Museum, for reading and criticizing earlier drafts of the manuscript, useful comments and much encouragement. Andrew N. Ostrovsky thanks the Danish Ministry of Education for a post-doctoral stipend which made the work in University of Copenhagen Library possible. The Otto Kinne Foundation is acknowledged for financial support.

Notes

1 N.N. Shunatova and A.N. Ostrovsky, 'Individual autozooidal behaviour and feeding in marine bryozoans', *Sarsia*, 86 (2001), 113-142.
2 N.N. Shunatova and A.N. Ostrovsky, 'Group autozooidal behaviour and chimneys in marine bryozoans', *Marine Biology*, 140 (2002), 503-518.
3 J.E. Winston, 'Polypide morphology and feeding behaviour in marine ectoprocts', *Bulletin of Marine Science*, 28(1) (1978), 1-31 (1).
4 A. Trembley, *Memoires pour servir a l'histoire d'un genre de polypes d'eau douce, a bras en forme de cornes*, 2 vols (Paris, Jean and Herman Verbeek, 1744), xvi, 324 (vol 2, 128-147, Plate 10, figures 8-9).
5 See for synonymy L.H. Hyman, *The invertebtares: smaller coelomate groups*, 5 (McGraw-Hill Book Company, New York, Toronto, London, 1959), 275-515.
6 H. Baker, *An attempt towards a natural history of the polype* (London, 1743), 248.
7 H. Baker, *Employment for the microscope* (London, 1753), xiv, 442.
8 Trembley, note 4.
9 Baker, note 7, p. 309.
10 See for further discussion Shunatova and Ostrovsky, note 1.
11 A.J. Rösel von Rosenhof, *Der monatlich-herausgegebenen Insecten-Belustigung*, 3 (Nürnberg, 1755), 624.
12 J. Ellis, *An essay towards a natural history of the corallines, and other marine productions of the like kind, commonly found on the coasts of Great Britain and Ireland* (London, 1755), 120.
13 L. Bomme, 'Bericht wegens een zonderling Zeeinsect, gevonden aan eenige Zeewieren, gevischt op het Strand van het Eiland', *Verhandelingen Uitgegeven door het Zeeuwsch Genootschap der Wetenschappen te Vlissingen*, 1 (1769), 394-402 (400).
14 Trembley, note 4.
15 L. Spallanzani, *Opuscules de physique, animale et végétale*, 2 vols (Geneve, 1777), vol 1, cxxiv, 255, vol 2, ii, 404.
16 F. Cavolini, *Memoire per servire alla storia de'polipi marini di Filippo Cavolini* (Napoli, 1785), 279.
17 L. Joliet, 'Contributions à l'histoire naturelle des Bryozoaires des côtes de France', *Archives de Zoologie expérimentale et générale*, 6 (1877), 193-304.
18 Cavolini, note 16, p. 244.
19 Rösel von Rosenhof, note 11.
20 W. Sharpey, 'Cilia', in *The cyclopaedia of anatomy and physiology,* edited by R.B. Todd (Sherwood, Gilbert and Piper, London, 1836), 606-638.
21 G.J. Allman, *A monograph of the fresh-water Polyzoa, including all the known species, both british and foreign* (Ray Society, London, 1856), 120.
22 J.C. Eichhorn, *Beiträge zur Naturgeschichte der kleinsten Wasserthiere in den Gewässern und um Danzig* (Danzig, 1776), 94.
23 J.G. Steinbuch, *Analecten neuer Beobachtungen und Untersuchungen für die Naturkunde* (Fürth, 1802), 135.
24 Baker, note 6, p. 22.
25 J.V. Thompson, 'On Polyzoa, a new animal discovered as an inhabitant of some zoophytes with a description of the newly instituted genera of *Pedicellaria* and *Vesicularia* and their species', *Zoological researches and illustrations*, 4(5) (1830), 89-102.

26 H.M.D. de Blainville, *Manuel d'Actinologie ou de Zoophytologie,* 2 vols (Paris, F. G. Levgrault, 1834), 694.
27 M.P. Gervais, 'Recherches sur les Polypes d'eau douce des genres *Plumatella*, *Cristatella* et *Paludicella*', *Annales des Sciences Naturelles,* 2 Serie, 7 (Masson, Paris, 1837), 74-93.
28 M.A. Nordmann, 'Recherches microscopiques sur l´anatomie et le développement du *Tendra zostericola*, espèce de polype de la section des Bryozoaires', *Annales des Sciences Naturelles*, 2 Serie, 11 (Masson, Paris, 1839), 185-191.
29 A.H. Hassall, 'Supplement to a catalogue of Irish zoophytes', *The Annals and Magazine of Natural History*, 7(44, 45) (1841), 276-87, 363-373.
30 P.J. van Beneden, 'Recherches sur l'organisation des *Laguncula* et l'histoire naturelle des différents polypes Bryozoaires qui habitent la côte d'Ostende', *Nouveaux Mémoires de l'Académie de Bruxelles*, 18 (1844), 1-29.
31 J. Reid, 'Anatomical and physiological observations on some Zoophytes', *The Annals and Magazine of Natural History*, 16 (1845), 385-400.
32 A. Hancock, 'On the anatomy of the freshwater Bryozoa, with descriptions of three new species', *Annals and Magazine of Natural History*, 5 (1850), 173-204.
33 J. Leidy, 'On the genus *Plumatella*', *Proceedings of the Academy of Natural Sciences of Philadelphia*, 5 (1850-1851), 261-262.
34 J. Leidy, 'On *Cristatella*', *Proceedings of the Academy of Natural Sciences of Philadelphia*, 5 (1850-1851), 265-266.
35 J. Leidy, 'On some American fresh-water Polyzoa', *Proceedings of the Academy of Natural Sciences of Philadelphia*, 5 (1850-1851), 320-322.
36 J. Leidy, 'On *Urnatella gracilis* and a new species of *Plumatella*', *Proceedings of the Academy of Natural Sciences of Philadelphia*, 7 (1854-1855), 191-192.
37 J. Leidy, 'Remarks on Polyzoa', *Proceedings of the Academy of Natural Sciences of Philadelphia*, 10 (1858), 188-190.
38 P.H. Gosse, 'Notes on some new or little-known marine animals', *The Annals and Magazine of Natural History*, 2 Serie, 16 (1855), 27-36.
39 G. Busk, 'Zoophytology', *Quarterly Journal of Microscopical Science*, 4 (1856), 308-312.
40 T.H. Huxley, 'Note on the reproductive organs of the cheilostome Polyzoa', *Quarterly Journal of Microscopical Science*, 4 (1856), 191-192.
41 T. Hincks, 'Note on the ovicells of the cheilostomatous Polyzoa', *Quarterly Journal of Microscopical Science*, New Series, 1 (1861), 278-281.
42 F. Müller, 'On the common nervous system (Kolonialnervensystem) of the Bryozoa (Polyzoa), exemplified in *Serialaria Coutinhii*, n. sp.', *Quarterly Journal of Microscopical Science*, New Series, 1 (1861), 300-305.
43 F.A. Smitt, 'Om Hafs-Bryozoernas utveckling och fettkroppar', *Öfversigt af Kongliga Vetenskapsakademien Förhandling*, 1 (1865), 5-50.
44 F.A. Smitt, 'Kritisk fürteckning öfver Skandinaviens Hafs-Bryozoer', *Öfversigt af Kongliga Vetenskapsakademien Förhandling*, 2 (1865), 115-533.
45 H. Reichert, 'Vergleichende anatomische Untersuchungen über *Zoobotryon pellucidus* (Ehrenberg)', *Abhandlungen der königlichen Akademie der Wissenschaften zu Berlin*, 2 (1869), 233-338.
46 E. Claparède, 'Beiträge zur Anatomie und Entwicklungsgeschichte der Seebryozoen', *Zeitschrift für Wissenschaftliche Zoologie*, 21 (1871), 138-174.

47 W. Repiachoff, 'Zur Entwickelungsgeschichte der *Tendra zostericola*', *Zeitschrift für wissenschaftliche Zoologie*, 25 (1875), 129-142.
48 W. Repiachoff, 'Zur Naturgeschichte der chilostomen Bryozoen', *Zeitschrift für wissenschaftliche Zoologie*, 26 (1876), 139-160.
49 E. Ehlers, '*Hypophorella expansa*. Ein Beitrag zur Kenntniss der minirenden Bryozoen', *Abhandlungen der physikalischen Classe der königlichen Gesellschaft der Wissenschaften zu Göttingen*, 21 (1876), 3-157.
50 Joliet, note 17.
51 W.J. Vigelius, 'Morphologische Untersuchungen über *Flustra membranaceo-truncata* Smitt', *Biologisches Centrablatt*, 23 (1884), 705-721.
52 A. Ostroumoff, 'Die Bryozoen der Bucht von Sebastopol. Vollständigere Ausgabe mit einem ganz neuen Theile über die Morphologie der Bryozoen', *Proceedings of the Society of Naturalists of Kazan*, 16(2) (1886), 1-124 (in Russian).
53 E. Pergens, 'Untersuchungen an Seebryozoen', *Zoologischer Anzeiger*, 12 (1889), 504-510, 526-533.
54 H. Proucho, 'Contribution à l'histoire des Bryozoaires', *Archives de Zoologie expérimentale et générale*, 10 (1892), 557-656.
55 Bomme, note 13.
56 R.E. Grant, 'Observations on the structure and nature of *Flustra*', *Edinburgh New Phylosophical Journal*, 3 (1827), 107-118, 337-342.
57 J.J. Lister, 'Some observations on the structure and functions of tubular and cellular polypi, and of Ascidiae', *Philosophical Transactions of the Royal Society of London*, 1 (1834), 365-388.
58 J.G. Dalyell, *Rare and remarkable animals of Scotland, represented from living subjects: with practical observations on their nature,* 2 vols (Van Voorst, London, 1847-1848), vol 1, xii, 268, vol 2, iv, 322.
59 Allman, note 21.
60 T. Hincks, '*A history of the British marine Polyzoa',* 2 vols (Van Voorst, London, 1880), 601.
61 J. Fleming, 'Observations on the *Sertularia cuscuta* Ellis', *Memoirs of the Wernerian Natural History Society*, 4(2) (1823), 485-491.
62 J. Fleming, '*A history of British animals*', (Edinburgh, London, 1828), xxiii, 565.
63 G. Johnston, '*A history of the British zoophytes*', (Edinburgh, London, Dublin, 1838), 341.
64 G. Johnston, '*A history of the British zoophytes*', 2nd edition, (Van Voorst, London, 1847), xiv, 488.
65 A. Hyatt, 'Observations on Polyzoa. Suborder Phylactolaemata', *Proceedings of the Essex Institute*, vols 4 and 5, (Salem, 1866-1868), iv, 103 (9).
66 H. Nitsche, *Beiträge zur Anatomie und Entwickelungsgeschichte der phylactolaemen Süsswasserbryozoen, insbesondere von* Alcyonella fungosa *Pall.* Akademisch Dissertation, (Grimm and Maass, Berlin, 1868), 57.
67 K. Kraepelin, 'Die Deutschen Süsswasser-Bryozoen. Anatomish-systematischer Teil', *Abhandlungen aus dem Gebiete der Naturwissenschaft herausgeben von dem Naturwissenschaftlicher Verein in Hamburg,* 10 (1887), 1-168.
68 A. Oka, 'Observations on fresh-water Polyzoa. (*Pectinatella gelatinosa*, nov. sp.)', *Journal of the College of Science, Imperial University of Japan, Tokyo*, 4 (1891), 89-150.
69 F.V. Raspail, 'Analyse du polypier de l'*Alcyonelle* des etangs Ferussac', *Bulletin des Sciences Mathematique*, 9 (1828), 138-141.

70 F.V. Raspail, 'Histoire naturelle de l'*Alcyonella fluviatile* (*A. stagnorum,* Lamk.) et de tous les genres voisins, consideres, soit sous le rapport de leur organisation et de leur identite specifique, sout sous le rapport physiologique de leurs tentacules avec les branchies des mollusques et des animalcules ou infusoires ou sermatiques', *Memoires de la Societe d'Histoire Naturelle de Paris*, 4 (1828), 75-165, 246.
71 B.-C. Dumortier, 'Recherches sur l'anatomie et la physiologie des polypiers composés d'eau douce', *Bulletins de l'Académie Royale des Sciences, Bruxelles*, 2 (1835), 421-454.
72 Grant, note 56.
73 Lister, note 57, p. 384, plate XII, figures 2, 5.
74 A.B. Farre, 'Observations on the minute structure of some of the higher forms of polypi, with views of a more natural arrangement of the class', *Philosophical Transactions of the Royal Society of London*, 1 (1837), 387-426.
75 Bomme, note 13.
76 Trembley, note 4.
77 Baker, note 6.
78 Baker, note 7.
79 Bomme, note 13.
80 Hincks, note 60, volume 1, p. xv.
81 Baker, note 6, p. 23.
82 Johnston, note 64, p. 258.
83 Farre, note 74, p. 408.
84 Dalyell, note 58.
85 Hassall, note 29, plate VI, figure 1.
86 Dalyell, note 58, volume 2, p. 281.
87 Allman, note 21.
88 Allman, note 21, Plate IX, figure 2.
89 Nitsche, note 66.
90 See also Kraepelin, note 67.
91 Hyatt, note 65, p. 9.
92 Kraepelin, note 67.
93 Nitsche, note 66, p. 27, plate III, figure 31.
94 S. Silbermann, 'Untersuchungen über den feineren Bau von *Alcyonidium mytili*', *Archiv für Naturgeschichte*, 72 (1906), 265-308.
95 Silbermann, note 94, 277-8, plate XIX, figure 13A-C.
96 G. Lutaud, 'Sur la ciliature du tentacule chez les Bryozoaires Chilostomes', *Archives de Zoologie expérimentale et générale (Notes et Revue)*, 92 (1955), 13-19.
97 A. Gerwerzhagen, 'Beiträge zur Kenntnis der Bryozooen. I. Das Nervensystem von *Cristatella mucedo* Cuv.', *Zeitschrift für Wissenschaftliche Zoologie*, 107 (1913), 309-345.
98 P. Brien, 'Classe des Bryozoaires', in *Traité de Zoologie*, edited by P.-P. Grasse, 5(2) (Masson et Sie, Paris, 1960), 1053-1335.
99 For further discussion see G. Lutaud, 'The bryozoan nervous system', in *The Biology of Bryozoans,* edited by R.M. Woollacott and R.L. Zimmer (Academic Press, New York, London, 1977), 377-410.
100 H. Mukai, K. Terakado and C.G. Reed, 'Bryozoa', in *Microscopic Anatomy of Invertebrates,* edited by F.W. Harrison, vol 13, (New York, Wiley-Liss, 1997), 45-206.
101 Hyatt, note 65, p. 9.

102 For terminology see J.E. Winston, note 3.
103 Hincks, note 60.
104 L. Calvet, 'Contribution à l'histoire naturelle des Bryozoaires Ectoproctes marins', *Travailes de l'Institut de Zoologie de l'Université de Montpellier,* Nouvelle Série, 8 (1900), 1-488 (4).
105 Baker, note 7.
106 Criticized in F. Borg, 'Studies on recent cyclostomatous Bryozoa', *Zoologiska Bidrag från Uppsala*, 10 (1926), 181-250.
107 F. Borg, 'On the structure of cyclostomatous Bryozoa', *Arkiv für Zoologie*, 15 (1923), 1-17 (8).
108 Borg, note 106.
109 Borg, note 106, p. 247.
110 Borg, note 106, p. 248.
111 See also F. Braem, '*Victorella sibogae* Harmer', *Zeitschrift für Morphologie und Ökologie der Tiere*, 36 (1940), 267-78.
112 L. Silén, 'On the division and movements of the alimentary canal of the Bryozoa', *Arkiv für Zoologie utgivet av Kongliga Svenska Vetenskapsakademien* 35, 3(12) (1944), 1-40.
113 Borg, note 106, p. 247.
114 E. Ries, 'Fütterungsversuche bei *Zoobotryon* (Bryozoa)', *Zeitschrift für Vergleichende Physiologie*, 23 (1936), 64-99.
115 See also Shunatova and Ostrovsky, note 1.
116 Borg, note 106.
117 E. Marcus, 'Beobachtungen und Versuche an lebenden Meersbryozoen', *Zoologische Jahrbücher Abteilung für Systematik, Ökologie und Geographie der Tiere*, 52 (1926), 1-102.
118 E. Marcus, 'Beobachtungen und Versuche an lebeden Sübwasserbryozoen', *Zoologische Jahrbücher Abteilung für Systematik, Ökologie und Geographie der Tiere*, 52 (1926), 279-350.
119 E. Marcus, 'Bryozoa', *Die Tierwelt der Nord und Ostsee*, 4 (7c) (1926), 1-100.
120 E. Marcus, 'Über *Lophopus crystallinus* (Pall.)', *Zoologische Jahrbücher Abteilung für Anatomie und Ontogenie der Tiere*, 58 (1934), 501-606.
121 E. Marcus, 'Sobre Bryozoa do Brasil', *Boletins da Faculdade de Filosofia, Ciências e Letras. Universidade de São Paulo XXII, Zoologia*, 5 (1941), 3-208.
122 Summarized in S.F. Harmer, 'Recent work on Polyzoa', *Proceedings of the Linnean Society of London*, 143 (1931), 113-68.
123 E. Marcus, note 117.
124 C.M. Brooks, 'Notes on the statoblasts and polypids of *Pectinatella magnifica*', *Proceedings of the Academy of Natural Sciences of Philadelphia*, 81 (1929), 427-441.
125 D. Atkins, 'The ciliary feeding mechanism of the Entoproct Polyzoa, and a comparison with that of the Ectoproct Polyzoa', *Quarterly Journal of Microscopical Science*, New Series, 75(299) (1932), 393-423.
126 Proucho, note 54.
127 See also F. Braem, 'Über die Querstreifung im Pharynx der gymnolämen Bryozoen und über den Bau des Munddarms', *Zeitschrift für Morphologie und Ökologie der Tiere*, 36 (1940), 668-676.
128 Marcus, note 121.
129 Borg, note 106.

130 C.J. Cori, 'Ordnung der Tentaculata. Bryozoa', in *Handbuch der Zoologie* III, 2(5), (Walter de Gruyter and Company, Berlin, 1941), 263-374, 375-503.

131 Cori, note 130, p. 409, figures 349, 366, 492.

132 H. Graupner, 'Zur Kenntnis der feineren Anatomie der Bryozoen (Nervensystem, Muskulatur, Stützmembran)', *Zeitschrift für Wissenschaftliche Zoologie*, 136 (1930), 38-77.

133 For comparison see J.B. Markham and J.S. Ryland, 'Morphometry and function in the lophophore of *Flustrellidra hispida* (Bryozoa)', in *Bryozoaires Actuels et Fossiles: Bryozoa Living and Fossil*, edited by F.P. Bigey, *Bulletin de la Société des Sciences Naturelles de l'Ouest de la France,* Mémoire Hors Série, 1 (1991), 237-242.

134 Lister, note 57.

135 Cori, note 130, figure 346.

136 Silén, note 112.

137 See also M.A. Best and J.P. Thorpe, 'Effects of particle concentration on clearance rate and feeding current velocity in the marine bryozoan *Flustrellidra hispida*', *Marine Biology*, 77 (1983), 85-92.

138 E. Hunter and R.N. Hughes, 'The effect of cell concentration on colony growth and feeding in the bryozoan *Celleporella hyalina*', *Journal of the Marine Biological Association of the United Kingdom*, New Series, 73 (1993), 321-331.

139 L. Silén, 'The main features of the development of the ovum, embryo and ooecium in the ooecioferous Bryozoa Gymnolaemata', *Arkiv für Zoologie*, 35A (1945), 1-34.

140 L. Silén, 'On the fertilization problem in the gymnolaematous Bryozoa', *Ophelia*, 3 (1966), 113-140.

141 Marcus, note 117.

142 J.S. Bullivant, 'Release of sperm by Bryozoa', *Ophelia*, 4 (1967), 139-142.

143 P.L. Cook, 'Observations on living Bryozoa', *Atti della Società Italiana di Scienze Naturali e del Museo Civico di Storia Naturale di Milano*, 108 (1968), 155-160.

144 D.P. Gordon, 'Zooidal dimorphism in the polyzoan *Hippopodinella adpressa* (Busk)', *Nature (London)*, 219 (1968), 633-634.

145 R. Ström, 'Sexual reproduction in a stoloniferous bryozoan, *Triticella koreni* (G. O. Sars)', *Zoologiska Bidrag från Uppsala*, 38 (1969), 113-128.

146 P.J. Chimonides and P.L. Cook, 'Observations on living colonies of *Selenaria* (Bryozoa, Cheilostomata). II', *Cahiers de Biologie Marine*, 22 (1981), 207-219.

147 P.E.J. Dyrynda and P.E. King, 'Gametogenesis in placental and non-placental ovicellate cheilostome Bryozoa', *Journal of Zoology (London)*, 200 (1983), 471-942.

148 Shunatova and Ostrovsky, note 1.

149 B. Werner, 'Das Prinzip des andlosen Schleimfilters beim Nahrungswerb wirbelleser Meerestiere', *Internationale Revue der gesamten Hydrobiologie und Hydrographie*, 44 (1959), 181-215.

150 Cori, note 130.

151 C.B. Jørgensen, *Biology of suspension feeding*, (Pergamon Press, Oxford, London, New York, Paris, 1966), xv, 357.

152 J.S. Bullivant, 'The method of feeding of Lophophorates (Bryozoa, Phoronida, Brachiopoda)', *New Zealand Journal of Marine and Freshwater Research*, 2(1) (1968) 135-46.

153 J.S. Bullivant, 'A revised classification of suspension feeders', *Tuatara*, 16 (1969), 151-60.

154 Lutaud, note 96.

155 See also G. Lutaud, 'L'innervation du lophophore chez le Bryozoaire chilostome *Electra*

pilosa (L.)', *Zeitschrift für Zellforschung und mikroskopische Anatomie*, 140 (1973.), 217-234.
156 L.W. Smith, 'Ultrastructure of the tentacles of *Flustrellidra hispida* (Fabricius)', in *Living and Fossil Bryozoa*, edited by G.P. Larwood (Academic Press, London, New York, 1973), 335-42.
157 Bullivant, note 152, p. 137.
158 For general discussion see also M.A. Sleigh and E. Aiello, 'The movement of water by cilia', *Acta Protozoologica*, 11 (1972), 265-277.
159 J.R. Blake and M.A. Sleigh, 'Hydromechanical aspects of ciliary propulsion', in *Swimming and Flying in Nature*, edited by T.Y.T. Wu, C.J. Brokaw and C. Brennen (Plenum Press, New York, 1975), 1, 185-209.
160 Bomme, note 13.
161 Marcus, note 117.
162 J.S. Bullivant, 'The rate of feeding of the bryozoan, *Zoobotryon verticullatum*', *New Zealand Journal of Marine and Freshwater Research*, 2(1) (1968b), 111-133.
163 G. Becker, 'Untersuchungen über den Darm und die Verdauung von Kamptozoen, Bryozoen und Phoroniden', *Zeitschrift für Morphologie und Ökologie der Tiere*, 33 (1938), 72-127.
164 D.P. Gordon, 'Microarchitecture and function of lophophore in the bryozoan *Cryptosula pallasiana*', *Marine Biology*, 27 (1974), 147-163.
165 Gordon, note 164, p. 151.
166 J.S. Ryland, *Bryozoans* (Hutchinson University Library, London, 1970) 175 (43).
167 T.H.J. Gilmour, 'Ciliation and function of the food-collecting and waste-rejecting organs of lophophorates', *Canadian Journal of Zoology*, 56 (1978), 2142-2155.
168 D.I. Rubenstein and M.A.R. Koehl, 'The mechanisms of filter feeding: some theoretical considerations', *American Naturalist*, 111 (1977), 981-994.
169 Gilmour, note 167, p. 2152.
170 Bullivant, note 152.
171 R.R. Strathmann, 'Function of lateral cilia in suspension feeding of lophophorates (Brachiopoda, Phoronida, Ectoprocta)', *Marine Biology*, 23 (1973), 129-136.
172 R.R. Strathmann, 'Cinefilms of particle capture by an induced local change of beat of lateral cilia of a bryozoan', *Journal of Experimental Marine Biology and Ecology*, 62 (1982), 225-236.
173 R.R. Strathmann, 'Comment on Dr. Gilmour's views on feeding by hemichordates and lophophorates', *Canadian Journal of Zoology*, 60 (1982), 3466-3468.
174 For discussion see Rubenstein and Koehl, note 168.
175 S. Vogel, *Life in moving fluids: the physiological biology of flow* (Princeton University Press, Princeton, NJ, 1994), xiii, 467.
176 M.W. Hart and R.R. Strathmann, 'Mechanisms and rates of suspension feeding', in *Ecology of Marine Invertebrate Larvae*, edited by L. McEdward, Marine Science Series, (CRC Press, Boca Raton, FL, 1995), 193-221.
177 C. Nielsen and J. Rostgaard, 'Structure and function of an entoproct tentacle with a discussion of ciliary feeding types', *Ophelia*, 15 (1976), 115-140.
178 For review see also C. Nielsen, 'Structure and function of metazoan ciliary bands and their phylogenetic significance', *Acta Zoologica*, 68 (1987), 205-262.
179 R.R. Strathmann, 'The feeding behaviour of planctotrophic echinoderm larvae: mechanisms,

regulation, and rates of suspension feeding', *Journal of Experimental Marine Biology and Ecology*, 6 (1971), 109-160.

180 R.R. Strathmann, T.L. Jahn and J.R.C. Fonseca, 'Suspension feeding by marine invertebrate larvae: clearance of particles by ciliated bands of a rotifer, pluteus and trochophore', *Biological Bulletin*, 142 (1972), 505-519.

181 Strathmann, note 171.

182 M.W. Hart, 'Deconstructing suspension feeders by analysis of film and video', *Invertebrate Biology*, 115 (1996), 185-190.

183 Hart, note 182.

184 Strathmann, note 171.

185 Strathmann, note 171.

186 Strathmann, note 172.

187 J.S. Ryland, 'Physiology and ecology of marine bryozoans', *Advances in Marine Biology*, 14 (1976), 285-443.

188 J.E. Winston, 'Feeding in marine bryozoans', in *Biology of Bryozoans*, edited by R.M. Woollacott and R.L. Zimmer (Academic Press, London, New York, 1977), 233-271.

189 Winston, note 3.

190 M. LaBarbera, 'Feeding currents and particle capture mechanisms in suspension feeding animals', *American Zoologist*, 24 (1984), 71-84.

191 F.K. McKinney and J.B.C. Jackson, *Bryozoan evolution* (University of Chicago Press, Chicago, London, 1989), 238.

192 M.W. Hart, 'Particle captures and the method of suspension feeding by echinoderm larvae', *Biological Bulletin*, 180 (1991), 12-27.

193 Hart, note 182.

194 Ryland, note 187.

195 Ryland, note 187, p. 327.

196 J.S. Ryland, 'Parameters of the lophophore in relation to population structure in a bryozoan community', in *Proceedings of the 9th European Marine Biology Symposium*, edited by H. Barnes (Aberdeen University Press, 1975), 363-393.

197 Ryland, note 187.

198 Ryland, note 196.

199 See also Cori, note 130.

200 For comparison see Markham and Ryland, note 133.

201 W.G. Sanderson, J.P. Thorpe and A. Clarke, 'Aspects of structure and function in the feeding of the Antarctic cheilostomate bryozoan, *Himantozoum antarcticum* (Calvet)', in *Proceedings of the 11th International Bryozoology Association Conference*, edited by A. Herrera Cubilla and J.B.C. Jackson (Smithsonian Tropical Research Institute, Panama, 2000), 355-364 (361).

202 Best and Thorpe, note 137.

203 Nielsen, note 178.

204 Sanderson, Thorpe and Clarke, note 201, p. 361.

205 See also M.A. Best and J.P. Thorpe, 'Feeding current interactions and competition for food among the bryozoan epiphytes of *Fucus serratus*', *Marine Biology*, 93 (1986), 371-375.

206 Best and Thorpe, note 137.

207 Borg, note 106.

208 Atkins, note 125.

209 Best and Thorpe, note 137, p. 91.
210 See for further discussion F.K. McKinney, 'Feeding and associated colonial morphology in marine bryozoans', *Critical Reviews in Aquatic Sciences*, 2(2) (1990), 255-280.
211 For comparison see Markham and Ryland, note 133.
212 Best and Thorpe, note 137.
213 Winston, note 188.
214 Winston, note 3.
215 J.E. Winston, 'Current-related morphology and behaviour in some Pacific coast bryozoans', in *Advances in Bryozoology*, edited by G.P. Larwood and M.B. Abbott, Systematics Association Special vol 13 (Academic Press, London, New York, 1979), 247-268.
216 Summarized in J.E. Winston, 'Feeding behaviour of modern bryozoans', in *Studies in Geology 5: Lophophorates. Notes for a Short Course*, edited by T.W. Broadhead (University of Tennessee, 1981), 1-21.
217 Winston, note 3.
218 Winston, note 188.
219 Bullivant, note 152.
220 Winston, note 3, p. 1.
221 Winston, note 3, p. 29.
222 M.H. Dick, 'Bryozoan behaviour in relation to autocleaning in *Holoporella brunea* (Hincks)', *Northwest Science*, 58(3) (1984), 195-207.
223 Winston, note 3, p. 13.
224 M.H. Dick, 'A proposed mechanism for chimney formation in encrusting bryozoan colonies', in *Bryozoa: Present and Past*, edited by J.R.P. Ross (Western Washington University, Bellingham, WA, 1987), 73-80 (74).
225 P.L. Cook, 'Bryozoa from Ghana', *Zoologische Wetenschappen Musee Royal l'Afrique centrale Tervuren, Belgique*, 238 (1985), 1-315.
226 Reviewed by McKinney, note 210.
227 Hunter and Hughes, note 138, p. 327, figure 2E.
228 Borg, note 106.
229 Strathmann, note 171.
230 Strathmann, note 172.
231 Hunter and Hughes, note 138.
232 H.U. Riisgård and P. Manríquez, 'Filter-feeding in fifteen marine ectoprocts (Bryozoa): particle capture and water pumping', *Marine Ecology Progress Series*, 154 (1997), 223-239.
233 C. Nielsen and H.U. Riisgård, 'Tentacle structure and filter-feeding in *Crisia eburnea* and other cyclostomatous bryozoans, with a review of upstream-collecting mechanisms', *Marine Ecology Progress Series*, 168 (1998), 163-186.
234 Rubenstein and Koehl, note 168.
235 LaBarbera, note 190.
236 J. Shimeta and P. Jumars, 'Physical mechanisms and rates of particle capture by suspension feeders', *Oceanography and Marine Biology Annual Review*, 29 (1991), 191-257.
237 Vogel, note 175.
238 Hart and Strathmann, note 176.
239 H.U. Riisgård and P.S. Larsen, 'Filter-feeding in marine macroinvertebrates: pump characteristics, modelling and energy cost', *Biological Review*, 70 (1996), 67-106.

240 H.U. Riisgård, P.S. Larsen and P.F. Nielsen, 'Particle capture in the mussel *Mytilus edulis*: the role of latero-frontal cirri', *Marine Biology*, 127 (1996), 259-266.
241 See for discussion also C.B. Jørgensen, 'A hydromechanical principle for particle retention in *Mytilus edulis* and other ciliary suspension feeders', *Marine Biology*, 61 (1981), 277-822.
242 C.B Jørgensen, 'Fluid mechanics of the mussel gill: the lateral cilia', *Marine Biology*, 70 (1982), 275-281.
243 C.B. Jørgensen, 'Fluid mechanical aspects of suspension feeding', *Marine Ecology Progress Series*, 11 (1983), 89-103.
244 J. Shimeta and M.A.R. Koehl, 'Mechanisms of particle selection by tentaculate suspension feeders during encounter, retention, and handling', *Journal of Experimental Marine Biology and Ecology*, 209 (1997), 47-73.
245 D. Wildish and D.D. Kristmanson, *Benthic suspension feeders and flow* (Cambridge University Press, 1997), xiii, 409.
246 Riisgård and Manríquez, note 232, p. 237.
247 For further discussion see also Shunatova and Ostrovsky, note 1.
248 Nielsen, note 178.
249 Nielsen and Riisgård, note 233.
250 Riisgård and Manríquez, note 232.
251 Best and Thorpe, note 205.
252 W.G. Sanderson and J.P. Thorpe, 'Effects of temperature on the feeding activity of some temperate intertidal Bryozoa', in *Bryozoans in Space and Time*, edited by D.P. Gordon, A.M. Smith and J.A. Grant-Mackie (National Institute of Water and Athmospheric Resarch Ltd, Wellington, 1996), 271-281.
253 See similar opinion in B. Okamura, 'Particle size and flow velocity induce an inferred switch in bryozoan suspension-feeding behavoiur', *Biological Bulletin*, 173 (1987), 222-229.
254 Shunatova and Ostrovsky, note 1.
255 Shunatova and Ostrovsky, note 1, p. 138.
256 I.I. Antipenko, *Feeding and behaviour of some fresh-water bryozoan species (Bryozoa, Phylactolaemata),* Unpublished MSc Thesis, (St. Petersburg State University, 1998), 112 (in Russian).
257 Bullivant, note 162.
258 N.R. Menon, 'Clearance rates of food suspension and food passage rates as a function of temperature in two North-Sea bryozoans', *Marine Biology*, 24 (1974), 65-67.
259 J. Dudley, 'Differential utilization of phytoplancton food resources by marine entoprocts', *Biological Bulletin*, 139(2) (1970), 420.
260 Ryland, note 196.
261 See also Winston, note 188.
262 Winston, note 3.
263 Winston, note 215.
264 D. Jebram, 'Interrelations of nutrition, food uptake and growth in bryozoans', in *Advances in Bryozoology*, edited by G.P. Larwood and M.B. Abbott (Academic Press, London, 1979), 121-140.
265 J.P. Thorpe, D.R.K. Clarke and M.A. Best, 'Natural variation in tentacle number in marine bryozoans and the possible effects of intraspecific and interspecific ecological competition for food', in *Bryozoa: Ordovician to Recent*, edited by C. Nielsen and G.P. Larwood (Olsen

and Olsen, Fredensborg, 1985), 319-327.
266 A. Herrera and J.B.C. Jackson, 'Environmental versus taxonomic variation in lophophore morphology of Panamanian cheilostome bryozoans', *Marine Biology*, 112 (1992), 99-106.
267 Best and Thorpe, note 137.
268 Best and Thorpe, note 205.
269 M.A. Best and J.P. Thorpe, 'Particle size, clearance rate and feeding efficiency in marine Bryozoa', in *Biology and Palaeobiology of Bryozoans*, edited by P.J. Hayward, J.S. Ryland and P.D. Taylor (Olsen and Olsen, Fredensborg, 1994*a*), 9-14.
270 W.G. Sanderson, J.P. Thorpe and A. Clarke, 'A preliminary study of feeding rates in the Antarctic cheilostomate bryozoan *Himantozoum antarcticum*', in *Biology and Palaeobiology of Bryozoans*, edited by P.J. Hayward, J.S. Ryland and P.D. Taylor (Olsen and Olsen, Fredensborg, 1994*a*), 167-172.
271 Sanderson and Thorpe, note 252.
272 Riisgård and Manríquez, note 232.
273 See also F.K. McKinney, M.R.A. Listokin and C.D. Phifer, 'Flow and polypide distribution in the cheilostome *Bugula* and their inference in *Archimedes*', *Lethaia*, 19 (1986), 81-93 (quoted in McKinney 1990).
274 Wildish and Kristmanson, note 245.
275 Reviewed in M.A. Best and J.P. Thorpe, 'The effect of suspended particulate matter (silt) on the feeding activity of the intertidal ctenostomate bryozoan *Flustrellidra hispida* (Fabricius)', in *Bryozoans in Space and Time*, edited by D.P. Gordon, A.M. Smith and J.A. Grant-Mackie (National Institute of Water and Athmospheric Resarch Ltd, Wellington, 1996), 39-45.
276 For discussion see also D.P. Gordon, A.G. Clark and J.F. Harper, 'Bryozoa', in *Animal Energetics*, edited by T.J. Pandian and F.J. Vernberg, volume 2 (London, Academic Press, 1987), 173-199.
277 Best and Thorpe, note 269.
278 Dudley, note 259.
279 Ryland, note 196.
280 J.S. Ryland, 'Physiology and ecology of marine bryozoans', *Advances in Marine Biology*, 14 (1976), 285-443.
281 Winston, note 188.
282 Winston, note 3, p. 1.
283 Winston, note 216.
284 Dudley, note 259.
285 Ryland, note 196.
286 Winston, note 188.
287 Winston, note 216.
288 McKinney and Jackson, note 191.
289 McKinney, note 210.
290 Herrera and Jackson, note 266.
291 Best and Thorpe, note 269.
292 Sanderson, Thorpe and Clarke, note 201, p. 361.
293 J.S. Porter, P. Bloor, B.A. Stokell and J.S. Ryland, 'Intra- and interspecific variation in tentacle number in the genus *Alcyonidium*', in *Proceedings of the 11th International Bryozoology Association Conference*, edited by A. Herrera Cubilla and J.B.C. Jackson

(Smithsonian Tropical Research Institute, Panama, 2000), 321-328 (328).

294 Sanderson, Thorpe and Clarke, note 200, p. 361.

295 Reviewed in O.D. Hunt, 'The food of the bottom fauna of the Plymouth fishing grounds', *Journal of the Marine Biological Association of the United Kingdom,* New Series, 13 (1925), 560-599.

296 Jørgensen, note 151.

297 Bullivant, note 162.

298 Winston, note 188.

299 Winston, note 3.

300 Winston, note 215.

301 D. Jebram, 1977. 'Experimental techniques and culture methods', in *The Biology of Bryozoans,* edited by R.M. Woollacott and R.L. Zimmer (Academic Press, New York, London, 1977), 273-305.

302 McKinney and Jackson, note 191, p. 238.

303 B. Okamura, 'Particle size, flow velocity and suspension feeding by erect bryozoans *Bugula neritina* and *Bugula stolonifera*', *Marine Biology*, 105 (1990), 33-35.

304 M.A. Best and J.P. Thorpe, 'The uptake of 14C radioisotope tracer from dissolved glycine and glucose by the marine bryozoan *Flustrellidra hispida*', in *Bryozoaires Actuels et Fossiles: Bryozoa Living and Fossil*, edited by F.P. Bigey, *Bulletin de la Société des Sciences Naturelles de l'Ouest de la France,* Mémoire Hors Série, 1 (1991), 13-23.

305 M.A. Best and J.P. Thorpe, 'An analysis of potential food sources available to intertidal bryozoans in Britain', in *Biology and Palaeobiology of Bryozoans*, edited by P.J. Hayward, J.S. Ryland and P.D. Taylor (Olsen and Olsen, Fredensborg, 1994), 1-8.

306 Best and Thorpe, note 275.

307 Best and Thorpe, note 305.

308 D.K.A. Barnes and A. Clarke, 'Seasonal variation in the feeding activity of four species of Antarctic Bryozoa in relation to environmental factors', *Journal of Experimental Marine Biology and Ecology*, 181 (1994), 117-133.

309 D.K.A. Barnes and A. Clarke, 'Seasonality of feeding ecology in polar suspension feeders', *Polar Biology*, 15 (1995), 335-340.

310 W.G. Sanderson, S.P. Hardling and J.P. Thorpe, 'An investigation of the effects of tidal fluctuations on the availability of seston as food for intertidal Bryozoa', in *Bryozoans in Space and Time*, edited by D.P. Gordon, A.M. Smith and J.A. Grant-Mackie (Wellington, National Institute of Water and Athmospheric Resarch Ltd, 1996), 259-269.

311 B. Okamura, note 253.

312 Marcus, note 117.

313 N.R. Menon, 'Heat tolerance, growth and regeneration in three North Sea bryozoans exposed to different constant temperatures', *Marine Biology*, 15 (1972), 1-11.

314 N.R. Menon, note 258.

315 Riisgård and Manríquez, note 233.

316 Sanderson and Thorpe, note 252.

317 B. Okamura, 'The effects of ambient flow velocity, colony size, and upstream colonies on the feeding success of Bryozoa. I. *Bugula stolonifera* Ryland, an arborescent species', *Journal of Experimental Marine Biology and Ecology*, 83 (1984), 179-193.

318 B. Okamura, 'The effects of ambient flow velocity, colony size, and upstream colonies on the feeding success of Bryozoa. II. *Conopeum reticulum* (Linnaeus), an encrusting species',

Journal of Experimental Marine Biology and Ecology, 89 (1985), 69-80.

319 B. Okamura, 'Seasonal changes in zooid size and feeding activity in epifaunal colonies of *Electra pilosa*', in *Bryozoa: Present and Past*, edited by J.R.P. Ross (Western Washington University, Bellingham, WA, 1987), 197-203.

320 Okamura, note 303.

321 Okamura, note 303.

322 B. Okamura, 'The influence of neighbors on the feeding of an epifaunal bryozoan', *Journal of Experimental Marine Biology and Ecology*, 120 (1988), 105-123.

323 B. Okamura, 'Microhabitat variation and patterns of colony growth and feeding in a marine bryozoan', *Ecology*, 73 (1992), 1502-1513.

324 L.W. Buss, 'Bryozoan overgrowth interactions – the interdependence of competition for space and food', *Nature*, 281(5731), (1980), 475-477.

325 Best and Thorpe, note 205.

326 M.A. Best and J.P. Thorpe, 'Effects of food particle concentration on feeding current velocity in six species of marine Bryozoa', *Marine Biology*, 93 (1986), 255-262.

327 See also L.W. Buss, 'Mechanisms of competition between *Onychocella* (Hastings) and *Antropora tincta* (Hastings) on an Eastern Pacific rocky shoreline', in *Recent and Fossil Bryozoa*, edited by G.P. Larwood and C. Nielsen (Olsen and Olsen, Fredensborg, 1981) 39-49.

328 L.W. Buss and J.B.C. Jackson, 'Planctonic food availability and suspended feeder abundance; evidence of in situ depletion', *Journal of Experimental Marine Biology and Ecology*, 49 (1981), 151-161.

329 Thorpe, Clarke and Best, note 265.

330 P.S. Larsen, S.S. Matlok and H.U. Riisgård, 'Bryozoan filter feeding in laminar wall layers: flume experiments and computer simulation', *Vie Milieu*, 48(4) (1998), 309-319.

331 J.E. Eckman and B. Okamura, 'A model of particle capture by bryozoans in turbulent flow: significance of colony form', *American Naturalist*, 152(6) (1998), 861-880.

332 J.W. Bishop and L.M. Bahr, 'Effects of colony size on feeding by *Lophopodella carteri* (Hyatt)', in *Animal Colonies: Development and Function through Time*, edited by R.S. Boardman, A.H. Cheetham and W.A. Oliver (Dowden, Hutchinson and Ross, Stroudsberg, PA, 1973), 433-437.

333 B. Okamura and L.A. Doolan, 'Patterns of suspension feeding in the freshwater bryozoan *Plumatella repens*', *Biological Bulletin*, 184 (1993), 52-56.

334 Winston, note 3, p. 1.

335 Okamura, note 319.

336 McKinney, note 210.

337 Riisgård and Manríquez, note 232.

338 Porter, Bloor, Stokell and Ryland, note 293.

339 Sanderson, Thorpe and Clarke, note 201, p. 361.

340 Shunatova and Ostrovsky, note 1.

341 See as examples Riisgård and Manríquez, note 232.

342 Nielsen and Riisgård, note 233.

343 Sanderson, Thorpe and Clarke, note 201, p. 361.

Alan Stanley Horowitz (1930–1999)

Joseph F. Pachut, Robert L. Anstey, N. Gary Lane, and J. Robert Dodd
Department of Geology, Indiana University-Purdue University at Indianapolis; Michigan State University, East Lansing; and Indiana University, Bloomington

Dr Alan Stanley Horowitz, 68, passed away on 18th February 1999 in Bloomington, Indiana, after a protracted battle with leukemia. He was curator of Paleontology, an emeritus Senior Research Scientist, and part-time Professor of Geology at Indiana University.

Alan was born on 12th June 1930 in Ashland, Kentucky, the son of Samuel and Irene Strouss Horowitz. He was an Eagle Scout and graduated Phi Beta Kappa from Washington and Lee University in 1952. In 1953, he completed a Masters degree on igneous rocks of the Nunatarssuaq area, northwest Greenland at The Ohio State University and followed that with a Doctoral degree on the fauna of the Late Mississippian Glen Dean Formation at Indiana University in 1957. His Greenland work was subsequently published in 1964 in a paper co-authored with A.T. Fernald.

After obtaining his Ph.D., Alan accepted a position as a research geologist with Marathon Oil Company in Denver, Colorado, where he worked from 1957 to 1965. At Marathon he served as palaeontologist with a team that pioneered the use of new techniques and approaches to the study of carbonate rocks. Alan returned to his alma mater, Indiana University, in 1966 to become Curator of Paleontology and he served in that capacity until his retirement in 1996. He willingly expanded his duties beyond research and curator of collections to include service on graduate research committees and teaching graduate courses. These extra contributions were formally recognized in 1980 with his appointment as research scientist and Professor of Geology (part-time).

An expert on the geology of Indiana and the south-central United States, Alan specialized in Late Mississippian (Chesterian) rocks and fossils. He was an internationally known authority on fossil bryozoans and blastoids, authoring or co-authoring 53 papers, 57 abstracts, 1 fieldtrip guidebook, 5 bulletins and monographs, and 1 book (*Introductory Petrography of Fossils*, co-authored with Paul E. Potter [1971]). He also made major contributions to palaeontology by compiling data from the published literature on various fossil groups. His research was supported by grants from The Petroleum Research Fund of the American Chemical Society and from The National Science Foundation. Additionally, Alan was a visiting professor at Aarhus University in Denmark in 1971-

Figure 1. Alan Stanley Horowitz

1972. He regularly attended meetings of the Geological Society of America, the International Carboniferous Congress, and the International Bryozoology Association.

Alan was a recognized bibliophile, with a library exceeding 3000 books. He could usually be found wearing his ever-present visor in a corner of his darkened office surrounded by shelves and stacks of books. In the field, he'd usually have his nose "to the rocks", wearing knee pads and blinders on his glasses to cut out peripheral distractions. Alan would always lend a sympathetic ear to both faculty and students when problems (academic, scientific, or personal) arose. Generations of geology graduate students benefited from his counsel whether or not his name appeared on the signature page of their dissertation.

Alan collaborated extensively with a number of fellow geologists and palaeontologists, including especially Joseph Pachut, Johnny Waters, John B. Droste, Paul Potter, and James Stratton. Alan's scientific legacy consists of numerous goals that he continuously communicated to his students and colleagues. First, look for everything in the fossil record, not just the large conspicuous fossils (e.g. robust trepostomes). Second, compile and archive the existing scientific knowledge concerning any organic group, and generate computer-based compilations that are as complete as possible. Third, analyze your data quantitatively and apply as many new and modern techniques as you can to old problems. Alan's influence can be seen in the scientific work of all the paleontologists with whom he has worked.

He is survived by two sisters, Mrs Carol Picker of Washington Crossing, Pennsylvania and Mrs Ann Itkin of Brooklyn, New York, four step-children, Mrs Paula Jarrett, Peter Perry, Michael Perry and John Perry, and 12 grandchildren. He was preceded in death by his wife, Lillian Perry Horowitz.

We have lost a major contributor to the study of bryozoans. However, Alan will be remembered even more for his friendship and genuine warmth toward his colleagues. If pressed for a single item that characterizes him it would have to be his optimistic view of life. He was a classic example of a person who professed "the glass is half-full" view of life, always able to perceive the positive aspects of an issue, a person, or a difficult or complicated situation. As we continue to attempt to develop a similar outlook (we tend to "make haste slowly"), his encouragement and assistance will be missed.

Selected bryozoan-related bibliography of A.S. Horowitz

A.S. Horowitz and T.G. Perry, 'Fauna of the Glen Dean Limestone (middle Chester) of Indiana and northern Kentucky', *Geological Society of America Bulletin*, 67, 12 (2) (1956), 1707-1708.

T.G. Perry and A.S. Horowitz, 'Bryozoans from the Glen Dean Limestone (Middle Chester) of southern Indiana and Kentucky', *Indiana Geological Survey Bulletin*, 26 (1963), 51 pp.

A.S. Horowitz, 'The ectoproct (bryozoan) genus *Actinotrypa* Ulrich', *Journal of Paleontology*, 42 (2) (1968), 356-373.

A.S. Horowitz, 'On nomenclatorial diversity in *Fistulipora* McCoy and allied genera', *Journal of Paleontology*, 44 (4) (1970), 776-778.

T.G. Perry, A.S. Horowitz and R.L. Anstey, 'Restudy of some Nicholson types of Palaeozoic bryozoans', in *Living and Fossil Bryozoa*, edited by G.P. Larwood (London, 1973), 429-436.

J.F. Stratton and A.S. Horowitz, 1976. Studies of the flow of water through models of *Polypora*', in *Bryozoa 1974*, edited by S. Pouyet, *Documents des Laboratoires de Géologie de la Faculté des Sciences de Lyon*, H.S. 3 (fasc. 2) (1976), 425-438.

A.S. Horowitz, 'Late Mississippian and early Pennsylvanian bryozoan faunas of Arkansas and Oklahoma: a review', *Oklahoma Geological Survey Guidebook*, 18 (1977), 101-105, 1 table.

J.F. Stratton and A.S. Horowitz, 'Astogenetic variability in a frond of *Polypora laevinodata* (Hall) (Bryozoa)', *Proceedings of the Indiana Academy of Science for 1976*, 86 (1977), 290-292.

J.F. Stratton and A.S. Horowitz, *Variability in seven Devonian species of* Polypora *M'Coy.* (Bloomington, Indiana, 1977), 23 pp.

J.F. Stratton and A.S. Horowitz, '*Polypora* M'Coy from the Devonian of southeastern Indiana', *Indiana Geological Survey Bulletin*, 56 (1978), 48 pp., 4 pls., 3 figs.

A.S. Horowitz, J.F. Stratton and J.A. Waters, 'Quantitative clustering applied to some polyporan data from the Middle Devonian of Indiana', in *Advances in Bryozoology*, edited by G.P. Larwood and M. B. Abbott, *Systematics Association Special Volume* 13, (London, 1979), 373-382.

J.F. Stratton and A.S. Horowitz, 'Two bryozoan faunules from the Late Pennsylvanian of Illinois', *Transactions of the Illinois State Academy of Science*, 80, supplement, (1987), 40.

S.M. Kelly and A.S. Horowitz, 'Growth forms and paleoecology of Mississippian bryozoans: critical application of Stach's 1936 model, eastern United States', in *Bryozoa: Present and Past*, edited by J.R.P. Ross (Bellingham, 1987), 137-144.

J.F. Pachut and A.S. Horowitz. 'Multivariate discrimination and classification of some North American Mississippian species of *Fistulipora* M'Coy', in *Bryozoa: Present and Past*, edited by J.R.P. Ross (Bellingham, 1987), 205-212.

A.S. Horowitz and J.F. Pachut, '*Strotopora* Ulrich and *Cliotrypa* Ulrich and Bassler (Cystoporida, Bryozoa)', *Geological Society of America Abstracts with Programs*, 20 (5) (1988), 349.

A.S. Horowitz, 'Extinctions', *M.A.P.S. [Mid-America Paleontology Society] Digest*, 11 (6) (1988), 3-6. [Reprinted *Bulletin of the Southern California Paleontological Society*, 23 (9-10), (1991), 88-92.]

A.S. Horowitz and J.F. Pachut, 'Estimate of Devonian bryozoan diversity', *Geological Society of America Abstracts with Programs*, 21 (4) (1989), 15.

A.S. Horowitz and J.F. Pachut, 'Specific, generic, and familial diversity of Devonian bryozoans', *Journal of Paleontology*, 67 (1) (1993), 42-52.

J.F. Pachut, R.L. Anstey and A.S. Horowitz, 'The H.A. Nicholson collection of Paleozoic stenolaemate bryozoans: comparison of cladistic and phenetic classifications', *Journal of Paleontology*, 68 (5) (1994), 978-994.

A.S. Horowitz and J.F. Pachut, 'Lyellian bryozoan percentages and the fossil record of the Recent bryozoan fauna', *Palaios*, 9 (1994), 500-505.

A.S. Horowitz, J.F. Pachut and R.L. Anstey, 'Devonian bryozoan diversity, extinctions, and originations', *Journal of Paleontology*, 70 (3) (1996), 373-380.

A.S. Horowitz and J.F. Pachut, 'Diversity of Cenozoic Bryozoa: a preliminary report', in *Bryozoans in Space and Time*, edited by D.P. Gordon, A.M. Smith, and J.A. Grant-Mackie (Wellington, 1996), 133-137.

A.S. Horowitz and J.F. Pachut, 'Permian bryozoan diversity, extinctions, and originations', *Geological Society of America Abstracts with Programs*, 29 (3) (1997).

A.S. Horowitz and J.F. Pachut, 'Bryozoan species diversity from the fossil record', in *Proceedings of the 11th International Bryozoology Association Conference*, edited by A. Herrera Cubilla and J.B.C. Jackson (Panama City, 2000), 245-248.

Thirty four years of bryozoology: the I.B.A. International Conferences (1968-2001)[1]

Salvador Reguant
Reial Acadèmia de Ciències i Arts de Barcelona, Departament d'Estratigrafia, Paleontologia i Geociències Marines, Facultat de Geologia, Universitat de Barcelona, Martí i Franquès s/n, 08028 Barcelona, Spain

1. Introduction

Annoscia and Cheetham[2] report that the International Bryozoology Association (I.B.A.) was founded after the first informal meeting called to discuss Bryozoa in Stockholm in 1965, while the first I.B.A. Conference was held in San Donato Milanese, near Milan, in 1968. Since then eleven conferences have been celebrated, the last of which was in Ireland. Most of the communications to these twelve conferences have been published in their respective Proceedings.

It seems reasonable to suggest that these Proceedings reflect the activity and research interests of bryozoologists around the world and offer a faithful picture of modern-day bryozoology. A thirty-four year period is sufficient I feel to provide scientometric information about the number of contributors, their country of origin and the main lines of research in bryozoology.

This paper presents the results drawn from a review of the eleven volumes of proceedings, corresponding to the 11 Conferences, with supplementary data on the 12th Conference provided by Patrick Wyse Jackson, convenor of that meeting. Taken together, the proceedings offer valuable insights into recent developments in bryozoology.

2. The I.B.A. Conferences

The following lists the twelve I.B.A. Conferences held to date, giving their location; date; the organizing institution; convenor(s);and bibliographic reference to the scientific proceedings:

1. San Donato Milanese, Italy; 12th-16th August, 1968; AGIP Direzione Mineraria. Enrico Annoscia; Annoscia, E. (ed.), 1968. Proceedings of the First International Conference on Bryozoa. *Atti della Societa di Scienze Naturalia del Museo Civico di Storia Naturale di Milano.* Volume 108, Milano.
2. Durham, England; 6th-16th September, 1971; University of Durham; Gilbert P. Larwood; Larwood, G.P. (ed.), 1973. *Living and Fossil Bryozoa.* Academic Press, London and New York.
3. Lyon, France; 2nd-11th September, 1974; University Claude Bernard; Louis David; Pouyet, S. (ed.), 1975. *Bryozoa 1974.* Proceedings of the Third Conference. International Bryozoology Association. *Documents des Laboratoires de Geologie de la Faculte des Sciences de Lyon*, HS 3 (fasc. 1, 2). Universite Claude Bernard, Lyon.
4. Woods Hole, Massachusetts, U.S.A.; 7th-17th September, 1977; Marine Biological Laboratory; Marie B. Abbott; Larwood, G.P. and Abbott, M.B. (eds), 1979. *Advances in Bryozoology.* Academic Press, London.
5. Durham, England; 1st-6th September, 1980; University of Durham; Gilbert P. Larwood; Larwood, G.P. and Nielsen, C. (eds), 1981. *Recent and Fossil Bryozoa.* Olsen & Olsen, Fredensborg.
6. Vienna, Austria; 18th-23rd July, 1983; University of Vienna; Norbert Vávra; Nielsen, C. and Larwood, G.P. (eds), 1985. *Bryozoa: Ordovician to Recent.* Olsen & Olsen, Fredensborg
7 Bellingham, Washington, U.S.A.; 4th-9th August, 1986; Western Washington University. June R.P. Ross; Ross, J.R.P. (ed.), 1987. *Bryozoa: Present and Past.* Western Washington University Press, Bellingham.
8 Paris, France; 17th-22nd August, 1989; Muséum National d'Histoire Naturelle and Université P & M. Curie; Françoise P. Bigey and Jean-Loup d'Hondt. Bigey, F.P. and J.L. d'Hondt (eds), 1991. *Bryozoaires Actuels et Fossiles.* Bulletin de la Societe Sciences Naturelles de la Ouest de France, Memoire H.S. 1. Nantes.
9 Swansea, Wales; 25th July-1st August, 1992; University of Swansea; John S. Ryland and P.J. Hayward; Hayward, P.J., Ryland, J.S. and Taylor, P.D. (eds) 1994. *Biology and Palaeobiology of Bryozoans.* Olsen & Olsen, Fredensborg.
10 Wellington, New Zealand; 30th January-3rd February 1995; NIWA; Dennis P. Gordon; Gordon, D.P., Smith, A.M. and Grant-Mackie, J.A. (eds), 1996. *Bryozoans in Space and Time.* NIWA, Wellington.
11 Panama City, Republic of Panama; 26th-31st January 1998; Jeremy B.C. Jackson; Herrera-Cubilla, A. and Jackson, J.B.C. (eds), 2000. *Proceedings of the 11th International Bryozoology Association Conference.* Smithsonian Tropical Research Institute, Panama.
12 Dublin, Ireland; 16th-21st July, 2001; Trinity College, Dublin; Patrick N. Wyse Jackson; Wyse Jackson, P.N., Buttler, C.J. and Spencer Jones, M.E. (eds), 2002. *Bryozoan Studies 2001.* A.A. Balkema, Rotterdam (scientific papers), and Wyse Jackson, P.N. and Spencer Jones, M.E. (eds) 2002. *Annals of Bryozoology.* International Bryozoology Association, Dublin (historical papers).

3. Participation

Participation at conferences has ranged from 49 at the first meeting to double that at the most recent meeting in Ireland. Table 1 shows the number of participants at each conference, the number of papers and posters presented, the number of papers published in the proceedings and the number of authors involved in the writing of those published papers. As can be seen there is no information available concerning the numbers of participants that attended some conferences.

Table 1. Number of papers, authors and participation at each I.B.A. Conference

Conference	Papers/posters presented	Papers Published	Number of authors of published papers	Number of delegates
1	45	45	46	49
2	65	58	64	?
3	55	54	54	67
4	53	38	53	65
5	50	35	44	?
6	53	39	56	?
7	75	40	54	92
8	75	55	78	85
9	77	45	65	74
10	74	43	72	73
11	78	45	68	66
12	114	70*	95	104

*published in two volumes; one containing scientific contributions and the other historical papers.

Table 2 shows the number of conferences that each author has contributed to either with a presentation of scientific findings or a discussion on some aspects of bryozoology.

Table 2. Number of published contributions presented at the I.B.A. conferences by individual authors (excludes data from Dublin)

Number of conferences	10	9	8	7	6	5	4	3	2	1
Number of contributors	3	4	2	6	9	7	15	40	42	184

The most frequent contributors are Professors Frank McKinney of Boone, Appalachian State University, and June R. P. Ross of Bellingham, Western Washington University, and Ehrhard Voigt from Hamburg University. They have all presented papers at 11 conferences.

As can be seen a total of 312 bryozoologists have authored papers presented at the first eleven IBA conferences. Some of them are collaborators on various aspects (biology or geology) of bryozoology, but they are not bryozoologists. However, this figure, 312 in the first 31 years, does not differ much from the number of active bryozoologists around the world. This is estimated at between 200 and 300, as reported in *Bryozoa*, the annual newsletters of the I.B.A. (for example, there were 244 members in 1986, and 238 in 1995).

Table 3 lists the number of contributors by country at each of the conferences (country populations in millions, according to *Encyclopaedia Britannica* 1991, are included in brackets) and the total number of contributors at each conference.

Only three African countries are represented, though only in Egypt is there a team of bryozoologists. One contributor from Algeria was a resident in Algeria at the time of the 1st and 6th conferences but now works in France. One of the handicaps of contemplating only the I.B.A. conferences is that there are some bryozooloogists who, for a variety of reasons, have never attended these conferences. Morocco, for example, is not present, but a number of monographs and papers have been published in recent years by Moroccan bryozoologists.

4. Research interests

According to the web page edited by Phil Bock: "Systematic List of Families of Bryozoa"[3] the general classification of the Phylum Bryozoa,containing exclusively the Ectoprocta, is as follows:

- Class Stenolaemata
  - Order Cyclostomata
  - Order (?) Hederellida
  - Order Cryptostomida
  - Order Cystoporida
  - Order Trepostomatida
  - Order Fenestrida
- Class Gymnolaemata
  - Order Ctenostomata
  - Order Cheilostomata
- Class Phylactolaemata

The Phylactolaemata, are exclusively a freshwater class, the body walls of which have no skeleton, though they produce resistant bodies called statoblasts. Such bodies have been reported as fossils from the Pleistocene and Upper Tertiary and also from the Cretaceous, although this is controversial.[4] It is also necessary to note that the number of Phylactolaemata families is very small compared with that of other bryozoan classes; there are only five in total.

Nearly all other Bryozoa have mineralized skeletons and their characteristics are essential for specific identification. McKinney & Jackson claim, "most descriptions of

Table 3. Number of contributors to the conference proceedings by country at each I.B.A. conference (country populations in millions shown in brackets)

Conference	**1**	**2**	**3**	**4**	**5**	**6**	**7**	**8**	**9**	**10**	**11**	**12***
EUROPE												
United Kingdom (57)	4	10	6	4	11	11	5	11	13	18	14	18
Ireland (4)								3	2	1	1	3
Italy (59)	3	2	3		1	6	2	10	4		5	3
Spain (39)			2				2	6	4			1
France (57)	5	4	10	6	4	5	4	5	5	2	3	4
Luxembourg (0.4)								2	2	2		2
Germany (79)	7	6	4	4	2	3	4	4	2	6	5	10
Austria (8)	1		1	1	1		1	4	2	1	3	2
Hungary												1
Sweden (8)	3	1	1		2		1					1
Denmark (5)		1	2			2	3	2	1		3	4
Norway (4)								2	1	1	1	2
The Netherlands (15)	1	2	1	1			1	1				1
Belgium (10)	1											
Czech Republic (16)								1				1
Slovakia (5)											1	
Croatia (4)												2
Romania (23)	1	1	1		1	2						
Poland (38)								1				1
USSR (286)	3	6	5		1							
AMERICA												
USA (249)	13	25	15	33	15	17	23	17	13	17	21	27
Canada (27)		1				1	1			1		
Jamaica (2.5)					1							
Panama (2.3)						1				2	2	1
Brazil (150)	1											1
Chile (13)	1			1				4	3	3	3	2
OCEANIA												
Australia (17)		1	1	2	2	1	1	1	3	5	3	6
New Zealand (3.5)			1				1	2	2	8	1	3
ASIA												
Japan (123)	1	2	2	1	1	2	2		4	2		3
South Korea (43)										1		
China (1,134)					2			1				
Taiwan (22)												2
India (685)		2				4			1			
Singapore (2.5)										1	1	
Iran (49)											1	
AFRICA												
Algeria (23)	1					1						1
Egypt (48)							2	1	3		1	1
South Africa (43)												1
TOTAL	**46**	**64**	**55**	**53**	**44**	**56**	**53**	**78**	**65**	**72**	**68**	**104**

* The figures for Dublin refer to participants, not all of whom contributed to the conference proceedings.

living bryozoans with calcified skeletons do not differ substantially from descriptions of fossil species in the same class. This is because the majority of taxonomists who work on living bryozoans have depended on skeletal characters for species recognition, to the virtual exclusion of soft organs and other attributes. The dominant focus has been on morphology of zooids. In addition to presence or absence of structures, a variety of linear measures of size, shape, skeletal construction, and types of heterozooids have been used for both living and fossil species".[5]

Hederellida (2 families), Cryptostomida (32 families), Cystoporida (13 families), Trepostomatida (33 families), and Fenestrida (13 families) are exclusively Palaeozoic; while Cheilostomata (130 families) is exclusively Meso-Cenozoic (including Recent). Cyclostomata (31 or 58 families according to different opinions) and Ctenostomata (41 families) are reported from the Palaeozoic until Recent.

Thus, the first classification of contributions to the I.B.A. Conferences is: general contributions on bryozoans; contributions on Palaeozoic bryozoans; contributions on Meso-Cenozoic (only fossils) bryozoans; contributions on Recent bryozoans; contributions on freshwater phylactolaemates. Table 4 shows the percentage of contributions in each category at each Conference.

Table 4. Percentage of contributions by topic at the I.B.A. Conferences

Conference	**1**	**2**	**3**	**4**	**5**	**6**	**7**	**8**	**9**	**10**	**11**	**12***
General	16	7	9	11	3	3	2	4	0	2	11	21
Palaeozoic	18	19	20	13	23	13	42	27	18	12	4	21
Meso-Cenozoic	44	21	30	29	31	21	17	16	22	26	20	20
Recent	22	45	43	50	43	56	35	47	51	42	48	26
Freshwater	4	7	4	8	6	5	2	4	9	7	15	12

* % figures refer to papers published in the scientific proceedings only and not the historical proceedings.

Research on living and fossil bryozoans gave rise to roughly the same number of contributions over the last 31 years, according to this data. At five conferences (1, 3, 5, 7 and 12) the contributions on fossil bryozoans exceeded those dealing with living specimens. In contrast, the number of contributions on recent Bryozoa was greater than that devoted to fossil bryozoans at the other seven conferences. It is also interesting to note the large number of contributions on Palaeozoic bryozoans at the 7th Conference.

5. Perspectives

The study of Bryozoa, as the study of any group of organisms, may be carried out from a number of perspectives. I analyzed the contributions at the first eleven I.B.A. Conferences according to the following classification:

- Bryozoology, including the history of bryozoan research.
- Biology, comprising anatomy, physiology, genetics, etc..
- Description of faunas and Catalogues of species in collections or in rock-stratigraphic units.
- Ecology and Paleoecology.
- Biogeography and Paleobiogeography.
- Evolution and related topics.
- Stratigraphy and Sedimentation.
- Applied use of bryozoans.
- Others (mainly descriptions of new methods).

Some contributions clearly belong to more than one perspective. On the other hand, the perspective to which each contribution belongs is, in some cases, difficult to decide, and is always somewhat subjective. Table 5 shows the results of this analysis. Biology, Description of faunas and Catalogues, and Ecology and Paleoecology are the main aspects tackled in bryozoan research.Their variation in percentages are shown in figure 3.

Table 5. Number of contributions on different perspectives of bryozoan research

Conference	**1**	**2**	**3**	**4**	**5**	**6**	**7**	**8**	**9**	**10**	**11**
Bryozoology	2	0	3	0	0	2	0	0	0	1	0
Biology	15	27	12	18	15	14	12	15	11	12	18
Description/ Catalogues	14	13	21	12	9	10	14	20	13	7	5
Ecology and palaeoecology	5	15	13	8	7	7	11	16	18	15	11
Palaeobiogeography and biogeography	2	1	2	0	5	3	2	0	3	5	4
Evolution	1	2	4	1	3	1	3	4	0	2	5
Stratigraphy and sedimentology	0	0	0	0	1	1	1	2	1	1	2
Applied use	3	0	0	0	0	0	0	1	0	1	0
Others	4	3	3	2	0	3	2	4	1	0	4

Clearly, the systematic study of newly collected material, including the description of new taxa, is normally published in monographs or papers that are longer than normal contributions to the I.B.A. Conference Proceedings. However, some new taxa are described in the twelve proceedings. Table 6 shows the new species and new genera included in such contributions. Also one new Cretaceous superfamily, seven new families (from Cretaceous to Recent),and two Palaeozoic subfamilies were established. For comparative purposes, between 1990 to 1992, the *Zoological Record* reported 379 new Paleogene species. As can be seen in Table 6, only 11 new Paleogene species are included in the I.B.A. Proceedings.

Table 6. Number of new species and genera (in brackets) established in the I.B.A. Proceedings

Conference	**1**	**2**	**3**	**4**	**5**	**6**	**7**	**8**	**9**	**10**	**11**	**12**
Recent		1	8(6)	4		2		19(10)	6(1)	7(1)		1(1)
Neogene	2	1										1(1)
Paleogene	2		1(1)				1(1)			(1)	7	
Cretaceous	2(2)	1	2	8(1)	2(1)		1	1(2)		(1)		(2)
Jurassic		2(1)							1			
Triassic					1(1)		(1)	1				
Palaeozoic										(1)		1

One perspective designated is the applied use of bryozoans. At the first Conference the application of the fossil bryozoan in oil research was included by the organizers as this conference was supported by a leading Italian oil company: AGIP Mineraria. This perspective virtually disappeared from all subsequent conferences. However, at the 8th and 10th Conferences, papers were delivered on the medical applications of substances produced by some bryozoan species. Indeed, the title of the Keynote Address by David J. Newman at the 10th Conference was particularly significant: “Bryostatin – from bryozoan to cancer drug”.

6. Conclusions

The analysis of contributions at the twelve I.B.A. conferences, covering a period of 34 years, is perhaps a rudimentary method for determining the evolution of recent bryozoology. However, the data, tables and figures presented here give a general perspective on the bulk of work being undertaken around the world, the number of bryozoologists and the countries from which they work, the aspects of bryozoology most frequently dealt with, and also the frequency of research in each geological period.

Notes

1 This paper is a slightly modified version of S. Reguant, ‘Thirty one years (1968-1998) of bryozoology: the I.B.A. international conferences’, *Memorias de la Real Academia de Ciencias y Artes de Barcelona* 58(4) (1999), 141-153. (Used with the permission of the Reial Academia). This present paper includes additional information about the 12th international conference held in Dublin in 2001.

2 E. Annoscia, ‘Conference Chairman Address’, *Atti della Società Italiana di Scienze Naturali*

e del Museo Civico di Storia Naturale di Milano, 108 (1968), 5-6; Cheetham, this volume
3 P. Bock, 'Systematic List of Families of Bryozoa', Webpage: http:// www.civgeo.rmit.edu.au/bryozoa/famsys.html (dated 2000)
4 R.S. Boardman, A.H. Cheetham, D.B. Blake, J. Utgaard, O.L. Karklins, P.L. Cook, P.A. Sandberg, G. Lutaud and T.S. Wood, 'Introduction, Order Cystoporata, Order Cryptostomata', in *Part G. Bryozoa. Revised. Volume 1. Treatise on Invertebrate Paleontology*, edited by R.C. Moore and R.A. Robison (Boulder, Colorado, and Lawrence, Kansas, The Geological Society of America Inc, and the University of Kansas, 1983).
5 F.K. McKinney and J.B.C. Jackson, *Bryozoan Evolution* (Chicago and London, University of Chicago Press, 1989).

Canu and Bassler

JoAnn Sanner
Department of Paleobiology, National Museum of Natural History, Smithsonian Institution, Washington, DC 20650-0121, USA

Raymond Smith Bassler (1878-1961) (Figures 1 and 2) officially joined the United States National Museum (USNM) as a preparator in 1902. From the age of 15, he had prepared Palaeozoic fossils for E.O. Ulrich in Cincinnati, Ohio, and when Ulrich came to the USNM in 1900 Bassler followed and continued unofficially in the same capacity. By 1904 Bassler had finished his Master's degree, had started on his Ph.D. (earned in 1905 from George Washington University), had become an assistant curator, and had published papers with Ulrich on Palaeozoic bryozoans as well as bryozoans from the Maryland Miocene.[1] At the same time he became an assistant professor of geology at George Washington University, a position from which he retired in 1948. In 1908 he became curator, and from 1929 to 1948 was head curator of the USNM's Department of Geology.[2] Although Bassler continued to collect Palaeozoic fossils and publish on them,[3] his interests turned more toward Cretaceous, Tertiary, Quaternary, and Recent bryozoans based on collections made by the U.S. Geological Survey, the U.S. Bureau of Fisheries vessel *Albatross*, and his own efforts. All of this material fueled his collaboration with Ferdinand Canu.

Ferdinand Canu (1863-1932) received his education at the Conservatoire des Arts et Métiers and the Musée d'Histoire Naturelle de Paris. For 33 years, until his retirement in 1914, he taught in the public schools of Paris. In 1890 he began the study of meteorology, geology, and palaeontology producing papers on the first two subjects in 1894 and 1895.[4] His first bryozoan publication came in 1897, and by his death he had over 100 publications on Cretaceous, Tertiary, Quaternary, and Recent bryozoans. He was awarded the French Legion of Honour[5] and in 1923, the Daniel Giraud Elliot Medal of the National Academy of Sciences of the United States for *North American Later Tertiary and Quaternary Bryozoa* which was coauthored with Bassler.[6]

Correspondence between Canu and Bassler began in 1905 with the usual exchange of specimens and reprints,[7] and by 1909 Bassler asked Canu to identify specimens for him. In agreeing, Canu explained that a fee was customary for this service but that specimens to complete his collections would be acceptable.[8] In 1912, after Canu completed notes on the Eocene of Wilmington, North Carolina, Bassler asked him to be senior author of a volume on American Tertiary Bryozoa because 'my knowledge of Tertiary Bryozoa is not

Figure 1 (left). Bassler at Mammoth Cave, Kentucky, 1890 (age 12)
Figure 2 (bottom). Bassler's first fossil collection, Cincinnati, Ohio, 1893

Figure 3. Bassler and his assistant Adelaide Quizenberry with the manuscript of Canu and Bassler's North American Early Tertiary Bryozoa *(c. 1919).*

very deep'.[9] Canu agreed enthusiastically and proposed amendments to Bassler's plan for the project.[10] It took more than four years to complete the manuscript of *North American Early Tertiary Bryozoa* (Figure 3) because of the constraints of World War I on communications and resources, and the inability of Canu to work full time on the project due to the 'civil mobilization' of the French people.[11] As their correspondence increased it took on a more personal note which included the exchange of photos of themselves and other bryozoan colleagues, trading stamps, postcards, political opinions, war news, and news of their families.[12] By the end of World War I they moved beyond their original agreement and began work on the Philippine bryozoans from the *Albatross* collections[13] and by 1925 they had completed ten publications.

Bassler expressed an interest in travelling to France and meeting Canu as early as 1914,[14] but the trip did not take place until the summer of 1926 (Figures 4 and 5). It was his first trip outside the United States and included not only France but also Germany (where he met Ehrhard Voigt, and Ernst and Eveline Marcus),[15] and Austria. Bassler stayed with Canu for the duration of his visit to France which seems to have cemented their friendship completely.[16] Bassler returned to France twice more[17] before Canu finally visited the United States in the summer of 1930 (Figures 6 and 7), where he spent the first week with Joseph A. Cushman and Bassler in New England and then on to Washington with Bassler for the rest of his stay.[18]

Canu's health began to deteriorate before his trip to the United States,[19] and by

Study on the morphology and biology of the Homo risor

M. Canu chasseur

M. Canu chanteur

M. Canu minéralogiste

M. Canu acteur

M. Canu architecte

M. Canu dessinateur

M. Canu acrobate

M. Canu musicien

M. Canu botaniste

M. Canu pêcheur
de baleines dans l'Ontario

M. Canu peintre

M. Canu brigand

Ruins of Humbugopolis
M. Canu archéologue

M. Canu galant
offre un bouquet de
Genista europea
au photographe

Figure 4 (above and facing page). Photographs taken of Canu at the Palace grounds, Versailles, by Madam Jean (Canu's housekeeper) [this page; bottom right - holding a spray of yellow Broom]; with titles by Canu. Series presented to Bassler on the occasion of his 1926 trip to Europe.

Figure 5. Canu and Bassler in Canu's study, Versailles, August, 1926.

September of that year he decided not to accept any new work since he and Bassler were still working on the Vincentown and Australian monographs and Bassler's duties as Head Curator were taking up more of his time.[20] Their close collaboration ended on 12th February 1932 when according to Bassler '…awakening with a slight headache, which

Figure 6 (right). Canu and Bassler studying the Calvert Formation (Miocene) at Chesapeake Beach, Maryland, USA (Summer, 1930)

Figure 7 (bottom). "A last farewell". Canu and Bassler at Union Station, Washington, D.C. (July 1930)

grew steadily worse, he [Canu] passed away from cerebral hemorrhage within a few hours. Thus came to a close our association in scientific work of almost a quarter of a century'.[21]

Acknowledgements

I wish to thank William Cox and his colleagues in the Office of the Smithsonian Institution Archives for access to Bassler's files and Ronald Lindsey of the Natural History Branch Library, Smithsonian Institution Libraries for help with Web searches.

Notes

All correspondence cited below is in the Smithsonian Institution Archives, record unit 7234, Ray S. Bassler Papers, 1875 to 1961 and undated.

1 K.E. Caster, 'Memorial to Ray S. Bassler', *Geological Society of America Bulletin*, 76 (1965), P167-74, P168-69.
2 http://www.siris.si.edu/; Smithsonian Institution Archives and Manuscript Catalog.
3 Caster, note 1, p. 169.
4 R.S. Bassler, attachment to letter to F. Canu dated 9th March 1928.
5 R.S. Bassler, letter to G. Canu, 16th August 1940.
6 http://www4.nationalacademies.org; recipients of the Daniel Giraud Elliot Medal.
7 F. Canu, postcard to R.S. Bassler, 17th February 1905.
8 F. Canu, letter to R.S. Bassler, 11th May 1909.
9 R.S. Bassler, letter to F. Canu, 30th August 1912.
10 F. Canu, letter to R.S. Bassler, 12th September 1912.
11 F. Canu and R.S. Bassler, 'North American Early Tertiary Bryozoa', *United States National Museum Bulletin*, 106 (1920), 1-879, 2.
12 R.S. Bassler, letter to F. Canu, 14th May 1914, and subsequent letters.
13 R.S. Bassler, letter to F. Canu, 12th November 1918; Bassler also describes the victory celebrations of the previous night in Washington.
14 R.S. Bassler, letter to F. Canu, 14th May 1914.
15 See Winston, figure 2, *this volume*.
16 R.S. Bassler, letter to F. Canu, 4th October 1926; after returning from Europe Bassler wrote '... you are no longer Monsieur Canu but Friend Canu for anybody can be a Monsieur but few a friend like you'.
17 R.S. Bassler, letter to G. Canu, 2nd April, 1932.
18 R.S. Bassler, letter to F. Canu, 5th May, 1930.
19 R.S. Bassler, letter to F. Canu, 30th January 1930.
20 R.S. Bassler, letter to F. Canu, 22nd September 1930.
21 F. Canu and R.S. Bassler, 'New Species of Tertiary Cheilostome Bryozoa from Victoria, Australia', *Smithsonian Miscellaneous Collections*, 93, 9 (1935), 1-54, 1.

North-eastern Pacific bryozoology and the American bryozoologists

Dorothy F. Soule
Hancock Institute for Marine Science, University of Southern California, Los Angeles, CA 90089-0371, USA

1. Early northeastern Pacific scientists

The earliest Bryozoa identifications from the Pacific coast of America were cursory descriptions found in the diverse, mostly paleontological, works of T.A. Conrad,[1] J.B. Trask[2] and W.M. Gabb and G.H. Horn,[3] each independent thinkers, perceptive, rugged individualists with disparate fields of interest, two of them physicians. The organizations that provided the earliest of them with their central motivating forces were the Philadelphia Academy of Natural Sciences, the Pacific Railroad Expedition to California and the Mexican Boundary (the Wilkes Expedition) at the time of the Gold Rush and California statehood, the John W. Audubon Expedition, and the US Army during the Civil War.

Figure 1. Timothy A. Conrad (1803-1877)

The first American Pacific descriptions were preceded by, or contemporaneous with, the descriptions of a few living species of bryozoans from Baja California by d'Orbigny in 1852 and 1853 and 13 from Mazatlán, Mexico, by George Busk in 1856 and 1857, mostly specimens scraped from the Reigan molluscan collections.

Studies by the British specialist Thomas Hincks in 1882 through 1884 on the "Polyzoa of the Queen Charlotte Islands" were based on collections provided by Dr G.W. Dawson of the Canadian Geological Survey. He provided descriptions of 95 species from British Columbia, many of them new, but as was the fashion of the times, descriptions were brief and illustrations rudimentary in most cases.

Later Dr Charles H. and Elsie O'Donoghue in 1923 to1926 published lists of about 210 species and varieties of bryozoans from the Vancouver Island region and Puget Sound, Washington, but with relatively few illustrations. Their collections were deposited at the Natural History Museum, London, but were only re-located in recent years for comparative studies.

2. Pioneering American scientists of the nineteenth century

2.1 Timothy A. Conrad (1803-1877)

Timothy Conrad (Figure 1), the eldest of ten children, was born near Trenton, New Jersey at his grandmother's home, but lived in Pennsylvania where his father was founder of a publishing firm. He was educated in private schools, was fluent in Latin, Greek and French, but did not study formally at a college or university. However, his parents' home was a center for discussion for learned friends who were interested in natural history, and as a boy he was president of a juvenile scientific society. Before he was 30, he was recognized as one of the foremost conchologists in the country and granted membership in the Philadelphia Academy of Natural Sciences.[4] His sister, Susan, hand-coloured the 17 plates in his first volume published in 1831.

At the Philadelphia Academy he met several wealthy amateur collectors who invited him to come to North and South Carolina and the new state of Alabama to look at the fossils on their extensive lands; he later travelled to Florida and Tennessee. Conrad, who was essentially a loner and never married, existed mostly on the good will of his wealthy patrons. Travelling on a pittance, by sea, on horseback, by wagon or walking, often in ill-health from exploring the coastal swamps unpleasant with a "miasma" of disease, he mapped the famous Claiborne (Alabama) sands, establishing the extent of Tertiary boundaries in the southeastern United States.

Conrad wrote poetry extensively while travelling. He was also involved in contentious disagreements with colleagues who published descriptions of specimens he collected and sent home before he could return.[5]

Conrad never ventured to California, but identified palaeontological specimens collected during the Charles Wilkes Expedition, also known as the Pacific Railroad Expedition to California and the Mexican Boundary, for which he received the great sum of $10. It is from this report[6] that we have the first California record of a bryozoan, one of the most recognizable of the California Pleistocene and Recent fauna, *Idmonea californica* (d'Orbigny, 1853), (= *Diaperoecia*, now *Diaperoforma*).

Conrad served in and then headed the New York State Geological Survey from 1837 to 1842, for which he was paid $1,500 per year, the only real salary he ever had. He invested in railroad stocks which fortuitously made his last years relatively comfortable. The New York survey was something of a fiasco, however, when district palaeontologists dumped their collections on him to identify, at which time he resigned.

Conrad apparently spent several intervals at the Smithsonian Institution but was not paid for those journeys. However, his impact in Washington was important, for in 1838 Conrad was one of those who helped found the now-prestigious American Association for the Advancement of Science (AAAS).

A reclusive figure, Conrad retired to live either at his brother's home in Philadelphia or at his modest retreat in Trenton, New Jersey, but continued to publish through the

Figure 2. John Boardman Trask (1824-1879)

Proceedings of the Philadelphia Academy of Natural Sciences. In 1870-1871 he served as volunteer assistant to the head of the North Carolina geological Survey, living with the family of the State Geologist in Raleigh. It was later reported by the family's small daughter that he subsisted on a diet of bran, biscuits and sour milk.

Conrad's publications totalled 197, including his poetry, and he was a prodigious letter writer, leaving an extensive correspondence with major national and international scientists and scientific institutions of his time. At his death, his small obituary remarked on his modest and unobtrusive life as a scholar, scarcely known outside his family, while he was known and honoured by the great and learned around the world. The only known picture of him is from a photograph poorly retouched by a member of his family.[7]

2.2 John Boardman Trask (1824-1879)

Trask (Figure 2) was born in Roxbury, Massachusetts, in 1823 or 1824[8] and came to California as a physician with the John Woodhouse Audubon party (son of the ornithologist, John J. Audubon), arriving in 1849, the year gold was discovered in California. During his travels he collected botanical specimens, particularly medicinal flora, and fossils. The party travelled from New York to Baltimore by boat, overland to Cairo (Illinois), down the Mississippi River to New Orleans, across Mexico on the Rio Grande, the Gila River and the Colorado River, and on to San Diego and San Francisco. The group was plagued by desertions, and also by cholera, which Trask was able to treat with some success although he was considered an eccentric, probably for his views on sanitation. He settled in San Francisco permanently.

Trask's education as a physician was typical of the times, wherein he apparently apprenticed to a physician, Dr John Adams McLean, in Norwalk, Connecticut, and then attended Yale College for a year, in 1846-1847. During that period he studied natural philosophy, had access to geological collections and studied mineralogy, botany and chemistry as well as the usual medical subjects taught at that time, also writing a medical thesis. Yale conferred an honorary doctorate on him in 1859.

Trask took the lead in establishing the California Academy of Sciences in San Francisco with seven others in 1853, and described Recent and fossil molluscans from the State in the first volume of the Academy's *Proceedings* in 1855-56.

Trask's small paper on nine new species of zoophytes from San Francisco Bay, five of which were bryozoans, was published by the California Academy.[9] He described the common *Crisia occidentalis*, *Scrupocellaria californica*, *Crisidia gracilis* (now *Filicrisia*; ? = *F. franciscana* (Robertson, 1910), *Menipea occidentalis* (now *Tricellaria*) and *Hippothoa amabilis*.

At a time when most arrivals in California were bent on profiting from the "Gold Rush", it is remarkable that these venturesome men, who were considered as "akin to cranks, pottering with shells and bugs and posies"[10] established the first Academy of Sciences west of the Mississippi in San Francisco. In 1858, Trask and Dr David Wooster established the first medical journal in California, which Trask edited for many years.

Trask began an unofficial geological survey of the Sacramento and San Joaquin valleys and the California coast between the 41st and 42nd degrees north latitude, after the State Legislature had failed to fund it; hence the disputed title given to Trask as the first State Geologist. Trask was initially interested in the mineralogy and mining areas of the State, and in underground water resources, but became entranced with the geological record of earthquakes. He calculated the extent of current earthquakes by using the telegraph to inquire where and at what time local telegraph offices felt the quakes,[11] and is thus considered one of the pioneers in California seismology.[12]

Trask served in the Civil War as a contract officer with the rank of Acting Assistant Surgeon, mainly at Alcatraz Island, the Presidio and Alameda County, California, and then signed up for duty in Washington, D.C. His treatment of patients with cholera and diarrhea was considered radical, as he experimented with doses of bismuth (precursors of modern digestive nostrums). His complaints about lack of camp sanitation standards and his insistence on discharging wounded and depressed soldiers rather than returning them to the lines led to increasing disagreements with the military and he requested cancellation of his contract.

Figure 3. William More Gabb (1839-1878)

Figure 4. George Henry Horn (1840-1897)

Trask returned to San Francisco to practice medicine and gradually disappeared from scientific circles until his death at the age of 55 in 1879, which went almost unremarked locally. However, he had been given recognition by various European and American scientific societies and had honorary degrees from Italian and German colleges, awarded for discoveries in organic chemistry, mineralogy, microscopy and medical botany.[13] The rigors of his extensive travels and explorations probably shortened his life.

Trask's collections and records deposited at the California Academy of Sciences were lost when the Academy building, donated by James Lick, was destroyed in the fires that followed the earthquake of 1906.[14] Lick is better known for the observatory named for him.

2.3 William More Gabb (1839-1878)

William Gabb (Figure 3) was born and raised in Philadelphia, Pennsylvania, and was acquainted early on with the Philadelphia Academy of Natural Sciences. He attended Philadelphia High School, from which he received a B.A. degree in 1857, a different concept of high school from modern standards. He then determined to specialize in geology and palaeontology, studying under James Hall in Albany, New York. He returned to his home in Philadelphia, but later spent some time at the Smithsonian Institution in Washington.

In 1861 Gabb was selected to serve as Paleontologist for the Geological Survey of

California; at 22 years of age he was considered to be the foremost authority on Cretaceous marine paleontology in America, with 22 papers already published. He would have known Timothy Conrad and his future collaborator, George H. Horn, at the Philadelphia Academy of Sciences[15] and may have known Horn in high school.

Gabb remained in California for six years, covering hundreds of miles on horseback with his companion, William H. Brewer. It was from this period that descriptions of six species of Bryozoa were published, *Semitubigera tuba* (now *Tubulipora*), *Crisia serrata* (now *Crisia serrulata*), *Phidolopora labiata* (? = *P. pacifica* (Robertson, 1910), *Idmonea californica* Conrad 1855 (=*Diaperoecia,* now *Diaperoforma*), and *Lichenopora californica* (Gabb and Horn, 1862), probably *Disporella californica* now. Geological results were published by the Geological Survey of California in 1864 and 1869.

Gabb went on to explore and map the peninsula of Baja California in 1867, travelling the length by mule and traversing its width 10 times. He combined research in geology, topography, natural history, palaeontology and ethnology in his reports. He then made several trips to Santo Domingo in the Caribbean, and explored Costa Rica, creating detailed maps of the lands he explored.

Gabb apparently fought "consumption" (tuberculosis) and "coastal fever" (possibly this was malaria), until pneumonia ended his life in Philadelphia at the young age of 47 years. He left a written legacy of 88 publications.[16]

2.4 George Henry Horn (1840-1897)

George H. Horn (Figure 4) was born in Philadelphia in 1840, and received his education at Central High School, where he received a Bachelor of Arts and Masters degrees. He then received his medical degree from the University of Pennsylvania in 1861. Horn enlisted in the U.S. Army as an Assistant Surgeon in 1863, and was sent to California, where he was promoted to Surgeon. He remained in the Army until 1866, spending time in Arizona and New Mexico as well as California. Interested in natural history in his youth, he became an expert in Coleoptera. He encouraged the foot soldiers in collecting wherever he was posted as Surgeon, thus procuring extensive collections and many rare specimens.[17]

On Horn's return to Philadelphia, he established an obstetrical and gynocological practice, and was considered an expert in difficult cases. He never married, working long hours on his entomological collections at night, and retaining a principal interest in the California fauna. He was long connected with the Philadelphia Academy of Natural Sciences where he served in various positions, and was a prominent member of the American Philosophical Society and the American Entomological Society.

Figure 5. Alice Robertson (1859-1922)

Horn's collaboration with William Gabb[18] on Bryozoa was probably as illustrator, at which he was considered skilled,[19] but not artistic.[20] He published about 240 papers, only six of which were not entomological. Most were monographic, and he briefly described about 240 genera and 1,550 species. He was always interested in evolutionary relationships.

Horn's early death at the age of 57 from apoplexy was attributed to the intense schedule he kept in his medical practice, coupled with the extensive work he put into his monographic studies of Coleoptera. The latter which made him a worldwide authority in this group.

3. Early twentieth century California scientists

3.1 Alice Robertson (1859-1922)

Alice Robertson, of Scottish ancestry, was a pioneer in America as a woman in science.[21] Records of her early life have not been found, other than that she was born in Philadelphia, according to *American Men of Science*.[22] We have no information as to when she went to California, but she received her B.S. in 1898 when she was almost 40 years old, her M.S. in 1899, and her Ph.D. in 1902, all from the University of California, Berkeley (UCB). She continued at Berkeley as Assistant in Hygiene for Women in 1902-1904, and Assistant in Zoology from 1904 to 1906. During her affiliation with Berkeley, she produced some of her best bryozoan research.[23]

Robertson left Berkeley in 1906 to become an instructor in Zoology at Wellesley College in Wellesley, Massachusetts, and served as an Associate Professor and Chair of the Zoology Department from 1909 until she left due to ill health in 1919 [reputed to have been from the worldwide influenza epidemic]. During her Wellesley years her publications continued from Berkeley.[24] Her research was painstaking and her illustrations accurate. At Wellesley she was regarded as an excellent and kindly teacher, with a hearty sense of humour and widely ranging interests in literature and politics.[25]

After a year's illness, Robertson spent some months at the University of Washington laboratory at Friday Harbor on Puget Sound. She then returned to the University of California, Berkeley, where she was associated with their Department of Public Health. She was also a zoology assistant with the Marine Biological Association of the San Diego's marine station, which was then directed by a UCB faculty member on a rotating basis (now UCB's Scripps Institute of Oceanography). Although her health was fragile, she continued her research and published a report on a collection of Bryozoa sent to her from the Bay of Bengal in 1921. She died unexpectedly in hospital following surgery, in 1922.

Robertson's publications on the Bryozoa of the west coast of North America from Alaska, and from Puget Sound to southern California[26] provided the only accurate descriptions of fauna with detailed illustrations prior to the monographs of Osburn.[27] Those of earlier investigators were limited in descriptions or lacked adequate illustrations. She made detailed embryological studies of the cyclostome genus *Crisia* for her dissertation.[28]

Unfortunately, Robertson's type specimens and collections were subsequently lost at the California Academy of Science in San Francisco. Untrained volunteer personnel placed all specimens labelled with the same species name, regardless of who had identified them or where the specimens were collected, in the same container to save space and money when new containers were purchased. The loss of her types has presented considerable difficulty for later investigators.

3.2 Raymond Carroll Osburn (1872-1955)

Raymond Osburn was born in the small town of Newark, Ohio. He was the first of the west coast Bryozoa research men who had not been born in the eastern United States and was not a geologist, palaeontologist or physician. He obtained his Bachelor of Science in 1898, his Master of Science in 1900, and an Honorary D.Sc. in 1949 from Ohio State University in Columbus.[29] His Ph.D. was awarded by Columbia University, New York in 1906 for work on the origin of verebrate limbs, particularly those of primitive sharks.

Osburn taught at Fargo College (North Dakota), at Barnard College in New York City, and at Connecticut College for Women before becoming Associate Director of the New York Aquarium under the famous director, Henry Fairfield Osborn. In 1917, he returned to Ohio State University, where he served as Chairman of Zoology and Entomology until

Figure 6. Raymond Carroll Osburn (1872-1955)

his retirement in 1942, and as summer Director of the Franz Theodore Stone Laboratory of Ohio State on Lake Erie, from its establishment in 1918 until 1937.[30]

Osburn had already retired from a distinguished career when he was invited in 1945 to come to the University of Southern California (USC) Allan Hancock Foundation as a Research Associate to study their extensive collections of eastern Pacific Bryozoa. Most of their collections were from 10 research cruises, with more than 1500 biological dredge stations and 2000 bottom samples, carried out by Captain G. Allan Hancock's ship *Velero III* in Oregon, California, Baja California, the Caribbean, Galapagos Islands and Peru. Although Hancock was a self-taught captain, licensed for any ship on any sea, he always employed a fulltime captain on all cruises. He later commissioned building of a research vessel, *Velero IV*, which made hundreds of research cruises in southern California before being decommissioned and taken to Alaska as a fishing vessel.

Hancock, a Los Angeles landowner and oil developer (the famous La Brea tar pits were on his ranch), had extensive agricultural operations in Mexico and had sailed often aboard his own freighter or his yachts. In the 1930s, he was persuaded by the California Academy of Sciences to undertake trawling for fishes, and a USC biology professor, Dr Irene McCulloch, asked if her graduate students might accompany them to collect invertebrates.

In 1939 the Captain completed a four-storey research building at USC to house his collections, and in 1945 purchased most of the defunct Boston Society of Natural History library, providing extensive literature that was crucial to Osburn's studies.

Osburn, with the aid of USC graduate students, also assembled a prodigious array of collections donated or on loan from various sources: the U.S. National Museum "Albatross' Fish Survey stations; the Alaska Crab Investigation; Prof. George E. and Nettie MacGinitie of the U.S. Navy's Arctic Marine Laboratory at Point Barrow, Alaska; the Pacific Biological Station at Nanaimo, British Columbia; the Oceanographic Laboratory (University of Washington) at Friday Harbor, Washington; Pacific Marine Station, Dillon Beach, California; Robertson's collections from the California Academy of Sciences; E. Alice Blagg's collection (unpublished) from Hopkins Marine Station (Stanford University), Pacific Grove, California; the Los Angeles County Museum of Natural History; Kerckhoff Marine Station (California Institute of Technology), Corona del Mar, California; Scripps Oceanographic Institution (University of California), La Jolla, California; and the Galtsoff Collection of the U.S. Fish and Wildlife Service from Panama. Most earlier collections were from shallow waters, making the *Velero* biological dredge samples particularly important.

Osburn retired for the second time in 1952, at the age of 80. The results of his studies were published in three volumes, *The Bryozoa of the Pacific Coast of America*,[31] for which he was awarded the Daniel Giraud Elliott Medal by the U.S. National Academy of Sciences in 1953, two years before his death.

Osburn's interests were wide ranging, from evolution, fisheries and insects to bryozoans. His early interest in Bryozoa began in summers spent at Woods Hole, Massachusetts, and publications covered Labrador, Newfoundland, Nova Scotia, the Canadian Arctic, Hudson Bay, and the Caribbean between 1912 and 1947 prior to undertaking the west coast studies.

A strong supporter of scientific societies, Osburn held many important posts. He was President of the American Microscopical Society, the American Limnological Society, the Ohio Academy of Sciences and the Ohio Chapter of Sigma Xi; he was Secretary of the American Fishery Society; Vice President of the Ecological Society of America; and active in may others. He also was for years the editor of the Bryozoa Section of *Biological Abstracts*.

Osburn's avocations were wide ranging as well; he and his wife, Mary Hubbell Osburn, whom he met while teaching in Fargo, were musicians, sometimes indulging friends and students in piano-violin duets, and he wrote lengthy poetic commentaries on all sorts of subjects. Although they had no children, they quite enjoyed the offspring of graduate students, being Uncle Ray and Aunt Mary to them, and the tall, slim gentleman dressed up as Santa for the annual Christmas party. John Soule was Osburn's last graduate student.

4. Latter twentieth century bryozoologists

4.1 John Dutcher Soule (1920-2001)

Born in Moline, Illinois, John Soule (Figure 7) graduated from Western Military Academy in Alton, Illinois and received a Bachelor of Arts from Miami University, Oxford, Ohio in 1942. He served four years in the US Army, as a medical field and laboratory officer in the Philippines and the occupation of Hiroshima-Kure, Japan. Soule came as a post-war graduate student to the University of Southern California (USC), where he met and studied under R.C. Osburn, receiving a Master of Arts in 1948 and a Ph.D. in 1952. He held a post-doctoral fellowship at the California Institute of Technology Kerckhoff Marine Laboratory in 1958 and 1959 under Dr. George Beadle.

Soule taught Histology and Pathology at the USC School of Dentistry for 41 years where he chaired the Histology Department for 14 years, was chair of Basic Sciences for many terms, and served on occasion as Assistant Dean for Admissions. He conducted research on histological stains, inventing Soule's Gold Chloride stain, studied comparative development of dentition in lower vertebrates, and did early studies on bioadhesives secreted by marine invertebrates including bryozoans, particularly as they applied to coral reef stabilization.[32]

Soule was the first bryozoan researcher in the USA to do extensive histological studies on post-larval development of Ctenostomata, discovering that the sequential development of musculature demonstrated the diphyletic nature of the group. Previously, the Ctenostomata had been separated into four divisions based mainly on adult anatomy or polypide development. He studied muscle sequences in six genera of Carnosa and 13 genera of Stolonifera, comparing them with anascan and ascophoran genera.[33] He wrote the Ctenostomata section in Osburn's monograph.[34]

After extensive histological studies on burrowing bryozoans[35] he became convinced that the Penetrantiidae were more aligned with the cheilostome line than with ctenostomes (*sensu stricto*), in keeping with the concept on the union of the two groups at the ordinal level expressed by Lars Silén, with whom the Soules consulted.[36] He coauthored several other papers on burrowing bryozoans including one on Cretaceous burrowers with E. Voigt.

Soule authored or co-authored over 100 publications on bryozoans and other topics. His research included zoogeographical and faunal studies of Bryozoa in California and Baja California. Using scanning electron microscopy, he and his wife, Dorothy, and others described new species of Hawaiian Smittinidae, Watersiporidae and Thalamoporellidae worldwide, new genera of tropical Cleidochasmatidae, Cribrilariidae, and other groups, and returned to faunal studies of the Santa Barbara Channel, California, with revisions of

Figure 7. John Dutcher Soule (1920-2001)

west coast species formerly attributed to Atlantic-Boreal or Mediterranean-Red Sea species.[37] They emphasized the generic and specific differences found in supposed cosmopolitan species, and the importance of recognizing true species diversity as it relates to species diversity calculations for environmental impact evaluations, especially in areas proposed for development.

Soule held joint appointments as Professor of Biological Sciences and Curator of Bryozoa of the Hancock Foundation, was a Research Associate of the American Museum of Natural History, The Santa Barbara Museum of Natural History and the Natural History Museum of Los Angeles County. He was a Fellow of the American Association for the Advancement of Science, the California Academy of Sciences and the Southern California Academy of Sciences. He served several terms as councilor and was a past president of the International Bryozoology Association and was an officer in several other scientific societies. He has been listed in *Who's Who in America, Who's Who in the West* and *American Men and Women of Science*. He was also the recipient of numerous teaching awards, teaching about 200 new dental students each year for 41 years. The Computer Learning Center at the USC Dental School is named for him.

Soule met his future wife when they were undergraduates at Miami University (Ohio), and they have two daughters, five grandchildren and one great grandson. They began collaborating on Bryozoa in 1961, although she had previously done illustrations for him. At the time of his death from Parkinson's syndrome in June 2001 at age 80, he had two joint papers in press and several in preparation.

Henry W. Chaney, now at the Santa Barbara Museum of Natural History, was his last doctoral student, and Soule served on doctoral committees of several USC graduate students who have published on bryozoans. These were Robert Woollacott of Harvard University, William Banta of American University in Washington, DC and Michael Temkin of St. Lawrence University in New York.

4.2 Christopher Gardiner Reed (1951-1990)

The first of the bryozoologists born on the west coast, Chris Reed was born in Seattle, Washington in 1951, and grew up on Vashon Island, Washington and in Milwaukee and Lake Oswego, Oregon, giving ample opportunity for his interest in wildlife of all sorts, including a pet skunk.[38]

Reed attended Washington State University for a short time and then transferred to the University of Washington, where he completed his Bachelors and Doctoral degrees. His dissertation was on the reproductive biology, larval morphology and metamorphosis of the ctenostome *Bowerbankia gracilis*. His postdoctoral work was with Robert M. Woollacott at Harvard University on the morphology and mechanics of metamorphosis of coronate bryozoan larvae.

Hired by Dartmouth University in 1982, he remained there until his untimely death in 1990. His career was distinguished by summer fellowships at Woods Hole, Massachusetts, and characterized by excellence in teaching. He was an exciting and dynamic lecturer, yet a kind and approachable person. Reed was awarded the Gross Taylor and Cornelia Pierce Williams Assistant Professorship chair for distinguished teaching, and in 1989 the J. Kenneth Huntington Memorial award as Dartmouth's outstanding teacher, when he was awarded tenure. His research was productive also, for he published 36 papers, some coauthored with R.M. Woollacott and R.L. Zimmer on bryozoan larvae and with others on brachiopod, nemertean and polycheate larvae.[39]

Zimmer stated that "Chris would be remembered for discoveries of intercoronal and juxtapapillary cells, the median band of mesoderm, and the dual polypide and cystid rudiments of bryozoan larvae; for his exciting studies on bryozoan metamorphosis including the elucidation of the roles of ciliary reversal, muscles and myoepithelial cells in the rapid morphogenetic movements of early metamorphosis; and for the elegance of his electron micrographs, the detail of his drawings, and the clarity of his writing."[40]

In 1986 Reed learned that he had cancer of the lymph system and he waged a courageous battle, with radiation, chemotherapy and an autologous bone marrow transplant that followed an unsuccessful international search for a bone marrow donor. He was married during this period to Paula Borden and had a daughter, Erica. Chris maintained a joy of life and sense of humour in spite of his ordeal, continuing to inspire (and tease) students, friends and colleagues throughout his short life span. He died shortly before his 39th birthday.

5. Contemporary bryozoologists

5.1 June R.P. Ross

In contrast to earlier bryozoan researchers, June Ross came to the western United States from Australia, by way of the east coast. June was born and raised in Australia. She received her Bachelor of Science in 1953, her Ph.D. in 1959, and a Doctor of Science in 1974, all from the University of Sydney in Australia.

June arrived in the USA in 1957 as a Research Fellow of the Peabody Museum at Yale University, where she was an American Association of University Women grantee and held the first of several National Science Foundation grants. From 1960 to 1965 she was a Research Associate at the University of Illinois, working on Middle Ordovician ectoprocts from New York State and Ordovician and Silurian ectoprocts from the central states. She then moved to Western Washington State University in Bellingham, Washington, serving there first as a Lecturer and as a Research Associate, becoming an Associate Professor in Biology in 1967 while studying Recent and fossil ectoproct phylogeny. She became a Professor in 1970, and is also a Research Associate of the University of Washington.

In addition to Biology and Zoology courses, June's teaching duties have included Evolution, Oceanography, History of Science, Electron Microscopy, Population Biology, and Human Populations and Natural Resources.

June has spent time at numerous research stations, including Friday Harbor Laboratory (University of Washington), the Lerner Laboratory in Bimini (American Museum of Natural History), the University of Georgia, Roscoff Marine laboratory in France, Scripps Institute of Oceanography in La Jolla (University of California), and in Japan and China as well as at industrial laboratories

June hosted thc 7th International Bryozoology Association Conference at Bellingham in 1986.[41] She was president of the IBA in 1992-1995.

June's scientific interests have ranged widely: from Palaeozoic to Recent bryozoans to foraminifera, fusulinids, corals and entomology; to plate tectonics, calcification, evolution, ecology and palaeoecology; and to sea level changes, sedimentology, biostratigraphy, biogeography and palaeobiogeography. The geographic coverage ranges through time virtually worldwide, from Pangea through the Tethys, to Asia, Australia, the United States, Canada, the Middle East, and Europe.

June's publications include 138 titles beginning in 1957, some co-authored with her husband, Charles A. Ross.[42] Other co-authors include Mathew Dick, who was her Masters student and is now on the faculty of Middlebury College, Middlebury, Vermont. She has written sections for Encyclopaedia Brittanica, McGraw Hill and Encyclopedia of Science and Technology.

June is listed in *Who's Who in the West*, *Who's Who in America and American Men and Women of Science*. Her other interests include community planning for low income housing, family planning and medical service, and the League of Women Voters.

5.2 Dorothy F. Soule

Born and raised in Lakewood, Ohio on Lake Erie, Dorothy Soule received her Bachelor of Arts in 1945 from Miami University, Oxford, Ohio, where she met John Soule when she was a freshman and he a senior zoology laboratory assistant. They married after her second year when he was in the US Army. They had two children while John attended Graduate School at the University of Southern California (USC) after World War II. She received her Master of Arts from Occidental College in Los Angeles in 1963 and her Ph.D. from Claremont Graduate University in Claremont, California in 1969 while their daughters were attending university. They have five grandchildren and one great grandson.

Dorothy taught embryology and comparative anatomy as an undergraduate at Miami University during World War II, when many faculty were drafted and married women were not asked their age. She moved to California when husband John shipped out to the Pacific theatre and became a Research Associate at USC in the Biochemistry Department of the Medical School. She later taught comparative anatomy and microbiology labs at Occidental College, zoology, invertebrate zoology and embryology at California State College, Los Angeles and physiology at California College of Medicine before returning to USC's Allan Hancock Foundation in 1965 to do field and laboratory research on tropical Pacific bryozoans with John Soule.

Dorothy has authored or co-authored over 150 publications, beginning in 1947. Her byrozoan papers have emphasized morphology and speciation, zoogeography of tropical oceans worldwide and temperate eastern Pacific fauna, post-larval development and natural and manmade environmental impacts on species diversity.[43]

Dorothy is a Research Professor of Marine Biology and Curator of Bryozoa, and was an Adjunct Professor of Environmental Engineering for 24 years. She served as Associate Director of the federal Sea Grant Program at USC for several terms and is on their advisory board. She is a Research Associate of the Natural History Museum of Los Angeles County and the Santa Barbara museum of Natural History, and is a Fellow of the Southern California Academy of Sciences.

In response to national and state legislation on environmental quality in 1970, she established Harbors Environmental Projects at USC, bringing together multidisciplinary teams of marine biologists, oceanographers, geologists, microbiologists and environmental chemists from USC and other local universities to assess the impacts of public and private coastal sewage and industrial waste discharges, fish processing waste management, oil spills, a tanker explosion, use of dispersants, and coastal zone planning and management.

Part of the Environmental Impact Report (State of California) or Environmental Impact Statement (Federal) process is to determine species diversities by various methods. This led the Soules to reevaluate so-called cosmopolitan species of bryozoans. Examination by scanning electron microscopy led to identification of multiple new genera and species, and restricted distribution for former cosmopolitan species. Consulting and research gave them opportunities to collect bryozoans widely in the Pacific Islands. Bryozoan publications have emphasized morphology elucidated by scanning electron microscopy, speciation and worldwide distribution of several tropical and subtropical families.

Dorothy established, contributed to, and edited 30 volumes *of Marine Studies of San Pedro Bay* on southern California marine environmental quality, and she is an editor of the *Irene McCulloch Foundation monographs* published with the Hancock Institute for Marine Studies. She served on the National Academy of Engineering Panel on Deep Sea Mining, the National Science Foundation Panel on Systematic Resources, the National Marine Fisheries Advisory Board, the Environmental Protection Agency Science Advisory Board Task Group on Marine Monitoring, the National Academy of Sciences-National Research Council Panel on Monitoring in Southern California, the Association of Systematics Collections Council on Environmental Quality, and chaired the National Marine Fisheries Anchovy Management Panel. She also served on several local Los Angeles Harbor and Marina del Rey planning and management task groups. She has been listed in *Who's Who in America* and *American Men and Women of Science.*

Dorothy is active as a speaker on marine biology, IndoPacific bio-geo-politics and El Niño/global warming for USC's Emeriti Center outreach program and serves on boards of several University and other organizations.

5.3 Russel L. Zimmer

Russel Zimmer is also a midwesterner, born in Springfield, Illinois in 1931. Russ graduated from Blackburn College at Carlinville, Illinois in chemistry and mathematics but moved west to earn his Masters of Science in Zoology at the University of Washington in Seattle in 1956 and his Doctorate in Zoology there in 1964. He then served two years in the U.S. Army at Fitzsimmons General Hospital in Denver, Colorado, doing experimental vascular surgery research.

Zimmer came to the University of Southern California (USC) in Los Angeles in 1960 as an Instructor, successively becoming an Assistant Professor, Associate Professor and Professor. From 1968 to 1976 he served as Director of USC's Catalina Marine Science Center on Santa Catalina Island (now the Philip K. Wrigley Marine Science Center). He subsequently served on many university and departmental committees and as Vice Chair and Chair of the Department of Biological Sciences. Most summers he returned as a Visiting Professor at the University of Washington laboratories at Friday Harbor on Puget Sound, and spent a year as Visiting Professor at Harvard University. Russ retired from teaching in 2001 but continues to be involved in research at Friday Harbor and in university activities.

Zimmer's dissertation and research concerned reproductive biology and development of phoronids, with interest in other lophophorates including bryozoans focusing on embryogenesis, metamorphosis, larval photoreceptors,[44] and on phylogenetic relationships. With his former graduate student, Robert M. Woollacott, Professor of Organismic and Evolutionary Biology at Harvard University, he published developmental and physiological studies.[45] Perhaps their most outstanding collaboration was in their *Biology of Bryozoans*[46] which was a compilation of articles by bryozoan biologists on the state of information at that time. He has contributed articles to a number of edited volumes on metamorphosis and invertebrates as well. Other former graduate students of bryozoans include William C. Banta, Professor of Biology at American University, and Michael Temkin of Saint Lawrence University. Other invertebrate (non-bryozoologist) students included Gary J. Brusca (now deceased), author of a prominent American invertebrate biology textbook, and diverse students who have published on diatoms, barnacles, polychaetes, abalone, cumaceans, copepods, phoronids and decapods.

Zimmer was honoured by his former students and colleagues at a symposium dedicated to him at the American Society of Zoology national meeting in December, 2001, organized by his last doctoral student, Scott Santagata, now a post doctoral fellow at the University of California, Los Angeles (UCLA).

Zimmer married Cheryl Hildebrand, whom he met at USC, in 1967; she was fluent in French and willing to translate obscure manuscripts. She cheerfully helped Russ in dealing with students, faculty and many visitors at the then-new Catalina Marine Science Center when the faculty facilities in those years were a collection of decrepit trailers. They have two sons born during the Catalina experience. The only public school was an hour's bus ride on a dirt mountain road to the little town of Avalon on Santa Catalina Island. The island is only about 30 miles from megalopolis Los Angeles by small boat or an adventurous helicopter ride, but far away in lifestyle.

5.4 Penny A. Morris

Penny Morris was born and educated in California. She holds a Bachelor of Arts from California State University, Los Angeles (1964), a Master of Arts in marine biology from San Francisco State University (1972), and a Ph.D. in palaeontology from the University of California, Berkeley (1975). Her undergraduate interest in bryozoan taxonomy came in a course taught by Dorothy Soule, for whom she continued to work on environmental projects at the University of Southern California between degrees. Dorothy also served on her dissertation committee at UC Berkeley under J. Wyatt Durham. Penny also worked as a Research Assistant in the Woods Hole Systematics-Ecology Program and at the University of Massachusetts, Amherst, in Botany. She participated in research and collecting cruises to the Mediterranean and to Baja California during these years.

Penny lived in Sumatra and other Asian countries where her husband, David Smith, was employed in the oil industry, and her two daughters were born abroad. On her return to the United States, she taught various biology, botany, geology and paleontology courses at North Harris College from 1984 to 1989, and from 1988 to the present at the University of Houston-Downtown.

Her research interests have ranged from the California *Celleporina* and *Hippothoa*[47] to bryozoan faunas in California,[48] Baja California, thermal vent fauna off Newfoundland, and Warrawoona Group fossils in Australia. Studies, including substrate relations with algae, hermit crabs, molluscs,[49] bacteria and rock, have led to research on organic biomarker molecules in specimens from Mars and the signals of organic matter in evaporates. The latter investigations resulted in NASA grants and included fieldwork in the Dead Sea area.

Penny is interested in minority education and research opportunities in Space science, and organized a community outreach programme funded by NASA in collaboration with Hispanic Serving group and a coalition of historically black universities.

5.5 Henry W. Chaney

Henry Chaney was born in the San Francisco area and raised in Santa Barbara, California. He receive a Bachelor of Arts from Claremont McKenna College, Claremont, California in 1976, a Masters of Science from the University of Oregon in Eugene, Oregon in 1977, and a Ph.D. at the University of Southern California in 1983. His major undergraduate professor was a former student of Dorothy Soule at California State University, Los Angeles, and he recommended that Chaney undertake doctoral studies with John and Dorothy at USC in Bryozoa although he was already known as a malacologist. His dissertation research was on histocompatability in *Thalamoporella californica* colonies.

After post-doctoral research positions at USC, Henry was appointed to the Howard-Berry Chair in Malacology at the Santa Barbara Museum of Natural History and is now also the Director of Collections and Research there. He oversaw construction and the equipping of a new building to house the invertebrate collections, and has implemented obtaining significant invertebrate collections from donors and from public agencies following environmental evaluations in California.

An expert SCUBA diver, Chaney has travelled widely in the tropical Pacific, western Australia, the Indian Ocean, Egypt and Saudi Arabia. He has published on tropical molluscs, and has been a co-author on Bryozoa with the John and Dorothy Soule.[50]

Chaney has served variously as President, Vice President and Treasurer of the Western Society of Malacologists, Secretary of the California Malacozoological Society and on other organization boards. He is also President of the Irene McCulloch Foundation and Director of the Terra Marine Research Foundation.

6. Acknowledgements

I would like to thank Susan S. Harrison for introducing me to internet search engines, and Jean Crampon of the Science and Engineering Library, University of Southern California, for obtaining some of the obscure documents that have contributed to this effort. Rick Hayduk assisted with graphics.

Notes

1. T.A. Conrad, 'A note on the Miocene and Post-Pliocene deposits of California with descriptions of two new fossil corals', *Proceedings of the Academy of Natural Sciences, Philadelphia*, 7 (1855), 441.
2 J.B. Trask, 'Nine new species of Zoophytes from the Bay of San Francisco and adjacent localities', *Proceedings of the California Academy of Sciences*, series 1, 1(2) (1857*b*), 100-102.
3 W.M. Gabb and G.H. Horn, 'Monograph on the fossil Polyzoa of the Secondary and Tertiary formations of North America', *Journal of the Academy of Sciences, Philadelphia*, new series, 5 (2) article 3 (1862), 111-179.
4 H.E. Wheeler, 'Timothy Abbott Conrad, with particular reference to his work in Alabama one hundred years ago', *Bulletins of American Paleontology*, 23 (77) (1935), 1-158.
5 Wheeler, note 4.
6 Conrad, note 1.
7 Wheeler, note 4.
8 R.E.C. Stearns, 'Dr. John B. Trask, a pioneer of science on the west coast', *Science*, n.s., 28 (712) (1908), 240-243; A.E. Leviton and M.L. Aldrich, 'John Boardman Trask: physician-geologist in California, 1850-1879', in *Frontiers of Geological Exploration of Western North America*, edited by A.E. Leviton, P.U. Rodda, E. Yochelson and M.L. Aldrich (Pacific Division American Association for the Advancement of Science, San Francisco, 1982), 37-69.

9 Trask, note 2.
10 Stearns, note 8.
11 J.B. Trask, 'On the direction and velocity of the earthquake in California, January 9, 1857', *Proceedings of the California Academy of Sciences*, series 1, 1(2) (1857*a*), p. 98.
12 Leviton and Aldrich, note 8.
13 A.W. Vogdes, 'A bibliographic sketch of Doctor John B. Trask', *San Diego Society of Natural History*, 1(2) (1907), 27-30.
14 Stearns, note 8.
15 Anon. 'Prof. W.M. Gabb', *Nature*, 18(454) (1878), 285.
16 Gale Group. *Dictionary of American Biography Base Set.* American Council of Learned Societies, 1928-1936. William More Gabb 1839-1878. Reproduced in *Biography Resource Center*, Farmington Hills, Michigan (2001).
17 Gale Group, *Dictionary of American Biography Base Set.* American Council of Learned Societies, 1928-1936. George Henry Horn 1840-1897. Reproduced in Biography Resource Center, Farmington Hills, Michigan (2001).
18 Gabb and Horn, note 3.
19 Anon. 'Scientific notes and news', *Science*, n.s, 6(153) (1897), 839-845; Anon. 'George Henry Horn', *Psyche*, 8(261) (1898), 159-160; Anon, 'Dr. George H. Horn', *Entomological News and Proceedings of the Entomological Section, Academy of Sciences, Philadelphia*, 18(1) (1898), 1-3.
20 J.B. Smith, 'George H. Horn', *Science*, n.s., 7(160) (1898), 73-77. 2001.
21 M.R.S. Creese, 'Alice Robertson', in *Ladies in the Laboratory? American and British Women in Science, 1800 –1900*, edited by M.R.S. Creese (The Scarecrow Press, Lanham, MD and London, 1998), 93.
22 J. McK. Cattell (ed), 'Robertson, Prof. Alice', in *American Men of Science*, 2nd edition (The Science Press, New York, 1910), 397.
23 A. Robertson, 'Papers from the Harriman Alaska Expedition 1900', *Proceedings of the Washington Academy of Sciences*, 2 (1900) 315-335; A. Robertson, 'Embryology and embryonic fission in the genus *Crisia*', *University of California Publications. Zoology*, 1(3) (1903), 115-156; A. Robertson, 'Non-incrusting chilostomatous Bryozoa of the west coast of North America', *University of California Publications. Zoology*, 2(5) (1905), 235-322.
24 A. Robertson, 'The incrusting chilostomatous Bryozoa of the west coast of North America', *University of California Publications, Zoology*, 4(5) (1908), 253-344; A. Robertson, 'The cyclostomatous Bryozoa of the west coast of North America', *University of California Publications, Zoology*, 6(12) (1910), 225-284.
25 S.C. Hart, 'In Memoriam', *Wellesley College News*, (1922), 8; H.G. Thomas, 'Dr. Alice Robertson – The Teacher', *The Wellesley Alumnae Quarterly*, 7 (1922), 23.
26 Robertson, note 23 (1900, 1905), note 24 (1908, 1910).
27 R.C. Osburn, *Bryozoa of the Pacific Coast of America Part 1, Cheilostomata - Anasca.* Allan Hancock Pacific Expeditions, 14(1) (University of Southern California, Los Angeles 1950), 1-270; R.C. Osburn, *Bryozoa of the Pacific Coast of America. Part 2, Cheilostomata – Ascophora.* Allan Hancock Pacific Expeditions, 14(2). (University of Southern California, Los Angeles 1952), 271-611; R.C. Osburn, *Bryozoa of the Pacific Coast of America. Part 3, Cyclostomata, Ctenostomata, Entoprocta and Addenda.* Allan Hancock Pacific Expeditions, 14(3). (University of Southern California, Los Angeles 1953), 613-841.
28 Robertson, note 23 (1903).

29 M.D. Rogick, 'Raymond Carroll Osburn', *Transactions of the American Microscopical Society*, 75(1) (1956), 143.
30 Rogick, note 29.
31 Osburn, note 27 (1950, 1952, 1953).
32 J.D. Soule, 'Histological and histochemical studies of the bryozoan-substrate interface', in *Living and Fossil Bryozoa*, edited by G.P. Larwood (Academic Press, London, 1973), 343-347; J.D. Soule and D.F. Soule, 'The bryozoan-coral interface on coral and coral reefs', *Proceedings of the Second International Symposium on Coral Reefs*, volume 1. (Great Barrier Reef Committee, Brisbane, Australia, 1974), 335-340; J.D. Soule and D.F. Soule, 'Fouling and bioadhesion: life strategies of bryozoans', in *Biology of Bryozoans*, edited by R.M. Woollacott and R.L. Zimmer (Academic Press, New York, 1977), 437-457.
33 J.D. Soule, 'Post-larval zoarial development on *Carnosa* (Bryozoa Ctenostomata)', *Bulletin of the Southern California Academy of Sciences*, 52(3) (1953*a*), 88-92; J.D. Soule, 'Post-larval development in relation to the classification of the Bryozoa Ctenostomata', *Bulletin of the Southern California Academy of Sciences*, 53(1) (1954), 13-34.
34 J.D. Soule, 'Suborder Ctenostomata', in R.C. Osburn, *Bryozoa of the Pacific Coast of America.* Pt. 3. Allan Hancock Pacific Expeditions, 14(3). (University of Southern California, Los Angeles, 1953*b*), 726-784.
35 J.D. Soule, 'Penetrantiidae and Immergentiidae from the Pacific, Bryozoa, Ctenostomata', *Transactions of the American Microscopical Society*, 69(4) (1950*a*), 359-367; J.D. Soule, 'A new species of *Terebripora* from the Pacific', *Journal of the Washington Academy of Sciences*, 40(11) (1950*b*), 378-381.
36 J.D. Soule and D.F. Soule, 'Systematics and biogeography of burrowing Bryozoans', *American Zoologist*, 9(3), edition 2 (1969), 791-802.
37 J.D. Soule and D.F. Soule, '*Spathipora*, its anatomy and phylogenetic affinities', in *Bryozoa – 1974*, edited by S. Pouyet (Lyon, France, Documents des Laboratoire de Géologie de la Faculté des Sciences de Lyon, HS3 1, 1975); J.D. Soule and D.F. Soule, 'Bryozoans from seagrass beds off the Arabian coast', *Bulletin of the Southern California Academy of Sciences,* 844(2) (1985), 67-75; J.D. Soule, D.F. Soule and H.W. Chaney, 'New tropical Pacific and Indian Ocean Cheilostomatidae (Cheilostomata, Ascophora): Systematics and biogeography', in *Bryozoaires Actuels et Fossiles: Bryozoa Living and Fossil*, edited by F.P. Bigey and J.-L. d'Hondt, Bulletin de la Société des Sciences Naturelles de l'Ouest de la France, Nantes. Mem. HIS (1991), 465-486.
38 R.L. Zimmer, 'Obituary. Christopher Gardiner Reed', in *Biology and Palaeobiology of Bryozoans*, edited by P.J. Hayward, J.S. Ryland and P.D. Taylor (Olsen and Olsen, Fredensborg, 1994), 233-234.
39 C.G. Reed and R.A. Cloney, 'The role of ciliary motility in the metamorphosis of a bryozoan', *American Zoology*, 20 (1980), 953; C.G. Reed and R.M. Woollacott, 'Mechanisms of rapid morphogenetic movements in the metamorphosis of the bryozoan *Bugula neritina* (Cheilostomata, Cellularoidea). 1. Attachment to the substratum', *Journal of Morphology*, 172 (1982), 335-348; C.G. Reed and R.M. Woollacott, 'Mechanisms of rapid morphogenetic movements movements in the metamorphosis of the bryozoan *Bugula neritina* (Cheilostomata, Cellularoidea) II. The role of dynamic assemblages of microfilaments in the pallial epithelium', *Journal of Morphology*, 177 (1983), 127-143; S.A. Stricker and C.G. Reed, 'The ontogeny of shell sectretion on *Terebratula transversa* (Brachiopoda: Articulata). I. Development of the mantle', *Journal of Morphology*, 183 (1985), 223-250; C.G. Reed, 'Phylogenetic implications

of the morphogenesis of the cystid epithelium during the metamorphosis of gymnolaemate bryozoans', in *Bryozoa: Present and Past*, edited by J.R.P. Ross (Western Washington University, Bellingham, 1987), 221-228; S.A. Stricker, C.G. Reed and R.L. Zimmer, 'The cyphonautes larva of the marine bryozoan *Membranipora membranacea*. I. General morphology, body wall and gut', *Canadian Journal of Zoology*, 66 (1988), 368-383.

40 Zimmer, note 38.

41 J.R.P. Ross (editor), *Bryozoa: Past and Present* (Western Washington University, Bellingham 1987).

42 J.R.P. Ross, 'Larger cryptosotome Bryozoa of the Ordovician and Silurian, Anticosti Island, Canada. Pt. 1', *Journal of Paleontology*, 34 (1960), 1056-1076; J.R.P. Ross, 'Lower Permian Bryozoa from Western Australia', *Palaeontology*, 6 (1963*a*), 70-82; J.R.P. Ross, 'Ordovician cryptostome Bryozoa, standard Chazyan Series, New York State and Vermont', *Geological Society of America, Bulletin*, 74 (1963*b*), 577-608; J.R.P. Ross, 'Champlanian Ectoprocta (Bryozoa), New York State', *Journal of Paleontology*, 41 (1967), 632-648; J.R.P. Ross, 'Calcification in the colonial stenolaemate Heteropora', in *Living and Fossil Bryozoa*, edited by G.P. Larwood (Academic Press, London, 1973), 475-483; J.R.P. Ross, 'Organization of calcified tissues in stenolaemate ectoprocts', in *Bryozoa – 1974*, edited by S. Pouyet (Lyon, France, Documents des Laboratoire de Géologie de la Faculté des Sciences de Lyon, HS3 1, 1975), 179-186; J.R.P. Ross, 'Biogeography of Carboniferous ectoproct Bryozoa', *Palaeontology*, 24 (1981), 313-341; Ross, note 41; J.R.P. Ross, and C.A. Ross, 'Pennsylvanian and Permian Bryozoa from northeast Greenland', *Meddelser om Gronland*, 167(7) (1962), 1-65; J.R.P. Ross and C.A. Ross, 'Bryozoan evolution and dispersal and Paleozoic sea-level fluctuations', in *Bryozoans in Space and Time*, edited by D.P. Gordon, A.M. Smith and J.A. Grant-Mackie (National Institute of Water and Atmospheric Research, Wellington, 1996), 243-258; J.R.P. Ross, M.A. Hamedi and A.J. Wright. 2000. 'Late Ordovician (Caradoc) bryozoans from the Kerman District, East-Central Iran', in *Proceedings of the 11th International Bryozoology Conference*, edited by A. Herrera Cubilla and J.B.C. Jackson (Smithsonian Tropical Research Institute, Republic of Panama, 2000), 329-335.

43 D.F. Soule, 'Morphogenesis of giant avicularia and ovicells in some Pacific Smittinidae', in *Living and Fossil Bryozoa*, edited by G.P. Larwood (Academic Press, London, 1973), 485-495; D.F. Soule and J.D. Soule, 'The Ectoprocta (Bryozoa) of Scammon's Lagoon, Baja California', *American Museum of Natural History Novitates*, 2199 (1964), 1-65; D.F. Soule and J.D. Soule, 'Ancestrulae and body wall morphogenesis of some Hawaiian and eastern Pacific Smittinidae (Bryozoa, Ectoprocta)', *Transactions of the American Microscopical Society*, 91(3) (1972), 251-260; D.F. Soule and. J.D. Soule, 'Morphology and speciation of Hawaiian and eastern Pacific Smittinidae (Bryozoa, Ectoprocta).', *American Museum of Natural History Bulletin*, 152(6) (1973), 365-440; D.F. Soule, J.D. Soule and H.W. Chaney, 'Some uncommon genera of Thalamoporellidaae: *Thairopora*, *Diploporella*, and new genera *Marsupioporella* and *Thalamotreptos*', in *Bryozoaires Actuels et Fossiles: Bryozoa Living and Fossil*, edited by F.P. Bigey and J.-L. d'Hondt, Bulletin de la Société des Sciences Naturelles de l'Ouest de la France, Nantes. Mem. HIS (1991), 447-464; D.F. Soule, J.D. Soule, and H.W. Chaney, 'The genus *Thalamoporella* worldwide (Bryozoa, Anasca): Morphology, evolution and speciation', *Irene McCulloch Foundation Monograph Series*, No. 1 (Los Angeles, Hancock Institute for Marine Studies, University of Southern California, 1992), 1-94; D.F. Soule, J.D. Soule, and H.W. Chaney, 'The Bryozoa', in *Taxonomic Atlas of the Santa Maria Basin and Western Santa Barbara Channel*, volume 13, edited by J.A. Blake, H.W. Chaney, P.H. Scott and A.L. Lissner

(Santa Barbara, Santa Barbara Museum of Natural History, 1995) 344 pp. Reprinted as *Irene McCulloch Foundation Monograph Series*, No. 2 (Los Angeles, Hancock Institute for Marine Studies, University of Southern California); D.F. Soule, J.D. Soule, and H.W. Chaney, 'New species of *Thalamoporella* (Bryozoa) with acute or subacute avicularium mandibles and review of known species worldwide', *Irene McCulloch Foundation Monograph Series*, No. 4 (Los Angeles, Hancock Institute for Marine Studies, University of Southern California, 1999), 1-57; D.F. Soule, J.D. Soule, H.W. Chaney and Morris, P.A., 'Phylum Bryozoa', in *Intertidal Invertebrates of the Central California Coast: Light and Smith's Manual*, 4th edition, edited by J.T. Carlton (University of California Press, Berkeley, 2001); D.F. Soule, J.D. Soule and Morris, P.A., 'Changing concepts in species diversity in the northeastern Pacific', in *Bryozoan Studies 2001*, edited by P.N. Wyse Jackson, C.J. Buttler and M.E. Spencer Jones (Balkema, Amsterdam, 2002).

44 R.L. Zimmer, 'What and how bryozoan larvae "see"' *American Zoologist*, 28 (1988), 141 A.

45 R.M. Woollacott and R.L. Zimmer, 'Attachment and metamorphosis of the cheiloctenostome bryozoan *Bugula neritina* (Linné)', *Journal of Morphology*, 134 (1971), 351-382; R.M. Woollacott and R.L. Zimmer, 'Origin and structure of the brood chamber in *Bugula neritina* (Bryozoa)', *Marine Biology*, 16 (1972*a*), 165-170; R.M. Woollacott and R.L. Zimmer, 'Fine structure of a potential photoreceptor in the larva of *Bugula neritina* (Bryozoa)', *Zeit. F. Zellforsch.*, 123 (1972*b*), 458-469; R.M. Woollacott and R.L. Zimmer, 'A simplified placenta-like system for the transport of extra-embryonic nutrients during embryogenesis of *Bugula neritina* (Bryozoa)', *Journal of Morphology*, 147 (1975), 355-378; R.M. Woollacott and R.L. Zimmer (editors), *Biology of Bryozoans* (Academic Press, New York, 1977); R.L. Zimmer and R.M. Woollacott, 'Mycoplasma-like organisms: occurrence with the larvae and adults of a marine bryozoan', *Science*, 220 (1983), 208-210.

46 R.L. Zimmer and R.M. Woollacott, 'Structure and classification of gymnolaemate larvae', in *Biology of Bryozoans*, edited by R.M. Woollacott and R.L. Zimmer (Academic Press, New York, 1977), 57-89; R.L. Zimmer and R.M. Woollacott, 'Metamorphosis, ancestrulae and coloniality in bryozoan life cycles', in *Biology of Bryozoans*, edited by R.M. Woollacott and R.L. Zimmer (Academic Press, New York, 1977), 91-142.

47 P.A. Morris, 'Middle Pliocene paleotemperature implications based on the bryozoan *Hippothoa* (Cheilostomata-Ascophora)', *Journal of Paleontology*, 50 (1976), 1143-1149; P.A. Morris, 'Pacific coast *Celleporina* Gray (1848): fossil and Recent', in *Advances in Bryozoology*, edited by G.P. Larwood and M.B. Abbott (Academic Press, London, 1979), 476-489; P.A. Morris, 'The bryozoan family Hippothoidae (Cheilostomat-Ascophora) with emphasis on the genus *Hippothoa*', *Allan Hancock Foundation Monograph Series*, 10 (University of Southern California, Los Angeles, 1980), 1-115.

48 J.D. Soule, D.F. Soule and P.A. Pinter, 'Phylum Bryozoa (Ectoprocta)', in *Intertidal Invertebrates of the Central California Coast, Light's Manual*, 3rd edition, edited by R.I. Smith and J.T. Carlton (University of California Press, Berkeley, 1975), 675-608; D.F. Soule *et al.*, note 43 (2001, 2002).

49 P.A. Morris, 'Bryozoan-algal associations on southern California waters', *Bulletin of the Southern California Academy of Sciences*, 658 (1969), 199-218; P.A. Morris, D.F. Soule and J.D. Soule, 'Bryozoans, hermit crabs and gastropods: life strategies can affect the fossil record', *Bulletin of the Southern California Academy of Sciences*, 88(2) (1989), 45-60.

50 H.W. Chaney, D.F. Soule and J.D. Soule, 'Systematics and Zoogeography of *Thalamoporella gothica* and its allied species (Bryozoa, Cheilostomata)', *Bulletin of Marine Science*, 45(2) (1989), 338-355; Soule *et al.*, note 43 (1992, 1995, 2001, 2002).

W.D. Lang, orthogenesis and the evolution of Cretaceous cribrimorph Bryozoa

Paul D. Taylor
Department of Palaeontology, The Natural History Museum, London SW7 5BD, UK

1. Introduction

Peter Bowler, in his aptly named book *The Eclipse of Darwinism*,[1] described the widespread scepticism around the beginning of the 20th Century among biologists who, although accepting the fact of evolution, questioned natural selection as the driving mechanism responsible for evolutionary change. Four alternatives to natural selection were identified by Bowler as being popular at this time: Theistic Evolution, Lamarckism, Orthogenesis, and the Mutation Theory. Orthogenesis was championed mainly by palaeontologists who perceived numerous apparent instances in the fossil record of parallel evolution following particular pathways (trends) that often led to morphologies which seemed to be of an increasingly non-adaptive nature and culminated in lineage extinction. Much of the evidence for orthogenesis came from the occurrence of fossils with grossly enlarged, thickened or elaborated shells and bones, features for which it was difficult to imagine any advantage to the organisms possessing them. Among the most complete theories proposed at the time to account for such 'overdevelopment' was that offered by W.D. Lang, an invertebrate palaeontologist working at the British Museum (Natural History) [BM(NH)]. Lang's theory sprang from his detailed morphological and systematic studies of Cretaceous cribrimorph bryozoans. His arguments, couched in emotive prose - lineages being 'doomed to extinction' as a result of individuals building their own 'tombs' through excessive calcification - are for the most part untenable in the light of our current understanding of evolutionary biology. Indeed, even at the time of publication Lang failed to attract significant numbers of followers who were willing to proselytise his theories.[2] Nevertheless, Lang's theories did find their way into student textbooks[3] still in use over half a century later.

With renewed interest in evolutionary trends,[4] it is opportune to review Lang's work on cribrimorph bryozoan evolution, not merely because it represents an imaginative attempt to interpret the fossil record, but also to highlight the fact that subsequent research on Cretaceous cribrimorphs has been so scant that we still lack the empirical data and phylogenetic framework needed to evaluate the pattern of cribrimorph evolution and test certain aspects of Lang's theories. After a biographical sketch of Lang, I discuss his thesis of cribrimorph evolution and assess his work in the context of modern evolutionary theory and advances in bryozoology. In addition to the primary literature, I have had access to 17 octavo fileboxes containing Lang's correspondence, official and personal, over a period of more than fifty years.[5]

2. W.D. Lang

William Dickson Lang was born on 29th December 1878 in the Punjab, India, where his father was a civil engineer working on the construction of the Jumna Canal.[6] The Lang family were firmly rooted in the Victorian upper middle-classes, with the army, the clergy[7] and the Indian Civil Service figuring prominently among the professions traditionally practised by his relatives and ancestors. Lang's father Edward endured worsening health while in India and retired from the Indian service the year after his son's birth, settling in Harrow in the northern suburbs of London with his family in 1879; however, he survived for only a year, leaving his widow Hebe to bring up William and the other children. W.D. Lang received his early education at Christ's Hospital, Hertford, before becoming a day pupil at Harrow, a leading bastion, then as now, of the English Public School tradition. An early interest in natural history was fuelled by family holidays to the coasts of Dorset, Yorkshire, Devon, Lincolnshire and North Wales. In later life, Lang subscribed to the idea that there are born naturalists,[8] of which he considered himself to be an example.[9] His leanings towards natural history were scarcely satisfied at Harrow where the emphasis was on the Classics. However, in 1898 he entered Pembroke College, Cambridge as a scholar to read the Natural Sciences Tripos with Zoology.

It was during 1898 that Lang first visited Charmouth in Dorset, a place he was to revisit almost annually thereafter in connection with his work on Jurassic biostratigraphy,[10] and which was to be his home[11] upon retirement 40 years later. Errol White[12] attributed Lang's failure to graduate from Cambridge with a First to a lack of supervision: there was only one science don at Pembroke and very few of Lang's fellow students read Natural Sciences. Lang's teachers at Cambridge included Sidney Harmer, destined to become Director of the British Museum (Natural History) and a leading authority on Recent Polyzoa (= Bryozoa), and Henry Woods, perhaps best known for his textbook on

Figure 1. An undated photograph of W.D. Lang taken sometime during the period of his employment at the British Museum (Natural History).

invertebrate palaeontology,[13] then in its second edition, and who encouraged Lang to apply for a position at the British Museum (Natural History). This post had become vacant on the retirement of Henry Woodward in 1901. Lang (Figure 1) was appointed in the face of strong competition and, on 1st October 1902, he entered the Department of Geology at the BM(NH) where he was to spend all of his working life. His remit covered fossil Protozoa, Coelenterata, Porifera, Polyzoa (Bryozoa) and miscellaneous smaller groups. Previously under the care of J.W. Gregory, these collections were in a poor curatorial condition when Lang arrived.

Most of Lang's efforts during his first few years at the BM(NH) were directed towards the registration and labelling of specimens - his diligence in carrying out this task is obvious from the large number of handwritten entries he made in the accession registers. The Keeper in these early years was Arthur Smith Woodward, a distinguished ichthyologist who, unfortunately, was later to be drawn into the Piltdown forgery. Lang's immediate superior was the echinoderm specialist Francis Arthur Bather who succeeded Smith Woodward as Keeper in 1924. Lang and Bather had very different personalities and backgrounds; they disagreed about various matters,[14] and Bather was a major critic[15] of his colleague's orthogenetic theories. Lang's career at the BM(NH) followed a common pattern: curation giving way to research which in turn was replaced by administration when Lang himself became Keeper in 1928, following a year's apprenticeship as Deputy Keeper. Stearn[16] remarked on the unusual working hours adopted by Lang who started very early in the morning and finished long before most of his colleagues, a regime which severely restricted any participation in scientific societies, made yet more difficult by his perceived physical weaknesses.[17] As Keeper, he ran the Department of Palaeontology along traditional and very formal lines, and is reputed[18] to have had a particular abhorrence of newspaper reporters who frequently arrived in search of a story.

The first[19] of Lang's papers devoted wholly or principally to bryozoans appeared in 1904. All of his early papers concerned Jurassic and Cretaceous cyclostomes, but within ten years his focus of interest had shifted to Cretaceous cheilostomes.[20] The most substantial of his publications on bryozoans were the two volumes of the *Catalogue of the fossil Bryozoa (Polyzoa) in the Department of Geology, British Museum (Natural History).*[21] Following on from three volumes published previously by J.W. Gregory, Lang's Catalogues dealt entirely with Cretaceous cribrimorph cheilostomes and went far beyond a simple listing of individual specimens present in the BM(NH) collection. Instead, they attempted a full monographic treatment of this group of bryozoans and included lengthy discussions of their morphology and evolution. Lang's final paper on bryozoans was published as early as 1925,[22] after which corals and Dorset geology dominated his scientific interests.[23]

In 1929 Lang was elected a Fellow of the Royal Society, but it is evident from the correspondence in his archives at The Natural History Museum (= BM(NH)) that he had been unsuccessfully proposed for election on several previous occasions.[24] He retired in 1938 on his 60th birthday and bade farewell to the BM(NH), never to return during the 28 years that remained of his life. His retirement was spent almost entirely in Charmouth where he was extremely active in various aspects of local natural history. For example, he became an authority on the Victorian fossil collector Mary Anning,[25] and wrote annual reports on natural history for the *Proceedings of the Dorset Natural History and Archaeological Society*. He was clearly held in great esteem as an expert in diverse areas of natural history, not only geology but also botany, entomology, ornithology and marine biology.

3. Orthogenesis

Bowler[26] provides a succinct history and philosophy of orthogenesis on which the brief account below is largely based. The term orthogenesis means 'linear evolution'. It was introduced by Wilhelm Haacke[27] who believed that the 'germ plasm' was structured in such a way that it predisposed variation to occur in particular directions, leading to linear patterns of evolution. The principal popularizer of orthogenesis was Theodor Eimer[28].

Eimer's studies of parallel evolution of colour patterns in butterflies,[29] which he believed had no function, led him to abandon an earlier belief in Lamarckism in favour of orthogenesis, a form of evolution which, in contrast to both Lamarckism and Darwinism, did not necessarily involve adaptation. Eimer went as far as to claim that "Orthogenesis is a universal law".[30]

Whereas Eimer and other biologists derived their evidence for non-adaptive parallel evolutionary trends entirely from living organisms without recourse to the fossil record, palaeontologists entering the fray had the advantage of being able to offer a time dimension to orthogenesis.

According to Gould:[31]

> The theory of orthogenesis became a touchstone for anti-Darwinian paleontologists, for it claimed that evolution proceeded in straight lines that natural selection could not regulate. Certain trends, once started, could not be stopped even if they led to extinction.

A recurrent theme among palaeontologists advocating orthogenesis was that the evolution of lineages ran in close parallel to the ontogeny of individuals: just as an ageing individual develops senile features and dies, so an ageing lineage evolves senile features and becomes extinct.[32] Alpheus Hyatt believed this to be the case for ammonoids, a group in which the appearance of supposedly senile species in the Cretaceous heralded their final extinction.[33] Two popular examples were commonly quoted in the palaeontological literature of orthogenetic evolution culminating in the appearance of morphological characters ultimately fatal to the lineage. The first was the progressive tightening of the coiling in shells of the Jurassic oyster *Gryphaea*, leading to individuals which could scarcely open their shells. The second was the evolution of gigantic antlers in *Megaloceros*, the Giant Irish 'Elk' of the Pleistocene.

Another widely cited instance of orthogenesis, the evolution of the horse, is very different from either *Gryphaea* or *Megaloceros* in that it does not involve the evolution of seemingly maladaptive features. This contrast illustrates just one of the many variations in the way that the term orthogenesis was applied. George Gaylord Simpson[34] remarked that orthogenesis:

> is usually employed not simply as the name of a phenomenon but also as the designation of some theory purporting to explain it.

The distinction made explicit in Simpson's statement is a particularly fundamental one between an evolutionary pattern and an evolutionary process.[35]

Orthogenesis in its various guises was given very serious consideration by palaeontologists during the early years of the 20th Century. Lull,[36] for example, regarded orthogenesis as an evolutionary theory worth entertaining either in opposition to natural selection or supplementary to it. He pointed out several lines of evidence for orthogenesis, particularly the widespread occurrence of parallel evolution, the incidence of 'over-specializations', and the apparent constitutional limitations on variation similar to those found by plant and animal breeders. As to the causes of orthogenesis, Lull noted that predetermination had been explained either by the existence of an intangible and unknown force within the organism or by the laws of organic growth.

Orthogenesis was discarded by most scientists, especially zoologists, as early as 1940. The following passage was written in 1937 by the President of the British Association for the Advancement of Science E.B. Poulton:[37]

> The appeal to Orthogenesis, or internal developmental force, as the motive cause of evolutionary progress has often been made – generally by palaeontologists rather than by observers of living forms. Any such belief in the potency of an internal tendency is, I think, open to the criticism made by Thisleton Dyer in his address to Section D at Bath in 1888: 'This appears to me much as if we explained the movement of a train from London to Bath by attributing to it a tendency to locomotion. Mr. Darwin lifted the whole matter out of the field of mere transcendental speculation by the theory of natural selection, a perfectly intelligible mechanism by which the result might be brought about. Science will always prefer a material *modus operandi* to anything so vague as the action of tendency.'.[38]

Julian Huxley also firmly rejected orthogenesis as a viable evolutionary mode in his classic text *Evolution. The Modern Synthesis*:[39]

> To sum up, the only important agency restricting the direction of evolutionary change is the historical one, leading to a purely apparent orthogenesis.

Criticism of orthogenesis was not confined to neontologists: the vertebrate palaeontologist Simpson[40] even went as far as to claim that:

> A dispassionate survey of many of the phenomena of orthogenesis, so called, strongly suggests that much of the rectilinearity of evolution is a product rather of the tendency of the minds of scientists to move in straight lines than of a tendency for nature to do so.

A late defence of orthogenesis appeared in 1940, written by the invertebrate palaeontologist Arthur E. Trueman.[41] Trueman upheld orthogenesis on the following grounds:

> the term orthogenesis has been so frequently misinterpreted also because at the present time it seems that a somewhat closer comparison between the results achieved by geneticists and the conclusions reached by palaeontologists can be made.
>
> The chief objections which have been made against it resulted from a belief that the term covers

> something mystical and akin to teleology, or that it involves predetermination; but I know of no palaeontologist who has put forward any such views. If they have spoken of an internal factor, an inherent tendency, or of "compulsion from within" (Lang, 1921, p. xviii) they have usually made it clear that some physiological or chemical factor may be involved.

Trueman drew a distinction between three terms that had become intertwined: 'trend', a tendency for parallel evolution to occur; 'orthogenesis', a trend driven by the action of some factor internal to the organism; and 'programme-evolution', a trend which was predetermined.

Despite the pleas of Trueman, orthogenesis had become inextricably linked with notions of predetermination and internal organismic control of evolution ("vitalism") which were totally out of tune with the emerging evolutionary consensus of Neodarwinism. Today, orthogenesis has effectively disappeared from the vocabulary of evolution, save for the occasional historical aside.

4. Orthogenesis and bryozoan evolution: Lang's thesis

As mentioned above, Lang began his research on bryozoans by studying the taxonomy of Mesozoic cyclostomes. His interest in evolution is first evident in 1905 with the publication of a scheme showing the proposed phylogeny of *Stomatopora* during the Early Jurassic.[42] However, Lang's orthogenetic ideas on bryozoan evolution did not emerge in print until 1916 by which time his systematic research had switched from cyclostomes to cribrimorph cheilostomes (Figure 2). The thesis he developed of Cretaceous cribrimorph evolution was complex and entailed the following principal propositions: (1) cribrimorphs evolved from membranimorph ancestors by fusion of marginal mural spines over the frontal membrane to produce a costate frontal shield (or secondary frontal wall); (2) cribrimorphs originated on at least 11 separate occasions from different membranimorph ancestors during the Cretaceous; (3) each distinct cribrimorph 'lineage' followed a trend through time of increasing elaboration and calcification of the zooid, such changes occurring in parallel in the different lineages; (4) these trends led to increasingly maladapted species and culminated in the extinction of the lineage; (5) from their initial appearance, each lineage was programmed to follow this evolutionary trajectory, with environmental change and natural selection playing no more than subsidiary roles; (6) calcium carbonate was secreted by the zooid primarily as a waste product but, through the action of natural selection, it was laid down in sites where it was of most use, or of least harm, to the bryozoan; (7) trends towards ever greater calcification resulted from the progressive removal over geological time of a physiological factor inhibiting calcium carbonate secretion until, finally, calcification occurred without inhibition, the bryozoan zooid was buried beneath its own waste products, and the lineage became extinct.

Lang's 1916 paper *Calcium carbonate and evolution in Polyzoa*[43] first expounded his

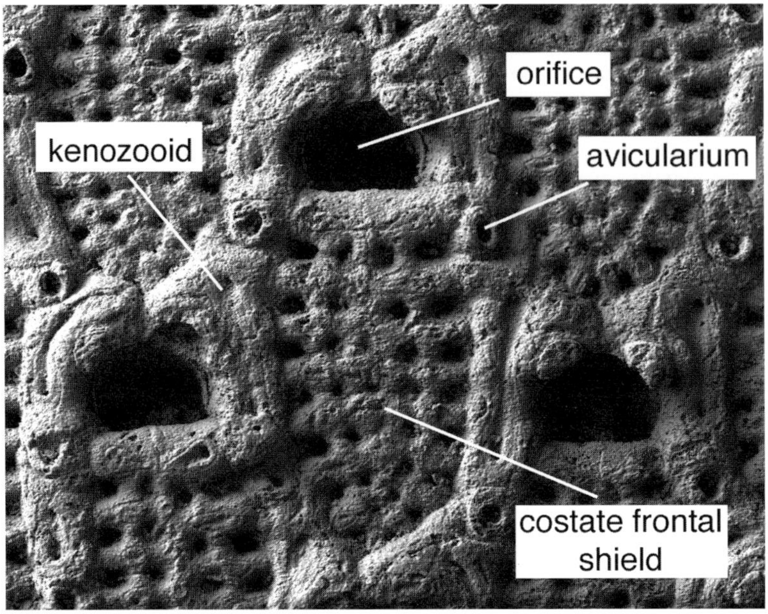

Figure 2. Scanning electron micrograph of a few zooids of the Cretaceous cribrimorph bryozoan Pelmatopora marsupitum *Lang, 1916 from the Campanian Chalk (*pilula *Zone), near Rottingdean Gap, Sussex, England; NHM D41056. The elongate autozooids have a semicircular distal orifice and a frontal shield consisting of costae with lateral fusions and pelmata (pores). Between the autozooids are smaller polymorphs - avicularia and kenozooids ('interoecial secondary tissue' of Lang). Magnification about x60.*

orthogenetic theories. Its starting point was a paper on trepostome bryozoans published the previous year by Cumings and Galloway.[44] This landmark paper on the skeleton of Palaeozoic trepostomes includes the following passage:[45]

> The thickening of the interzooecial walls, due to the development of the cingulum, is often very great … . There is an actual reduction in the size of the zooecial chamber. … We believe that this extreme development of secondary deposits is a senile feature, analogous to the great thickening of brachiopod shells and the shells of Mollusca in old age.

While Cumings and Galloway were concerned with the ontogeny of individual zooids, Lang[46] extended the idea of calcification and senility onto phylogeny and evolutionary time:

> Cumings and Davenport [sic for Galloway] attribute the secondary thickness of the walls to senility of the individual, at least they say it accompanies senility. We claim, further, that it marks the senility of the lineage to which the individual belongs

Lang's willingness to make such an exact analogy between ontogeny and phylogeny must be seen in the light of the widespread acceptance of Haeckel's Biogenic Law by palaeontologists at this time – if the ontogeny of modern organisms accurately recapitulates their phylogeny, is it not reasonable to infer the reciprocal, that phylogeny passes through equivalent stages to ontogeny leading to senility and ultimately to the death (extinction) of the lineage?[47] The notion that the precipitation of calcium carbonate in invertebrates in general evolved as an excretory function that had the unfortunate consequence of dooming the lineage to extinction is explicit in the following quotation:[48]

> Say that some metabolic process, such as one involved in nitrogenous excretion, resulted in the precipitation of calcium carbonate in the tissues or upon the surface of a marine organism - Mollusc, Brachiopod, or Polyzoan, and was turned to useful account as affording a supporting or protecting skeleton or shell; that this production of calcium carbonate became increasingly constitutional so that the mere need for a skeleton or shell was more than met; that the process could not be arrested or countered in all the organisms that had acquired it, and in these the disposal of superfluous calcium carbonate became a pressing problem; that, finally, those organisms that found no way out of the difficulty were doomed to extinction under a mass of calcium carbonate of their own making.

Referring specifically to cheilostome bryozoans, Lang[49] stated:

> The most hopeful of the Cheilostomes are those which have a skeleton of chitin only. When once calcium carbonate has begun to be deposited the whole lineage is doomed to a more or less stereotyped sequence of calcification until, in the end, it becomes extinguished under its superfluity of skeleton. .... The general phylogenetic future of ... 'Cribrimorphs' - can be predicted with certainty, though the details vary in every lineage. Further calcium carbonate is laid down in connection with (1) the spines that form the intraterminal frontal wall - the costae; (2) those that surround the aperture - the 'apertural spines' (considered with the first pair of costae, those which bound the aperture proximally, and, fused, form the 'apertural bar'); and (3) the extraterminal frontal wall, finally filling up the interzooecial spaces; and the process may take place in one or more of these directions simultaneously in any lineage. ... It is inconceivable that Steginomorphs [an advanced, heavily calcified group of cribrimorphs] can have any evolutionary future. In extreme cases they present externally a thick crust of calcium carbonate pierced here and there by holes representing apertures and avicularia, that tend to become smaller and more choked as more calcium carbonate is laid down. So the cingulum of Trepostomes marks the senility of the race, and the Palaeozoic Polyzoa fell victim to the same disease as those Cretaceous forms which, owing to their calcareous skeletons, have been preserved to us to study.

The evolutionary themes explicit here are parallel evolution and the predictable, predetermined evolutionary trajectorics followed by the lineages that drove them to extinction.

Lang's first paper on cribrimorph systematics was also published during 1916.[50] Here he remarked that:[51]

> The evolutionary aim in the development of all families appears to be the disposal to the best advantage or least detriment to the organism of superfluous Calcium Carbonate.

Three years later he published a larger work dealing with the systematics and evolution of one particular subfamily, the Pelmatoporinae.[52] Once again, Lang expounded the notion of increasingly uncontrolled formation of a calcium carbonate skeleton in cribrimorph zooids, propelling the lineage towards a self-imposed extinction. At most, natural selection served only to determine the locations where the calcium carbonate was deposited by the zooid, not the fact of its secretion or the quantity produced:[53]

> The evolutionary aim, then, behind the elaboration of the skeleton in the Cretaceous Cribrimorphs is the disposal of an increasing secretion of Calcium Carbonate where it will be least in the way; and it is mainly deposited, independently, in three positions, namely, on the intraterminal frontal wall [=costate frontal shield], so as further to solidify it; around the aperture, so as to build up a secondary aperture; and in the interoecial depressions. To explain the increasing complexity of aperture and intraterminal front-wall thus, is not, however, to deny to the organism any power of elaborating its skeleton to a useful end. It only claims that this secretion is unavoidably there to be disposed of, and, if the organism can employ it usefully, it is by so much the gainer; but sooner or later it will fail to cope with the abundance of its secretion, and the race will perish.

Two further threads in Lang's thesis are mentioned in this paper. The first is his conviction that all of the Cretaceous (including Danian) cribrimorph families became extinct before the end of that period and are in consequence not closely related to extant cribrimorph genera (e.g., *Cribrilina*) which must have evolved independently. The second is that the mural spines present in certain membranimorphs, including the ancestors of cribrimorphs, evolved as a response to the need to secrete calcium carbonate. More primitive, non-spinose membranimorphs fared less well according to Lang:[54]

> The Cribrimorph, then, only puts off the evil day. That it does put it off with some success seems likely when the more primitive Membranimorphs are considered. In a colony of these it is common to find individuals completely sealed up [i.e. with closure plates], with the scar of the aperture showing on the calcareous covering of the oecium. It is probable that this is the only answer that the Membranimorph individual can give to the demands of its deranged metabolism - it simply deposits Calcium Carbonate over its whole surface, building its own tomb, and thus experiencing the final doom of the race. The race, however, saved itself from such a crudely immediate ending by the capacity it acquired for limiting the areas of super-secretion to definite spots along the termen; and thus terminal spines arose, which, by further increase of size and by inter-fusion, form the Cribrimorph intraterminal front-wall.

A second paper published in the same year (1919) dealt with the morphology and evolution of a single subfamily, the Kelestominae.[55] The theme of excessive calcification during cribrimorph evolution is again discussed:

> it became a pressing problem in the organism's bionomy where to dispose of its increasing superfluity of calcium carbonate so as least to interfere with its normal functions. There is no evidence that the Polyzoa ever gained control over this derangement of metabolism, or that they ever learned to counter it by resorption; but a good deal to show that these calcareous lineages are doomed to ultimate extinction under their masses of superabundant skeletal tissue.

The first part of the cribrimorph *Catalogue of Cretaceous Bryozoa*[56] afforded Lang the opportunity to explain his thesis of cribrimorph orthogenesis more fully. In order to avoid any misinterpretation, it is worthwhile quoting some of the key passages from Lang's lengthy exposition:

> when a Cretaceous [cribrimorph] lineage can be followed to any considerable length, the amount of calcium carbonate in the skeleton becomes so pronounced that, not only does the skeletal structure become secondarily simple by the piling up of calcareous matter, so that further evolution is inconceivable, but the very life-processes of the organism appear to be in danger of obstruction, so constricted and tunnel-like become all the apertures in the skeleton, by which the organism communicates with its environment.[57]

> It is usual to consider a calcareous shell as an adaptation for protection. But to regard it (and its chitinous predecessor) as primarily a fortuitous metabolic product, is not to deny the shell a protective function. Rather, it is to emphasise the inevitability of the shell; its subsequent adaptation to protective purposes, though in keeping with organic behaviour, and to be expected from our knowledge of organisms, is yet a secondary production.

> Extravagances of growth ... have been referred to as being caused by a loss of equilibrium ... . In Vertebrates, it has been suggested that the original condition of equilibrium is controlled by those substances called hormones, contained in the secretions of the ductless glands, and that hormones act as inhibiting agents limiting the growth that otherwise would continue indefinitely.

> environment may provide the first stimulus whereby inhibitions are ultimately removed, and an expression-point reached in the evolution of a lineage.

> calcareous matter ... is always piled up according to the same architectural plan - a plan repeated in each lineage

> given a Membranimorph with well-developed terminal spines, it may be said with certainty that, either the lineage will go no further than a specialised Membranimorph, or, if it has a further history, the spines will arch over and form a Cribrimorph front-wall ... that this will ... become more and more solidified; that, if further evolution takes place, secondary calcium carbonate will be laid down either in the interoecial valleys, or in connection with the aperture, or in both places simultaneously; finally, if the lineage persists still longer, the areas of the secondary tissue will grow above the level of the intraterminal front-wall, and, spreading laterally, will fuse one with another to form a secondary (really tertiary) front-wall.

> Now, while it is conceivable that the environment might so affect an organism as to throw out of gear its normal metabolism, it seems improbable that it should be directly responsible for

causing independent groups of lineages of Cheilostome Polyzoa to run through a similar evolutionary history..' Such a phenomenon points rather to a compulsion from within - to a tendency in the ancestral form which becomes actual as its evolution is worked out in the offspring. ... Such predetermined evolution is very like what has been termed Orthogenesis by Eimer, and my friend Dr. F. L. Kitchin has spoken of it as Programme-Evolution.

potentiality .... is the quality inherent in a radical organism whereby all its descendant modifications develop according to a common general plan; its expression is controlled by inhibiting agents, which may ultimately be removed by environmental means; this release is apt to occur periodically, and to result in the realisation or actualising of potentiality pent up and hitherto repressed

Lang's summary states:

An organism, then, is regarded as a synthesis of structure and function, and as possessing tendencies* for structural development
* The question of the origin of these tendencies is not touched upon. But my friend Dr. Kitchin, who agrees with me in recognizing their importance, considers that they are not present *ab initio*, but that they arise as evolution proceeds, and that the environment has a share in their origin.

As for the role of natural selection, Lang made his position clear:

in so far as Natural Selection is present, its action is rather a negative one - insuring that the superfluous calcareous matter shall be laid down where it is least in the organism's way, than that the structure it builds should be directly useful.

By now Lang's detailed and comprehensive research on cribrimorph systematics was complete and he was able to classify Cretaceous species into 11 families: Myagroporidae, Otoporidae, Ctenoporidae, Thoracoporidae, Taractoporidae, Lagynoporidae, Andrioporidae, Calpidoporidae, Disheloporidae, Rhacheoporidae and Pelmatoporidae.[58] These families were largely founded on 'the nature of the secondary intraterminal front-wall, and of the costae which compose it'.[59] According to Lang:

it may be supposed that they have independently evolved from as many Membranimorph stocks. It is also possible that some of the sub-families comprised in a single family really had an independent evolution from Membranimorph ancestors, and, consequently, should be regarded as independent families.

Lang repeated his doubt that post-Cretaceous cribrimorphs were descended from Cretaceous cribrimorphs, all or most of which belong to lineages that became extinct during the Cretaceous:

if the majority of Cretaceous lineages are found to end in forms not only incapable of further evolution, but apparently becoming extinct under the products of their secretory activity, it

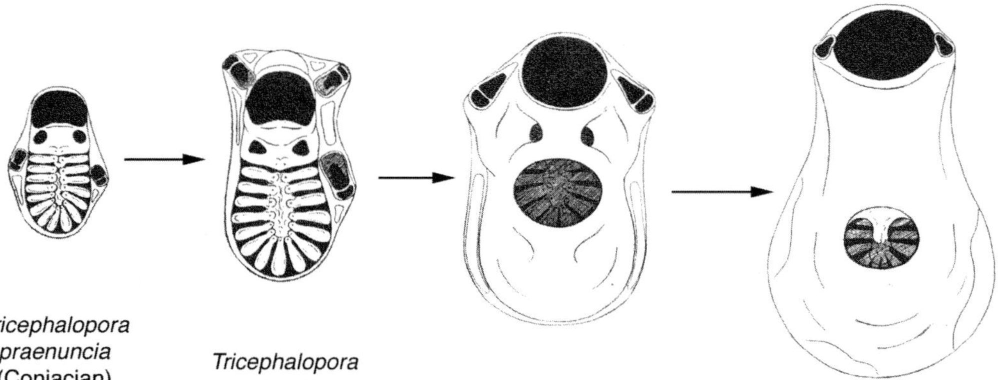

Figure 3. Zooids from four species constituting the central lineage in Lang's evolutionary tree of the Tricephaloporinae.[60] *The two Coniacian species have exposed costate frontal shields. The costate shield is largely overgrown by a tertiary frontal wall in the two Danian species, remaining visible through a small window. Further skeletal elaboration in* Haplocephalopora uniceps *takes the form of a long peristome.*

> follows that few, if any Cretaceous genera persisted into Tertiary times; and that the ancestors of Tertiary and Recent forms must be sought in the chitinous Cretaceous species which have left no remains

The two volumes of Lang's *Catalogue* contain numerous evolutionary trees expressing inferred phylogenetic relationships between species placed in stratigraphical order. Both anagenetic and cladogenetic evolutionary patterns are depicted in these trees, i.e. Cretaceous cribrimorph evolution according to Lang consisted of a mixture of ancestor-descendant transitions without branching of the lineage (chronospeciation), and splitting of the lineage into two (speciation *sensu stricto*). As an example of the evolutionary changes considered to occur in cribrimorphs by Lang, Figure 3 uses Lang's own drawings of four of the six species forming one of his inferred lineages.

Whereas Lang's thesis was devised at a time when orthogenetic theories were very much in vogue, during the next few decades the climate changed from one of widespread acceptance to almost universal refutation of orthogenesis. An early critic was Lang's colleague F.A. Bather who wrote of Lang in 1920:[61]

> He speaks of living matter as if it were the over-pumped inner-tube of a bicycle tyre, 'tense with potentiality, curbed by inhibitions' [of the cover] and 'periodically breaking out as inhibitions are removed' [by broken glass]. A race acquires the lime habit or the drink habit, and, casting

off all restraint, rushes with accelerated velocity down the easy slope to perdition.

A melancholy picture! But is it true? The facts in the case of the Cretaceous Polyzoa are not disputed, but they can be interpreted as a reaction of the organism to the continued abundance of lime-salts in the sea-water. If a race became choked with lime, this perhaps was because it could not keep pace with its environment. Instead of 'irresistible momentum' from within, we may speak of irresistible pressure from without.

Lang ceased publishing on evolution during the 1930s. His later publications and documents demonstrate, however, a continuing scepticism of Neodarwinism.[62] While still maintaining that the fossil record revealed constrained, parallel evolutionary trends, his correspondence shows that he came to regard seeking explanations for these trends as lying outside the realm of science:

To my mind, a Natural Science at most can demonstrate a chain of proximate causes and effects, which is rather a matter of description than explanation. Explanations belong (as I think) to the realm of Metaphysics, and depend upon one's philosophical background and outlook on matters such as the nature of time, and ultimately upon ones religious views. ... I hold that Darwin's theory of Natural Selection is logically bound to be true up to a point; but in any one instance the range of deviation possible to a character may vary from a little to a great deal before the axe of Nat. Selc. Cuts off the straggler; and it needs much much more than his theory to account for the origin of species.[63]

5. Scientific influences on Lang

It seems likely that Lang's orthogenetic theory of cribrimorph evolution originated sometime around 1910 and was quite probably complete by the outbreak of the First World War. There is some evidence in his correspondence and publications of the major influences on his work during this critical time. Among a small number of letters from the American palaeontologist E.R. Cumings of Indiana University is one dated 13th February 1911 which reveals the tide of feeling against Darwinian evolution, especially the Weismann Dogma, as well as the support for the Biogenic Law, and hints at the existence of a schism between biologists and palaeontologists interested in evolution:

I feel now that Hyatt is soon to come into his own, The experimentalists are about to prove the inheritance of acquired characteristics; and, as for the law of acceleration, it is based on such an array of paleontologic facts that it will very easily take care of itself. Beecher used to say that he "hoped Weismann would live to see the error of his ways" — and I guess he has. Zoologists are altogether too prone to neglect the geologic history of organisms. Your results from the Bryozoa are worth tons of Weismannian speculation.

Two letters in 1914 from the Geological Survey palaeontologist F.L. Kitchin (1870-1934), who was working on Jurassic *Gryphaea*, are also significant in that they discuss parallel evolution and orthogenesis in terms of internal predisposition and racial senility:

The "*Gryphaea*" style of shell must of course represent only a particular evolutionary phase repeated again and again in different ostrean stocks. ... It is clear that this morphic trend - reduction of attached stage *plus* increasing arcuation, prepared for rapid genetic decline; the end result was so often the inordinate thickening and foliaceous growth of degeneracy ... I have some views (good or bad) on limited potentialities and predispositions, on the bringing of these to particular expression by causes apart from adaptive selection or response, and on genetic old-age being reached more as a result of internal than external conditions.[64]

It is clear that the expression of a predisposition (inherited in common by separate genetic series) to evolve along certain definite lines has often been deferred for a very long time ("latent or potential homology" of Osborn). In other cases there has been some sudden evolutionary efflorescence, as in many ammonite genera, resulting in several or many little separate genetic lines which at once go through the same programme, evolving the successive transformation-stages in certain characters in just the same manner and order (though not necessarily at the same rate), thus, surely, showing the immediate results of the same predisposition bequeathed by the common ancestors.[65]

Lang's supposition that there existed physiological (hormonal) controls on skeletal growth which if removed would lead to excessive and potentially maladaptive growth of the skeleton may have originated from, or was at least reinforced by, a short and speculative paper published by Arthur Dendy in 1912.[66]

Another strong influence on Lang was Sidney S. Buckman, well-known for his meticulously detailed work on Jurassic biostratigraphy, who visited the British Museum (Natural History) during the early part of Lang's career there to work on the brachiopod collections. Lang himself wrote:[67]

Contact with Buckman was most stimulating. ... Buckman's enthusiastic, though often blind, acceptance of "biogenetic law" as applied to ammonites by Alpheus Hyatt, according as it did, with the work of many palaeontologists at that time, attracted me strongly for its direct appeal to serial change of form. At that time I was very much alive to the implications of "recapitulation", as we palaeontologists generally termed the biogenetic law, and I already saw that the principle could not be applied blindly, as Buckman apparently applied it, but that its expression was continually modified and masked by environmental conditions.

6. A current perspective on Lang's thesis

In a rare critique of Lang's work published during the past 25 years, the evolutionary palaeobiologist Thomas J. M. Schopf concluded:[68]

In the modern view, Lang's general observation of increasing complexity in calcification, with subsequent extinction of these lineages, is confirmed (Larwood, 1962, 1969). But we would neither interpret this, nor the trends noted by Voigt (1939), as being due to a vital force gone berserk, for which no other evidence has been uncovered in this or any other group of organisms.

Schopf's dismissal of a vitalistic force driving orthogenetic trends would be supported by all but the most heterodox of modern evolutionary biologists. However, as noted above (p. 281), Lang's thesis was complex and it is worth examining individually each of its seven main propositions:

(1) *Cribrimorphs evolved from membranimorph ancestors by fusion of marginal mural spines over the frontal membrane to produce a calcareous frontal shield (or secondary frontal wall).* Larwood[69] affirmed the origin of cribrimorphs from such an ancestral group, his model showing how initially erect costae in a membranimorph could become prostrate, lie over the frontal membrane as in myagromorphs,[70] and then fuse at their inner, distal ends in cribrimorphs proper (Figure 4). The membranimorph ancestors of cribrimorphs envisaged by both Larwood and Lang[71] would nowadays be regarded as pseudomalacostegans belonging to the 'grade' Family Calloporidae. There is as yet no published rigorous (i.e. quantitative cladistic) analysis of cribrimorph relationships that might support (or refute) a nesting of cribrimorphs within the Calloporidae. Nevertheless, this certainly seems likely to be true on the basis of both morphological and stratigraphical evidence.[72]

(2) *Cribrimorphs originated on at least 11 separate occasions from different membranimorph ancestors during the Cretaceous.* Lang's claim that Cretaceous cribrimorphs evolved polyphyletically has not been evaluated cladistically. However, the fact that these supposedly independent lineages have a considerable number of advanced characters in common suggests that the outcome of a cladistic analysis would be unlikely to support polyphyly. It is revealing that Lang was able to construct a hierarchical key[73] for the identification of his families and subfamilies of Cretaceous cribrimorphs: the fact that the families in this key are nested within as many as four levels gives an indication of how much homoplasy would need to have occurred in order for the families to have been derived independently from non-cribrimorph ancestors.

(3) *Cribrimorph lineages followed parallel trends through time of increasing elaboration and calcification of the zooids.* Once again, the lack of a cladistic analysis hampers adequate testing of this proposition. Larwood's comments[74] on the evolution of individual genera belonging to one subfamily (Pelmatoporinae) suggest that some genera (e.g. *Pelmatopora*) increased in complexity through time, whereas others did the just the opposite (e.g. *Aeolopora*). Nevertheless, heavily calcified cribrimorphs with tertiary frontal walls did not appear until relatively late in the Cretaceous; indeed, they are particularly characteristic of the terminal Maastrichtian stage.[75]

(4) *Evolutionary trends led to increasingly maladapted species and culminated in the extinction of the lineage.* Some of the more advanced cribrimorphs from the Upper

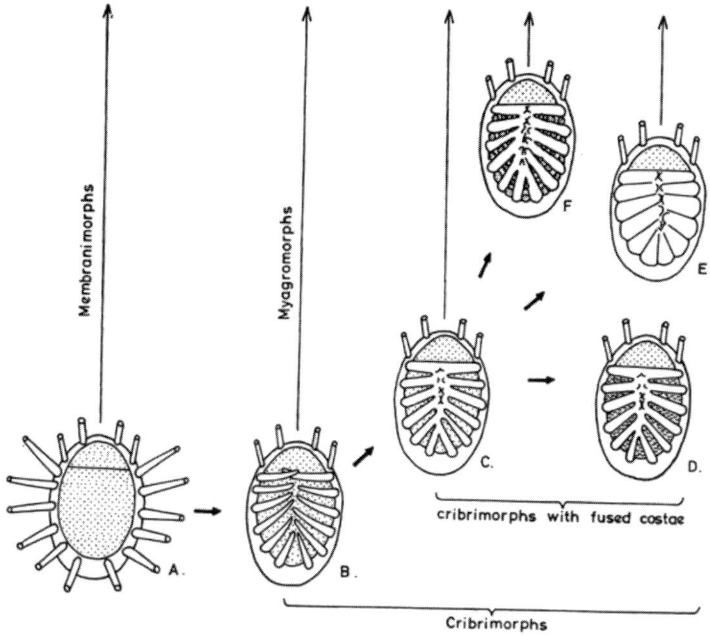

Figure 4. The origin of cribrimorphs from a membranimorph (calloporid) ancestor as envisaged by Larwood.[76]

Cretaceous exhibit 'bizarre' and complex skeletal morphologies which to Lang were of no or doubtful adaptive value. In some cases, particular skeletal structures appeared to impede the normal functioning of the zooid, as in the oral avicularia that almost occlude the orifice in *Ichnopora denticulata*,[77] leaving little space for the protrusion of the tentacle crown. At the time of Lang's research almost nothing was known about predation on bryozoans, but it is now clear that a wide range of small predators attack individual bryozoan zooids,[78] and that structures such as avicularia, spines and thickened frontal shields can be reasonably interpreted as adaptations against such predators.[79] Schopf[80] proposed that the overarching spines forming the frontal shield in cribrimorphs functioned in defence, and it has been suggested that progressive trends towards increased calcification in late Mesozoic bryozoans may have been driven by escalating predation pressure.[81] However, it remains to be demonstrated whether any trends in Cretaceous cribrimorphs are driven and require an explanation, or are simply passive and reflect increasing variance and the inevitable drift of morphology from the left-wall of simplicity.[82]

As Schopf noted, specializations of the type found in the most derived Cretaceous cribrimorphs may have made them particularly vulnerable to environmental changes, likely accounting for some of the many extinctions of Lang's 'senile' lineages. Recent research on the origins of ascophoran cheilostomes by Dennis P. Gordon[83] raises another possibility. Ascophoran cheilostomes, the dominant marine bryozoans at the present day, are a polyphyletic grade grouping that first appeared in the Late Cretaceous. The ancestry of at least some ascophorans undoubtedly lies within the cribrimorphs, with one or more cribrimorph-ascophoran transitions entailing lateral expansion of kenozooids over the costate frontal shield, to form an umbonuloid ascophoran frontal shield, coupled with reduction and eventual loss of the overgrown costate frontal shield. Gordon and Voigt[84] wrote:

> It is ironic that W.D. Lang perceived the so-called interzooecial secondary tissue (i.e., adventitious kenozooecia) as an orthogenetic liability – "Lineages which secrete calcium carbonate are ... doomed lineages" (Lang 1921, p. xxv). Far from heralding 'doomed lineages', the skeletal modifications of late Cretaceous pelmatoporine cribrilinids appear to have been a major innovation ... contributing to one of the greatest flourishings of bryozoan diversity in geological history.

If correct, this view means that one or more of the apparent extinctions of cribrimorph 'lineages' identified by Lang was in reality a taxonomic pseudoextinction, the result of transition from a cribrimorph grade to an ascophoran grade of organization.

(5) *From their initial appearance, each lineage was programmed to follow a particular evolutionary trajectory, with environmental change and natural selection playing no more subsidiary roles.* As noted already, these views are totally out of keeping with current evolutionary orthodoxy.

(6) *Calcium carbonate was secreted by the zooid primarily as a waste product but, through the action of natural selection, it was deposited in sites where it was of most use, or of least harm, to the bryozoan.* Although little is known about the physiology of skeleton formation (biomineralization) in bryozoans, the idea that the skeleton is primarily a waste product is unconvincing. Excretion in living bryozoans is accomplished by cyclical degeneration of the polypide (tentacles and gut) producing a brown body of waste material which is typically disposed of in the faeces of the newly regenerated polypide. There is no reason to suppose that this mechanism could not have been employed in Cretaceous cribrimorphs for the disposal of any excess calcium carbonate as an inert residue together with nitrogenous material and the products of cellular breakdown. Furthermore, bryozoans as a group include a spectrum of species, from those with entirely unmineralized organic skeletons to others with minor amounts of biomineralization to yet others with massively calcified skeletons, all seemingly representing successful survival for at least 450 million years of geological time. Highly calcified skeletons are not a necessary consequence of excretory processes in bryozoans.

(7) *Trends towards ever greater calcification resulted from the progressive removal*

over geological time of a physiological factor inhibiting calcium carbonate secretion until, finally, calcification occurred without inhibition, the bryozoan zooid was buried beneath its own waste products, and the lineage became extinct. Even in the paucity of information about processes of biomineralization in bryozoans, it is evident that bryozoans exert a high degree of control over the formation of their skeletons. This is apparent from the complexity and precision of the skeletons they are capable of building, an observation particularly true of cribrimorphs. At the finest scales, bryozoan skeletons are characterised by ultrastructural fabrics with specific orientations and shapes of calcite or aragonite crystallites.[85] Biomineralization in all studied bryozoans falls within the 'biologically controlled' category of Lowenstam and Weiner,[86] and it is difficult to imagine progressive, initially non-fatal loss of this control as envisaged by Lang.

7. Conclusions

W.D. Lang was an earnest, knowledgeable and philosophical naturalist. His orthogenetic thesis of cribrimorph evolution was a bold attempt to explain the evolutionary patterns he perceived in this group of bryozoans which radiated to a high diversity during the Late Cretaceous. It was devised at a time when natural selection was under attack, especially from palaeontologists, many of whom favoured orthogenesis as an alternative. In Lang's case it is unclear to what extent his strong religious beliefs led him to reject natural selection, with its large component of chance, and to opt for orthogenesis, which may have been more palatable to him, especially in its more extreme variant with factors internal to the organism powering predetermined evolutionary trends. Later in his life, however, Lang came to believe that seeking explanations for evolution fell within the remit of metaphysics rather than science.

A comprehensive evaluation of Lang's orthogenetic theories of Cretaceous cribrimorph evolution cannot be undertaken until a phylogenetic framework has been established through restudy of cribrimorph morphology[87] coupled with cladistic analysis. For the most part, the processes proposed by Lang have no place in modern evolutionary theory but would be difficult to falsify conclusively. However, the evolutionary patterns he inferred are both testable and worthy of future study. Such research would be of specific importance in clarifying the origin of ascophoran cheilostomes, and of more general interest in understanding evolutionary trends and the evolution of complexity.

8. Acknowledgements

I am grateful to Dennis Gordon, Andrew Smith, Phil Palmer, David Lewis and Mark Wilson for their critical readings of the manuscript.

Notes

1 P.J. Bowler, *The Eclipse of Darwinism* (Baltimore and London, 1983).
2 Arthur Trueman provided a rare exception, describing Lang's work as stimulating and brilliant. A.E. Trueman, 'The meaning of orthogenesis', *Transactions of the Geological Society of Glasgow*, 20 (1940), 77-95.
3 H.H. Swinnerton, *Outlines of Palaeontology* (Edward Arnold, 1947).
4 For example, K.J. McNamara (editor), *Evolutionary Trends* (Tucson, 1990); S.J. Gould, *Life's Grandeur* (London, 1996); D.W. McShea, 'Metazoan complexity and evolution: is there a trend?' *Evolution*, 50 (1996), 477-492; J. Alroy, 'Understanding the dynamics of trends within evolving lineages', *Paleobiology*, 26 (2000), 319-329.
5 Lang's correspondence files are archived in the Earth Sciences Library at the NHM, London.
6 E.I. White, 'William Dickson Lang', *Biographical Memoirs of Fellows of the Royal Society*, 12 (1966), 366-386. Two other biographies, both untitled, of Lang are known: H.D.T. [Henry Dighton Thomas], *Proceedings of the Geological Society of London*, 1636 (1967), 202-203; M.A.A., *Proceedings of the Geologists' Association*, 78 (1967): 387-389.
7 Lang himself was a staunch Christian, see W.D. Lang, 'Human Origin and Christian Doctrine', Nature, 136 (1935), 168-170.
8 M.K. Howarth, personal communication, May 2001.
9 A copy of a letter written by Lang to fellow Harrovian Sir Reginald Spence in September 1960 in the Lang correspondence archive at the NHM notes that 'In 1897 a visiting Master arrived to teach Biology as a "special subject" for 2 hours a week. Being a born Naturalist, I took to biology as to no normal school subject, ..".
10 C.P. Palmer, 'William Dickson Lang: his Liassic work appraised', *Proceedings of the Dorset Natural History and Archaeological Society*, in press.
11 Lang's retirement house in Charmouth was appropriately called 'Lias Lea'.
12 White, note 6, p. 374.
13 H. Woods, *Palaeontology Invertebrate* (Cambridge, 1893).
14 Lang, for example, criticized Bather's philosophy of how fossils should be exhibited to the public. W.D. Lang, 'Palaeontology and the public: a curator's aspirations.', *Proceedings of the Geologists' Association* 41 (1930), 175-179.
15 F.A. Bather, 'Fossils and life', *Report of the British Association for the Advancement of Science*, 88th Meeting (1920), 61-86.
16 W. T. Stearn, *The Natural History Museum at South Kensington*, p. 240 (London, 1981)
17 Lang was unfit to serve in the armed forces during the First World War and instead undertook war work which led to an entomological publication, *A handbook of British mosquitoes* (London, 1920).
18 G.F. Elliott, personal communication, *c.* 1980.
19 W.D. Lang, 'The Jurassic forms of the 'genera' *Stomatopora* and *Proboscina*', *Geological Magazine*, Decade 5, 2 (1904), 315-322.
20 Lang followed the practice then prevailing and included the Danian within the Cretaceous whereas it is now recognized as the first stage of the succeeding Paleocene. Some of the families which Lang regarded as becoming extinct during the Cretaceous can therefore now be seen to have survived into the Tertiary.
21 W.D. Lang, *Catalogue of the fossil Bryozoa (Polyzoa) in the Department of Geology, British Museum (Natural History). The Cretaceous Bryozoa (Polyzoa). Volume III. The cribrimorphs.*

– *Part I.* (London, 1921); *Catalogue of the fossil Bryozoa (Polyzoa) in the Department of Geology, British Museum (Natural History). The Cretaceous Bryozoa (Polyzoa). Volume IV. The cribrimorphs. – Part II.* (London, 1922).

22 W.D. Lang, 'Polyzoa' in L.R. Cox, 'The fauna of the basal shell-bed of the Portland Stone, Isle of Portland', *Proceedings of the Dorset Field Club*, 46 (1925), 52-55.

23 Lang summarised his approach to palaeontology in a short *Nature* 'birthday' article from which the following passage is taken: 'Palaeontological investigation, therefore, should I think, be carried out along the lines of bed-by-bed collecting; of tracing lineages; of noting growth-stages; and of observing trends and the evolutionary history of individual characters; and, whether investigating Palaeozoic corals, Cretaceous Polyzoa, or the faunal succession of the Dorset Lias, I have followed, and continue to be guided by, these methods of research.' W.D. Lang, *Nature*, 128 (1931), 1085.

24 For example, a congratulatory letter from W.B.R. King dated 23 February 1929 remarks 'Better late than never.'

25 See C. Tickell, *Mary Anning of Lyme Regis* (Lyme Regis, 1996).

26 Bowler, note 1.

27 W. Haacke, *Gestaltung und Vererbung: Ein Entwickelungsmechanik der Organismisen* (Leipzig, 1893)

28 G.H.T. Eimer, *On Orthogenesis and the impotence of Natural Selection in Species-Formation* (Chicago, 1898).

29 Lang discussed Eimer's butterfly work as an example of orthogenesis but, perhaps surprisingly, sought to explain the evolutionary trends postulated in these butterflies by Eimer as resulting from Natural Selection. W.D. Lang, 'Evolution: a resultant.', *Proceedings of the Geologists' Association*, 34 (1923), 7-20.

30 G.H.T. Eimer, *On Orthogenesis and the impotence of Natural Selection in Species-Formation* (Chicago, 1898), 21.

31 S.J. Gould, *Ever since Darwin* (New York, 1977), 84.

32 This idea bears an obvious relationship to the 'Biogenic Law' which states that ontogeny recapitulates phylogeny, and results from the tendency for evolutionary novelties to be added to the terminal developmental stages of individuals; here, however, phylogeny is considered to progress through stages paralleling those seen in ontogeny (e.g., youth, senility and death/ extinction).

33 According to Lang himself (1919, 61): 'A large part of the evolution of Jurassic and Cretaceous Ammonites is concerned with their catagenesis or decline.' This decline is manifested by the appearance of such features as uncoiling, sutural simplification, and a reversion to unornamented shells. W.D. Lang, 'The evolution of ammonites. A demonstration given at the British Museum (Natural History)', *Proceedings of the Geologists' Association* 30 (1919), 49-65.

34 G.G. Simpson, *Tempo and Mode in Evolution* (New York, 1944).

35 Plate (1913) coined the term 'orthevolution' to describe the evolutionary pattern of linear (or rectilinear) trends, and used the terms 'orthoselection' and 'orthogenesis' respectively for what might now be seen as Darwinian and non-Darwinian evolutionary processes. Confusion between pattern and process is also evident in two other terms - rectilinear evolution and programme-evolution – which were used as alternatives to orthogenesis, the former applying to a pattern and the latter to a process. L. Plate, *Selektionsprinzip und Probleme der Artbildung* (Leipzig, 1913).

36 R.S. Lull, *Organic Evolution* (Revised edition, New York, 1929).

37 E.B. Poulton, 'The history of evolutionary thought as recorded in meetings of the British Association', *British Association for the Advancement of Science, Nottingham*, 1937, 17-18.
38 Lang's response to Poulton was published the following year: '..the doctrine of trends (and this is where it has been misunderstood) does not explain; it describes: it is not a matter of speculation, but of observation. I suppose that, whatever Eimer exactly meant by Orthogenesis, he would have included the phenomenon of Trends under that term; yet Sir Edward Poulton, in his Presidential Address to the British Association last September, opposed Orthogenesis both to Darwinism and, by inference, to Lamarckism. .... [a trend] is only relevant to Darwinism in what it takes as given, namely, limited variation in definite directions, instead of unlimited variation in all directions.' W.D. Lang, 'Some further considerations on trends in corals', *Proceedings of the Geologists' Association*, 49 (1938), 149.
39 J. Huxley, *Evolution. The Modern Synthesis* (London, 1942).
40 Simpson, note 34, p. 164.
41 Trueman, note 2.
42 W.D. Lang, '*Stomatopora antiqua*, Haime, and its related Liassic forms', *Geological Magazine*, New Series, Decade 5, 2 (1905), 258-268.
43 W.D. Lang, 'Calcium carbonate and evolution in Polyzoa', *Geological Magazine*, Decade 6, 3 (1916), 73-77.
44 E.R. Cumings and J.J. Galloway, 'Studies of the morphology and histology of the Trepostomata or monticuliporoids', *Bulletin of the Geological Society of America*, 26 (1915), 349-374.
45 Cumings and Galloway, note 44, p. 362.
46 Lang, note 43, p. 75.
47 This idea can be traced back to Alpheus Hyatt. According to Gould "Hyatt believed that the sequence of new stages added to a lineage during the course of phylogeny runs parallel to the stages of an individual's ontogeny." S.J. Gould, *Ontogeny and Phylogeny* (Cambridge, Massachusetts, 1977), 93.
48 Lang, note 43, p. 75.
49 Lang, note 43, pp. 76-77.
50 W.D. Lang, 'A revision of the "cribrimorph" Cretaceous Polyzoa', *Annals and Magazine of Natural History*, Series 8, 18 (1916), 81-112.
51 Lang, note 50, p. 82.
52 W.D. Lang, 'The Pelmatoporinae, an essay on the evolution of a group of Cretaceous Polyzoa.' *Philosophical Transactions of the Royal Society of London*, Series B, 209 (1919), 191-228.
53 Lang, note 52, pp. 195-196.
54 Lang, note 52, p. 196.
55 W.D. Lang, 'The Kelostominae', *Quaterly Journal of the Geological Society of London*, 74 (1919), 204-220.
56 W.D. Lang, *Catalogue of the fossil Bryozoa (Polyzoa) in the Department of Geology, British Museum (Natural History). The Cretaceous Bryozoa (Polyzoa). Volume III. The cribrimorphs. – Part I.* (London, 1921);
57 Lang here added a cautionary footnote : 'It is true that not many lineages can be followed to such an extreme development. But it could hardly be otherwise with the fragmentary phylogenies at our disposal. What is significant is that when term links with term into a lineage, progress is *in that direction.*'
58 In the Discussion at the end of Lang's paper on the Kelestominae, Sidney F. Harmer, the leading British authority on Recent bryozoans at the time and a former teacher and now colleague of

Lang at the BM(NH), questioned the large number of families and subfamilies erected by Lang, believing that most could be accommodated within the extant family Cribrilindae. W.D. Lang (note 55), 219.

59 Harmer, note 58, p. xxxviii.

60 W.D. Lang, *Catalogue of the fossil Bryozoa (Polyzoa) in the Department of Geology, British Museum (Natural History). The Cretaceous Bryozoa (Polyzoa). Volume IV. The cribrimorphs. – Part II.* (London, 1922).

61 F.A. Bather (note 14), 78-79. Also highly critical of Lang's theories was R.M. Brydone who had first hand experience of Cretaceous cribrimorphs; R.M. Brydone, *Further notes on new or imperfectly known Chalk Polyzoa* (London, 1929).

62 'I sometimes wonder whether the gene is made to carry more burdens than it can reasonably bear'. Brydone, note 61, p. 377.

63 W.D. Lang, draft of a letter written to A. Arber, 23rd August 1945.

64 F.L. Kitchin, letter to W.D. Lang dated 3rd March 1914.

65 F.L. Kitchin, letter to W.D. Lang dated 6th March 1914.

66 A. Dendy, 'Momentum in evolution', *Report of the 80th Meeting of the British Association, Portsmouth 1911* (1912), 277-280.

67 W.D. Lang unpublished autobiography quoted at length by E.I. White (note 6).

68 T.J.M. Schopf, 'Patterns and themes of evolution among the Bryozoa', in *Patterns of Evolution*, edited by A. Hallam (Amsterdam, 1977), 159-207.

69 G.P. Larwood, 'Frontal calcification and its function in some Cretaceous and Recent cribrimorph and other cheilostome Bryozoa', *Bulletin of the British Museum (Natural History)*, (Zoology Series), 18 (1969), 171-182.

70 Further discussion of myagromorphs can be found in G.P. Larwood, 'Form and evolution of Cretaceous myagromorph Bryozoa', in *Bryozoa: Ordovician to Recent*, edited by C. Nielsen and G.P. Larwood (Fredensborg, 1985), 169-174. A further stage in the origin of cribrimorphs, not discussed by Lang or Larwood, is the loss of articulation at the spine bases.

71 e.g., Lang, note 56, pp. xxxvi-xxxviii.

72 D.P. Gordon, 'Towards a phylogeny of cheilostomes – morphological models of frontal wall/shield evolution', in *Proceedings of the 11th International Bryozoology Association Conference*, edited by A. Herrera Cubilla and J.B.C. Jackson (Balboa, Republic of Panama, 2000), 17-37.

73 Lang, note 56, pp. 4-6.

74 G.P. Larwood, 'The morphology and systematics of some Cretaceous cribrimorph Polyzoa (Pelmatoporinae)', *Bulletin of the British Museum (Natural History)*, (Geology Series), 6 (1962), 1-285.

75 G.P. Larwood, 'Colonial integration in Cretaceous cribrimorph Bryozoa', in *Advances in Bryozoology,* edited by G.P. Larwood and M.B. Abbott (London, 1979), 503-520. E. Voigt, 'The Bryozoa of the Cretaceous-Tertiary boundary', in *Bryozoa: Ordovician to Recent*, edited by C. Nielsen and G.P. Larwood (Fredensborg, 1985), 329-342.

76 Larwood, note 69, figure 1.

77 Larwood, note 76, text-figure 7c.

78 J.S.Ryland, 'Physiology and ecology of marine bryozoans', *Advances in Marine Biology*, 14 (1976), 285-443.

79 See discussion and references in P.D.Taylor, 'Bryozoans', in *Functional morphology of the invertebrate skeleton*, edited by E. Savazzi (Chichester, 1999), 623-646.

80 Schopf, note 68, pp. 188-189.

81 G.P. Larwood and P.D. Taylor, 'Mesozoic bryozoan evolution: response to increasing predation pressure?, in *Recent and Fossil Bryozoa*, edited by G.P. Larwood and C. Nielsen (Fredensborg, 1981), 312-313. F.K. McKinney and J.B.C. Jackson, *Bryozoan Evolution*. (Boston, 1989).

82 D.W. McShea, 'Mechanisms of large-scale evolutionary trends', *Evolution*, 48 (1994), 1747-1763.

83 D.P. Gordon and E. Voigt, 'The kenozooidal origin of the ascophorine hypostegal coelom and associated frontal shield', in *Bryozoans in Space and Time*, edited by D.P. Gordon, A.M. Smith and J. Grant-Mackie (Wellington, 1996), 89-107. D.P. Gordon, 'Towards a phylogeny of cheilostomes – morphological models of frontal wall/shield evolution', in *Proceedings of the 11th International Bryozoology Association Conference*, edited by A. Herrera Cubilla and J.B.C. Jackson (Balboa, Republic of Panama, 2000), 17-37. Note that defining an ascophoran cheilostome is not straightforward; at least some Cretaceous 'cribrimorphs' probably had proximal outpocketings of the frontal membrane forming a functional ascus, conventionally treated as the defining character of an ascophoran (D.P. Gordon, personal communication, June 2001).

84 Gordon and Voigt, note 83, p. 105.

85 e.g. M.J. Weedon and P.D. Taylor, 'Skeletal ultrastructure in primitive cheilostome bryozoans', in *Proceedings of the 11th International Bryozoology Association Conference*, edited by A. Herrera Cubilla and J.B.C. Jackson (Balboa Republic of Panama, 2000), 400-411.

86 H.A. Lowenstam and S. Weiner, *On Biomineralization* (Oxford, 1989).

87 This was begun but not completed by G.P. Larwood (note 75).

Eliza Catherine Jelly (28th September 1829 – 3rd November 1914): pioneer female bryozoologist

Hugh S. Torrens*[*] and Judith E. Winston[†]
[*]*University of Keele, Keele ST5 5BG, UK*
[†]*Virginia Museum of Natural History, Martinsville, VA 24112, USA*

1. Introduction

We should first note the great difficulty in studying the history of women, as one of us realised when investigating the career of the English fossil collector and dealer Mary Anning (1799-1847).[1] The real problems which such women scientists faced during their careers are nicely demonstrated by the treatment meted out to another of the first women in English science, who wrote a pioneering *Catalogue* of the fossils of Wiltshire in 1831. This is the Wiltshire-based geologist and palaeontologist Etheldred Benett (1775-1845). After she donated a copy of this in person to the British Museum library, she wrote thus in 1836 of her treatment at the hands of males:-

> They have made me a member of the Imperial Natural History Society of Moscow and have sent me a Diploma of the appointment, but it is provoking that no one will believe that a Lady could write such a trifling thing [as her *Catalogue*] - in this Diploma I am called Domin*um* Etheldred*um* Benett and Mr Lyell [Charles Lyell 1797-1875] told me that he had been written to by foreigners to know if *Miss* Benett was not a Gentleman, as of course I sent them all 'with Miss Benett's compliments': nor was this error confined to Foreigners. I had two letters the

> same day from the British Museum, one thanking me for fossils [collecting which was a respectably female activity] as Miss Benett, the other for my book as Etheldred Benett Esq., though I gave the Book into the hands of one of the Librarians and told him that it was written by myself, so you can see that scientific people in general have a very low opinion of the abilities of my sex.[2]

We shall find Eliza Jelly facing similar problems as a student of zoology 52 years later (see Figure 1).

Born ten years after Queen Victoria (1819-1901), Eliza Catherine Jelly lived through the entire Victorian era and into the Edwardian - from gaslight to electric lighting, from the horse-drawn carriage to the motor car - yet, largely because of her sex, many details of her life remain both enigmatic and symptomatic, of the silence afforded to the many women who were already working in science. We have been able to uncover a good deal about where, and with whom, she lived, but little of the how or why of her life and particularly why exactly she became a bryozoologist. What we do know of her character and motivation comes from her own writings, the preface to her book and her extant letters to colleagues, especially the series written from May 1895 to Sir Sidney Frederic Harmer (1862-1950), while he was superintendent of the University Museum of Zoology at Cambridge University, now preserved in the Bryozoa Section at the Natural History Museum, London. As much as we can, we hope to let Miss Jelly speak for herself.

2. Parentage

Eliza Catherine Jelly was born in Bath, Somerset, England, on 28th September 1829.[3] But only her christening, on 11th November 1829, was recorded in the registers of St. Swithin's, the Walcot parish church near the house at 13 Walcot Parade, where her parents then lived. The situation with her only brother, Harry Richardson Jelly (1831-1894), was different. His birth on 21st March 1831, was carefully recorded in these same registers, along with his baptism, when he was later christened in the same church.

Her father, Harry Jelly (1800-1843), was an Anglican clergyman and Oxford graduate. Her mother, Eliza Cave (*c.* 1798-1860), came, like her paternal great-grandfather, from a family of Bath builders. Her parents were married at Walcot by licence on 1st January 1829. Like many clergymen of the time, Harry Jelly was also a naturalist, with strong interests in palaeontology.[4] During the nineteenth century natural history was considered a most desirable hobby for a clergyman. Not only might hiking through the countryside keep him active and healthy, but searching for the plants, fossils, or insects that were his special interest allowed him many opportunities to interact in a casual and friendly way with his parishioners, increasing his ability to serve as their spiritual advisor.[5]

Harry Jelly had been orphaned in August 1801 at the age of less than one by the death of his father Thomas Jelly junior (*c.* 1763?-1801), an attorney in Bath. The same fate soon attended Harry's later close friend John Phillips (1800-1874), orphaned in 1808, who was to become Professor of Geology at Oxford University.[6] Phillips later referred to Harry Jelly as "one of my earliest friends... and a student in the school of Richardson and Smith".[7] Phillips had, by 1813, been sent to a school at nearby Holt in Wiltshire. This had been recommended to his uncle, William Smith (1769-1839), the pioneer of stratigraphy and mineral prospecting, by Smith's friend the naturalist parson, Rev. Benjamin Richardson (1758-1832) after whom Eliza's brother was to be named. From this school, in 1814 Phillips "passed to a most pleasant interlude by accepting a twelve-months invitation to the home of my ever-honoured friend, the Rev. Benjamin Richardson of Farleigh Castle, near Bath", where he joined Harry Jelly.[8]

3. Jelly ancestry

Eliza's great grandfather, Thomas Jelly senior (*fl.* 1752-died 1781), was a carpenter, builder and architect in Bath who was one of the creators of that fashionable spa city from 1752.[9] He married Mary née Smith (1725-1790),[10] then of Bradford-on Avon, Wiltshire, at Bath's St Michael church on 25th September 1760.[11] The close link between the Jelly and Richardson families arose from this marriage, because Benjamin Richardson's wife Ann was a daughter of another marriage,[12] now of Richard Whatly (1709-1782), also of Bradford-on-Avon, to Mary Smith's younger sister, Elinor Smith (1725-1786). These two sisters were daughters of Richard Smith, and his wife Elizabeth and baptised on 25th July 1725 and 26th February 1726 at St Martin in the Fields, London.[13] Harry Jelly was thus, not Richardson's nephew as claimed,[14] but "a distant relation" of Mrs Richardson by marriage.[15]

Benjamin Richardson had married Ann Whatly in 1785.[16] Benjamin Richardson's will[17] left everything to his widow Ann Richardson (1763-1847) and was witnessed by Harry Jelly's first cousin Selina Jelly (born 1789) and by Harry's brother-in-law A.G. Barretté. Ann Richardson's will[18] equally mentions her many Jelly relations, including Harry Jelly and his first cousin John Gresley Jelly (1790-1859), the father of Selina.

Thomas Jelly senior had had three children, Mary and two brothers, John Jelly (1761-1813)[19] and Thomas Jelly junior (*c.* 1763?-1801). John Jelly provides the first evidence of any Jelly family interest in natural history. After entering Lincoln's Inn in London from 1782 and marrying well,[20] he became a money lender and solicitor in Bath. Here he published in 1794 *Hortus Bathonienis or a Catalogue of the Plants cultivated in the Subscription Botanic Garden at Bath.*[21] But on 7th February 1795 a Commission of Bankruptcy was issued against him[22] and he was forced to sell his house and property, including the Botanic Garden he had created.[23] He had also collected material for a "*Flora Bathonica*, which, from [these] unforeseen circumstances, was never published".[24]

John's younger brother, Thomas Jelly junior, was equally unfortunate. He was baptised in 1773 but this seems certain to have been a late baptism. He was probably born about 1763. He too became a solicitor in Bath, but died in August 1801 supposedly aged 27 [probably 37], leaving a widow and six children[25] including Eliza Jelly's father. Such were the sad circumstances of Eliza Jelly's grandfather and great uncle.

4. Eliza's father

Harry Jelly was orphaned at an age of less than one. He is first heard of when he and John Phillips spent time together living with the Richardsons at Farleigh Hungerford rectory, Somerset in 1814 and 1815, before Phillips left for London on 21 November.[26] Certainly Harry Jelly was still living with the Richardsons between 1818 and 1821.[27] He next matriculated at Oxford University (St Alban Hall) aged 27 in 1828. He was awarded his B.A. degree in 1832 and his MA degree in 1835. He next became curate at Farleigh Hungerford in 1831 and 1832,[28] serving as assistant to its rector Benjamin Richardson, the man who had inspired him long before this to start collecting fossils. He is first recorded donating some of these, Chalk, fossils from Wiltshire and Transition beds [Silurian] fossils from Gloucestershire, to the Bath Literary and Philosophical Institute in 1826.[29] He was duly listed as a geological donor to that Institute in 1829 in William Lonsdale's MSS catalogue.[30] This overriding interest was one on which he also published some papers, as well as some valuable biographical notices.[31]

His only published volume, of *Sermons* published by subscription, shows that he next became assistant curate at Walcot, St Swithin's church, where Eliza had been baptised. He took up this position on 18th September 1832 at the annual salary of £100.[32] His final position was as Minister or perpetual curate of Trinity Church, Bath. Sadly he had by now become an invalid, suffering from consumption (tuberculosis). From his letters to John Phillips after 1836,[33] a little of his movements can be gathered. In 1836 he was living at 12 Claremont Place, Bath. By August 1837 he was at Redland, near Bristol and again visiting there in August 1838, when he writes that his "long illness [had] caused me to be ordered down to Penzance [Cornwall] last winter [between January and June 1838] as a means of escape from the severity of the winter". While he was there Harry Jelly collected corals and other fossils from the South Devon coast for Phillips.[34] In the same letter Jelly adds:-

> I am thinking of a longer journey... for I am seriously proposing a voyage this autumn [from October 1838] to Jamaica. I have a[n elder] brother[35] settled there in the practice of physic and he strongly urges my making a sea voyage as the best means of recovering my health.

This journey Jelly certainly made, as he donated specimens from Jamaica to the Geological Society of London in September 1839 (see Note 31).

His published volume of sermons is dated from Penzance March 1840. This shows that he had returned to Cornwall after Jamaica, having been forced by ill-health to give up his Trinity Church position in Bath. In February 1841 he was still in Penzance living at 3 Morrab Place. But his ill-health had failed to respond to either the West Country or the West Indies and on 18th March 1843 he died at Down Cottage, West Clifton, near Bristol aged 41, and was buried at St John the Evangelist church, Redland, Bristol.[36] Phillips recorded in an April 1843 letter how fine Jelly's personal collection of fossils had been, but all trace of it has since been lost.[37] Jelly's 1838 will, leaving effects under £6,000, was proved on 13th July 1843 in the Prerogative Court of Canterbury. This left all his estate to his widow Eliza, now of Westbury-on-Trym, Bristol, who died 8th September 1860 aged 70 at 11 Lockyer Street, Plymouth, Devon.[38] After her demise his estate was to be shared equally between his two children, Harry Richardson and Eliza Catherine since they had long reached the age of 21.

5. Eliza's natural history background

Under such a family's, and in particular such a father's, influence, it is easy to see how Eliza Catherine was drawn to the study of natural history. But for her, living invertebrates, especially molluscs and bryozoans, and plants, particularly algae and liverworts, were finally to prove more attractive than fossils. The Jelly family lived in the Bath and Bristol area until Eliza was 13, with visits to the sea coast at Penzance, in Cornwall, in hope of restoring their father's health. It is almost certain that Eliza's first experiences with bryozoans came while collecting on the seashores there. The family seems to have remained in Devon until 1860 where, as we have seen, Eliza's mother was living, in Plymouth, when she died.

The Jelly's were never wealthy, and the opportunities open to their children were limited by this fact. Harry Richardson Jelly did briefly attend one of England's best (or at least now most expensive!) schools, Eton College, in Berkshire from January 1845 at the age of 13 where his tutor was W.L. Evans but he appears to have very soon left, in the summer of 1846.[39] Probably funds collected for his education after his father's early demise had been exhausted. He next matriculated at Oxford (Brasenose College) in 1849 aged 18. He graduated B.A. in 1854 and now followed in his father's footsteps by becoming an Anglican clergyman. He was ordained at Winchester between 1854 and 1861, when, now as Jellie to confuse historians and genealogists, he married Alice Maria Loveless at Barnstaple in Devon.[40] He later held a large number of church positions, in Teignmouth, Devon 1863-4; Alberbury, Shropshire 1864-5; Alverstoke, Hampshire 1869-1871; Haddenham, Cambridgeshire 1871-75; Ealing [London] 1875-76; London 1876-77 and he was finally appointed chaplain to the Fulham and Hampstead hospitals in London, from 1883 until his death on 2nd September 1894 at 91 Tubbs Road, Willesden, Middlesex aged 63.[41] He seems never to have taken any interest in natural history, unlike his father and his sister.

6. The Stewart family

The 1871 census entry for Haddenham, Cambridgeshire records Harry Richardson Jellie's eldest son living there. He had been named Thomas Stewart Jellie, and was born in 1863 at Teignmouth, Devon.[42] This link with the Stewart family (just as with the Richardson family earlier) was to prove highly significant for Eliza Catherine's later life and work. No such advanced schooling as was given to her brother was available to her. The first women's college at Oxford was not founded until 1879, so she could not follow her brother there. However, despite her opportunities for formal education being so limited, her family background, plus the strong Victorian belief in self education seems to have enabled Eliza to pursue her scientific interests. Collecting natural history specimens and microscopy were both mid-Victorian passions. Eliza may have gravitated toward bryozoans as a good subject for microscopy. Well-brought up Victorian women were often discouraged from anything that resembled useful or gainful employment, but study of minute bryozoan specimens under the microscope was probably considered 'useless' enough to be acceptable, on the same level as 'desirable' female accomplishments like needlework, watercolour painting and making decorative objects from shells and seaweed collected at the seashore.

Eliza never married. Perhaps with no dowry it would have been difficult for her to have done so. After her mother's death in 1860, she lived in friends' homes, later working as a "Governess" and then as a "Lady's Companion". These were two of the few acceptable options by which well-bred young women without families could support themselves. For much of the decade from 1861 to 1870 she seems to have lived in the household of Colonel William Stewart (*c.* 1795-1865), after whom her brother's eldest son had been named. Stewart was a retired Madras Army colonel, who lived at Eldon Villa, in Lower Redland Road, Bristol. In the 1861 census entry for Eldon Villa, Eliza Catherine is listed living there as "an unmarried friend, without occupation aged 31".[43] It is possible that the Stewarts, who came from either Liverpool (the husband) or East India (the wife), were friends or relatives of Eliza's mother's family.

William Stewart died on 31st December 1865 at Eldon Villa and in his will, leaving effects under £8,000, he left Eliza Catherine Jelly £400.[44] Stewart's wife, Isabella neé Bowser (*c.* 1801-1879), continued to live at Eldon Villa until after 1870 when she moved to 17 Brock street, Bath where she died on 2nd December 1879.[45] By May 1870 Eliza was instead based on the Wirral peninsula in Cheshire.

7. Eliza's interests in natural history

Eliza's serious interests in natural history had certainly already started while she was living at Eldon Villa. Her first, and only, scientific journal publication was a list of the land and freshwater molluscs of Bristol, published when she was 38 from this address, uniquely under the new spelling of their name which her brother now used, Jellie.[46] She soon however reverted to the spelling Jelly. Her whereabouts between 1870 and 1880 are partly revealed by five of her letters to the botanist Rev. Edward Adolphus Holmes (1809-1886) F.L.S. 1834[47] preserved in the archives of the Linnean Society, London.[48]

In the first, of 12th May 1870, Eliza writes from Dee House, Parkgate, a sea-side resort near Neston in Cheshire, about the local natural history. She first discusses a moss she then called *Dicranella fallax*, noting Dr Robert Braithwaite (1824-1917)'s comments on this species. She had found it there in "a deepish ditch, down close to the water & hidden by grass etc [where it was] no easy matter, especially for a lady to hunt it up". Braithwaite specifically referred to her discovery of this species here in his own work, with another moss from Teignmouth which she must have collected while living in Devon.[49] This letter continues by knowledgeably discussing lichens she had also found there. Her next two letters, 16th and 23rd July 1870, are from 4 Peel Terrace, New Brighton, a larger resort at the north east corner of the Wirral peninsula. She had been in North Wales in the interim. These discuss other lichens and the alga *Conferva Brownii* which she had sent Holmes from Mousehole cave, Cornwall, "the only English habitat" the previous year. She also promised to send Holmes a carte de visite photograph of herself and to put him in touch with a female algologist in Jersey and others in Scotland.

Her fourth letter, of 24th October 1870, is from Albion Street in New Brighton and announces that she has now been busy, and successfully, collecting zoophytes on the local shores, but was not sure if Holmes took any interest in them. She asks if he had ever found the hydroid *Plumularia myriophyllum* on the south coast. She also asks if he had heard from the Cornish cryptogamist William Curnow (*c*. 1809-1887) from whom she "can get no answer to her letters". Her final surviving letter, of 22nd July 1871, comes from "The Oval, Addiscombe, Surrey" thus giving us her new permanent address. She now sends Holmes, a letter from Alexander Croall (1809-1885) about a sea-weed she had queried with him. She also returns Holmes' zoophytes, "one of which I cannot name as it does not belong to the families I am studying. I believe it is one of the Gorgoniadae" [i.e. sea fans]. This provides the first evidence we have that Eliza was now collecting and studying Bryozoa, having already clearly mastered to a considerable extent sea-weeds, algae, lichens and mosses.

Eliza Jelly is listed in the 1871 Census as a governess, i.e. a female teacher in a private household, living at 1 Pretender Villas, Oval Road, Addiscombe, Croydon, a few miles due south of London. Her earlier references to south coast specimens record her own new accessibility to the Sussex coast by train, whence came many of the Chalk fossils with which she later supplied G.R. Vine. She shared this Croydon house with three other women, one of whom was another lodger, Edith Williams, (*c*. 1837-1899), the third daughter of William Williams, a Cardiff brewer.[50] She will reappear a lot in Eliza's story.

In 1880 Eliza was still busy collecting. The first surviving specimen we can associate with her is her Cornish seaweed, *Nitophyllium gmelini* Lam. "with tetrapores from Penzance, May 1880", now preserved in the Dr Philip Brooks Mason (1842-1903) collection at Bolton Museum, Lancashire.[51] Eliza's presence in Cornwall is confirmed by the 1881 Census which records her sharing 5 Florence Place, Falmouth, Cornwall with five other unmarried women, including one born in North America.[52] Two more can be identified:

(1) Rebecca S[mith] Fox (1823-1901), a daughter of Joseph Fox (1792-1861) of the Quaker Fox family long resident there,[53] which included a number of American Consuls based at Falmouth and
(2) Edith Williams, who had been with her in 1871 at Croydon.

A problem is created by Augustus Hamilton (1854-1913) in New Zealand who later reported that "in 1880 [he] commenced a correspondence with Miss E.C. Jelly who was then residing at Bristol".[54] It seems most likely that Hamilton was simply wrong about when their correspondence actually started.

8. Eliza Jelly as bryozoologist

Eliza kept MSS "Notes on Polyzoa" which now survive as a single bound volume in the Harmer library at The Natural History Museum, London (hereafter NHM). These include the record of "*Diastopora saruiensis* Norman, Hastings, 1878, E.C.J." which seems to be her first specific bryozoan record. The same volume preserves two letters to her from the London surgeon and polyzoologist George Busk (1807-1886) of 32 Harley Street.[55] The first, dated 7th July 1882, identifies species of *Idmonea* and *Catenicella* for her. The second dated 26th July 1885 tries to do the same for species of *Membranipora*.

In 1880 the British Association for the Advancement of Science (= BAAS) had reappointed Peter Martin Duncan (1824-1891) and George Robert Vine senior (1825-1893), a bryozoologist and stay and corset maker in Sheffield, as a committee to report on British Fossil Polyzoa with Vine as its secretary and chief operative.[56] Vine's work, with his son G.R. Vine junior, is reviewed elsewhere in the present volume.[57] The third *Report* of this Committee was issued in 1883. It now comprised two members both from Sheffield, Henry Clifton Sorby (1826-1908) and Vine. It asked for contributions from "any student of British or Foreign Fossil Polyzoa" and included records from the local cabinet of a Miss Gatty.[58] This was the Sheffield bryozoologist, Horatia Katherine Frances Gatty, later Mrs Thomas Bainbridge Eden (1844-1945),[59] who was a daughter of the famous children's and natural history writer Margaret Gatty (1809-1873), who had earlier made her own special study of Bryozoa.[60]

Vine and Sorby's fourth *Report* was issued in 1884. It covered British Cretaceous Polyzoa and now thanked "Miss E.C. Jelly for lending specimens and furnishing the minute details necessary for this Report". Eliza Jelly seems to have been introduced to Vine by Miss Gatty, since this same *Report* also ended by thanking Miss Gatty for allowing her collection of Polyzoa to be examined. Eliza Jelly's Chalk Polyzoa recorded here included two provisionally new species from Beachy Head, Hastings, Sussex and other specimens from Haldon Hill, Devon.[61] The fullest tribute to Eliza Jelly's collaboration with Vine came in his fifth and final *Report*. This thanked both Jelly and Gatty for the loan of fossil specimens and noted that Vine had:-

> only intended to give the history and classification of Cyclostomatous Polyzoa... because of the limited time [he had available but] in making my wants known to my ever constant friend - Miss E.C. Jelly - her answer was prompt and welcome: 'The Cheilostomata *must* be done, and you can command my services to any extent in the work'. Of her labours I have gladly availed myself.[62]

From this it seems likely that Vine and Jelly had been collaborating since the start of this BAAS project. But only an undated 25 page alphabetical "Index to [recent] Polyzoa by E.C.J." now survives in Vine's archive at the National Museum of Wales, Cardiff.[63]

It seems certain that Eliza Jelly's first recorded donations of Bryozoa to museums, made to Sheffield Museum in February 1883, July 1885 and March 1886, came as a result of this same collaboration. Her donations comprise mainly microscope slides of material from around the world especially Australia, New Zealand and Great Britain (particularly Hastings and the south coast of England). The 1883 donation was of 74 slides from Britain, including Shetland and Guernsey, while the 1885/6 donation was of 177 foreign specimens in two parts. Her address at the time of these donations was not recorded. A single 1886 Jamaican specimen (E 86-114) came from a collector with the initials E.J.S.

Eliza made another donation in February 1883 to Liverpool Museum, whose records list 72 specimens of identified, and only British, Polyzoa, arranged according to Rev. Thomas Hincks (1818-1899)'s *History of the British Marine Polyzoa* of 1880, many from Devon, also others from Hastings, Guernsey and all round Britain as in her above first Sheffield donation. Again no address for Miss Jelly seems to have been recorded with this donation. Hincks' books provided another catalyst to Miss Jelly's specialist bryozoological activities. They too were in contact and some of her Bryozoa are now in the Hincks collection at the Natural History Museum, London. Hincks had been forced to retire from the Unitarian church in 1869 and he soon retired to Clifton where they could have been in close, if brief, proximity.[64]

According to a note by Dr Anna B. Hastings preserved at the Natural History Museum:-

> Miss Jelly was evidently in communication with a wide circle of polyzoologists and made a practice of distributing samples of collections sent to her. These she mounted and labelled before sending them out, with the result that various important collections contain slides

prepared by Miss J., and identically mounted and labelled. Samples of the same material can be found in several collections in this way. Examples of [such] twin slides [can be seen] in the Hincks and [John] Bracebridge Wilson [1828-1895][65] collections in the [Nat. Hist. Museum]... Much of the material described by Hincks in his *Annals [and Magazine] of Natural History* series [1880-1893] on "Contributions towards a general history..." were provided by Miss Jelly.[66]

Other such material is in the Arthur William Waters (1946-1929) collection at Manchester Museum.[67] These two were certainly in contact by April 1881.[68]

It is clear that Eliza was, from about 1880, corresponding with people all over the world about the study and collection of bryozoans. Miss Jelly was thus an important collector and distributor of bryozoan specimens, as well as being their bibliographer. Although we have no information that Eliza ever travelled abroad herself, her letters show that she corresponded with most of the bryozoan workers, both fossil and Recent, of the day. As the Belgian palaeontologist Edward Pergens reported of her in 1894 " qui bon nombre de bryozoologistes doivent la possession d'une grande partie de leurs matériaux".[69] She convinced them and other overseas correspondents to collect for her and to send her specimens. She described no new species herself, but sorted, identified, and mounted and labelled the material she amassed. Those specimens she considered might represent new taxa she passed on to colleagues like Waters for formal description:-

part of the material forming the subject of the present communication I received in exchange from Miss E. C. Jelly, in a small test tube, already washed out of the clay; and on two subsequent occasions she has kindly lent me a number of slides from her collection,[70]

or Randolph Kirkpatrick (*c*. 1863-1950)[71]

Miss E. C. Jelly recently sent to the Natural History Museum an interesting species of *Retepora* from Port Western [Australia], belonging to an undescribed species,[72]

or Augustus Hamilton

in 1880, when living at Petane, near Napier [New Zealand], I commenced correspondence with Miss E. C. Jelly..., on the subject of our New Zealand Bryozoa, and from that time until quite recently I have been forwarding parcels for recent and fossil species from different parts of New Zealand... Through Miss Jelly the specimens I collected have been passed on to the principal authorities on the various groups. Several of the new species have been described in English scientific publications and others are still awaiting description.[73]

In 1887 Eliza Jelly was elected a Fellow of the Royal Microscopical Society (one of the few scientific societies which then "admitted" women) when her address was given as "Hatchlands", now in Hatchlands Road, Red Hill, Surrey.[74]

9. Change of address

This was in a wealthier neighbourhood 30 kms (20 miles) due south of London, a little farther out of London to the south west of her previous Croydon address.[75] Hatchlands, which still stands, is the next house east of the Forester's Arms public house, at Shaw's corner, between Red Hill and Reigate. Eliza had clearly moved here between 1881 and 1887. Red Hill was equally convenient for London by train, and still allowed good access to London libraries. She remained a Fellow until 1897 when, as we shall see, she retired from such work.

She must have moved to Red Hill to be the single companion and caregiver to her old friend Edith Williams (*c.* 1837-1899), third daughter of a Cardiff brewer with whom, as shown above, she had been living at Croydon in 1871 and at Falmouth in 1881. Edith had first moved after 1881, when they shared their Falmouth home, to Apple Tree Farm, Winscombe near Weston-super-Mare, Somerset, maybe also with Eliza Jelly. But since Edith was to die of breast cancer, it seems likely that Eliza now agreed to share the costs, and the painful duties of house sharing for Edith Williams at Red Hill. The 1891 Census shows them living together there, with Edith Williams as head of the household, "living on her own means".[76] When Edith Williams died at Hatchlands on 1st December 1899 she left her entire estate, of £2,276, to her "dear friend, Eliza Catherine Jelly"[77] and they share a double grave in Chart Lane cemetery, Reigate (see Figure 4).

In 1887 and 1888 Eliza gave two further bryozoan collections to Oxford University,[78] the first was again British material (which survives today) and the second Australian, "156 slides of foreign polyozoa" (but of which no trace has been found recently).[79] Two letters from her survive at Oxford with these donations, the first, of July 28th [1888], reports that she had sent:-

> a series of mounted Polyzoa (foreign) as a supplement to the British collection I had previously given... I should like to hear that they are [now] in your possession. I do not suppose my name in connection with the previous set has been kept as E.C. Jelly, Esq. although the letter I received from some member of the family of Professor [Henry Nottidge] Moseley [1844-1891 - saying he was ill] was so addressed.[80]

The age old problems for women working in science clearly persisted into late Victorian times!

10. *The Synonymic Catalogue of the Recent Marine Bryozoa*

Sometime during the 1880s she began her book, *A Synonymic Catalogue of the Recent Marine Bryozoa*. This must have been inspired by the work she had done for Vine before 1885. His last "*Report on Fossil Polyzoa*" (see above) had noted that:-

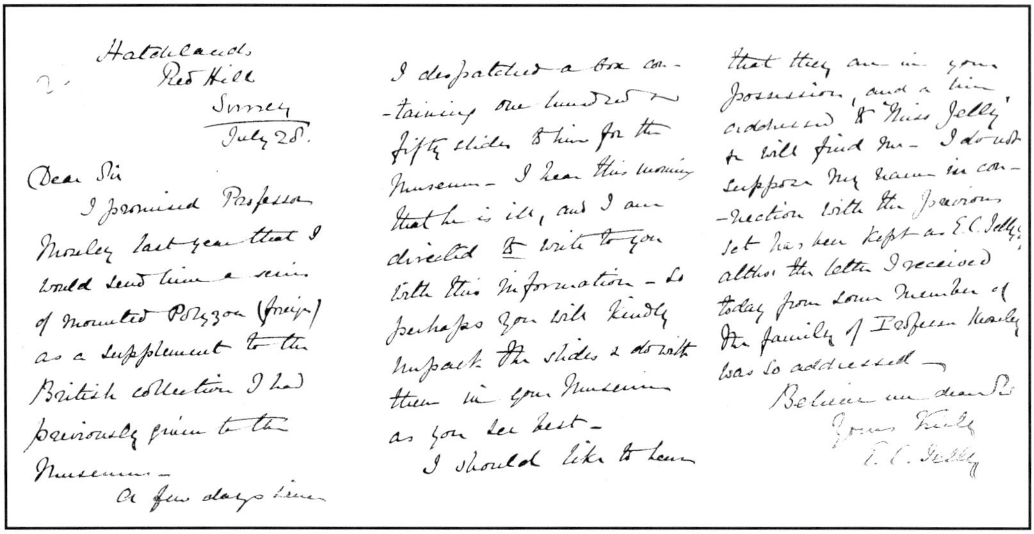
Hatchlands, Red Hill Surrey July 28.

Dear Sir

I promised Professor Moseley last year that I would send him a series of mounted Polyzoa (foreign) as a supplement to the British collection I had previously given to the Museum — A few days since I despatched a box containing one hundred & fifty slides to him for the Museum — I hear this morning that he is ill, and I am directed to write to you with this information — so perhaps you will kindly unpack the slides & do with them in your Museum as you see best — I should like to learn that they are in your possession, and a line addressed 'to Miss Jelly' will find me — I do not suppose my name in connection with the previous set has been kept as E.C. Jelly; although the letter I received today from some member of the family of Professor Moseley was so addressed —

Believe me dear Sir

Yours truly

E.C. Jelly

Figure 1. A sample of Eliza Jelly's handwriting. This 28th July [1888] letter was addressed to an unknown member of the Oxford University Museum staff (OUM archives).

> I owe to her the compilations of many of the elaborate lists given below [including one - listing Manzoni's Italian Pliocene Bryozoa. As Vine did] not have these papers by me, I can only give this list as supplied by Miss E.C. Jelly.[81]

Eliza's importance to bryozoan history lies first and foremost in this book, which consists of an alphabetical listing of names and synonyms for all the 1,696 Recent species known up to the time of the publication date given on its title page - 1889. In fact, since reviews only appeared from August 1890 and its Preface is dated "Red Hill, December 1889", it was clearly not published until 1890. The *Catalogue* was begun as a personal effort, partly to help her in organizing her own collection:-

> the study of the Bryozoa having been carried on by many naturalists in many parts of the world, it has followed as a natural result that more than one name has been given to a single species ... as also, that one name has fallen to the lot of two or three different species.

It was based not just on the published literature, but on determinations of valid taxa by herself and other experts. As she said in her preface:-

> two chief ideas have been kept in view. The first was to collect in a workable form all the names of recent Bryozoa that have been published, so far as books and scattered papers could be obtained... The second idea was to reduce the synonymy to something like 'fact,' including in such synonymy only the fossil forms belonging to recent species... With the great variety of opinions on much connected with the subject of synonymy, it would be impossible to adopt that of all writers, but the lines I worked upon are those of Mr. [Thomas] Hincks and Mr. [Arthur

William] Waters, for recent and fossil respectively, venturing to differ from these known authorities only on rare occasions.[82]

According to her preface, fellow bryozoologist Arthur Waters had encouraged her to publish the *Catalogue*, "had it not been for the kindly encouragement of Mr. Waters, my spirit would have failed me many times",[83] but she had had to raise the money for its publication herself. According to her letter dated 2nd March 1889, probably to Professor Moseley at Oxford:-

I have been at work for several years on a Catalogue of all published recent Polyzoa with a full synonymy, a work which has never yet been attempted, and which Mr Hi[n]cks in his letter accompanying my application to the Royal Society for a grant of money said would fill a vacant place in the literature of the subject.

The letter asks Moseley to recommend her *Prospectus* to the notice of the Oxford Museum Committee

and ends

I have but a small chance of ever seeing it published, and this would be a little disappointing after the heavy work and expense it has entailed for years... I have personally examined every reference.[84]

She seems certain to have received a grant from the Royal Society, probably of £100, towards printing costs.[85] She also wrote to bryozoan workers and museums around the world asking them to subscribe for copies to make publication possible, and she apparently received financial help from microscopist friends, especially from Charles Nathaniel Peal (*c.* 1832-1898) of Ealing, West London,[86] of whom she wrote:-

To Mr. Peal I owe everything connected with the publication; he undertook the seeing it through all its financial difficulties, and has proved himself a very real friend in bringing my work to a practical result.[87]

The book was published by Dulau and Co., booksellers and publishers, in Soho Square, London. Sadly this publisher's premises were destroyed, and their then Director Leslie Chaundy killed, in an air raid during the Second World War,[88] so we cannot discover how many copies of the book were printed. It was one of a number of such synoptic or bibliographic guides which Dulau published from 1888 to 1892, like those on Foraminifera by C.D. Sherborn 1888,[89] on British Fossil Vertebrata by A.S. Woodward and Sherborn 1890,[90] or on British Jurassic Gasteropoda by W.H. Hudleston and E. Wilson in 1892.[91] The first of these books had been earlier refused a grant of £100 from the Royal Society towards costs of printing, because its later famous bibliographer author, Charles Davies Sherborn (1861-1942), had written "a savage criticism of a foreigner" who had published

A CATALOGUE

OF THE

PUBLISHED SPECIES of RECENT POLYZOA

WITH A FULL SYNONYMY,

BY

E. C. JELLY, F.R.M.S.

The above work, which will be published as soon as a sufficient number of Subscribers has been obtained, gives *all published names* of recent Polyzoa, with *recent and fossil synonyms in full.* The main lines followed are those of Hincks and Waters, the two British authorities on recent and fossil Polyzoa respectively. No attempt at making such a Catalogue has hitherto been made, and in addition to showing all synonymy, it enables the future student to avoid the use of specific names which have been already appropriated.

An early reply on the accompanying form will be a favour.

Figure 2. Her 1889 Prospectus *- preserved at the Oxford University Museum of Natural History (OUM archives).*

an earlier quite inadequate foram bibliography.[92] Sherborn's bibliography was only published thanks to the enterprise of Frederick Justen (died 1906 - managing director of Dulau and Co.) who must have been equally involved with Eliza Jelly's book. Sherborn later paid these fine tributes to him:-

> Justen was a great and true friend and later on offered me a partnership in [that] firm. He had the most extraordinary knowledge of natural science literature and was a dear and amiable man..., who practically made the [Natural History Museum, London library] collection.[93]

The following review of Eliza's book appeared in August 1890 in the *Journal of Royal Microscopical Society*[94] probably soon after the book was published:-

> *Synoptic Catalogue of Recent Marine Bryozoa.** — Miss E. C. Jelly is to be congratulated on the completion of this most valuable work. Though it is not to be expected that her fellow-workers will in all cases agree with her in her views of synonymous names, they are all greatly indebted to her for the long and patient labour necessary for a work of this kind; in many cases a single species occupies more than a page and sometimes nearly two. When a recent form is also known in the fossil state, references to the fossil specimens are given. Only a systematist will fully appreciate the service Miss Jelly has performed for students of the Marine Bryozoa. *8vo, London, 1889, 322 pp.

In October 1890 *Nature* also gave it an enthusiastic review[95] saying:-

> It was a work that all students of the Polyzoa should be grateful for. It supplies an undoubted want and will greatly facilitate the investigation of the large and interesting class with which it deals..., the work ... bears the marks of careful and conscientious labour.... The value of such works depends entirely on the care and minute accuracy with which they are compiled. Miss Jelly's *Catalogue* affords abundant proof that these qualities have not been wanting in her case. The book is handsomely got up, and printed in a type which, so far as clearness in concerned, leaves nothing to be desired. Miss Jelly is to be congratulated on the completion of a very onerous task.

A third review was published in 1890. It regarded it as an "exceedingly useful book" which was almost too conscientiously complete. It had to point out a few faults, but then noted only one single incorrect synonym. It finally complained that the book was about Polyzoa, not Bryozoa, but concluded that "the author deserves the gratitude of all students" of that group.[96]

Later systematists certainly appreciated her efforts and put Miss Jelly's *Catalogue*, as it came to be called, to use:-

> Miss Jelly has long been known as an enthusiastic student of the Bryozoa, and as one whose intimate acquaintance with the literature of the subject gives authority to the excellent synonymic catalogue of the recent marine Bryozoa, published by her in 1889.[97]

Even today her work is a good starting point for taxonomic research into Recent species described up to the time of its publication, which comprise about a quarter of those known today.

The proceeds from the book were important to her financially, as her correspondence shows:-

> Mrs. Dymond tells me you seem disappointed at not having a copy of my *Catalogue* in the Museum library, but my dear Sir your museum did not subscribe to the work. I myself sent you prospectus and subscription forms for filling in when I forwarded to other Australian workers, but your museum committee took no notice of it, and I was not in a position to give copies anywhere.[98]

Eliza may also have depended on the sale of collections and specimens to augment her income. However, she clearly also collected for herself and for the joy of seeing and owning new specimens, as another letter to the curator of the Australian Museum shows:-

> Dear Sir: A letter from my old friend Mrs. Dymond, now in Sydney, tells me of her interview with yourself about a species of *Retepora* I asked her to try and get for me. It is very difficult to get the Bryozoa comprehended by anyone who does not know them, and I told her that Mr. Brazier's children called it the 'Red Coral'. It is really the *Retepora phoenicea* of Busk, of which I wanted to get a few full grown, fairly perfect specimens. I have never seen anything but fragments of it and I wanted a whole specimen if it could be had.[99]

Despite her lack of involvement in formal taxonomic publication she had clear opinions on species' identities, taxonomic characters, and methodology. Some of them are detailed in her *Catalogue*:-

> Waters' places *Cellepora rota* of MacGillivray as a synonym to this species, but MacGillivray insists on their being two quite distinct forms. It is possible that Mr. Waters has had imperfect specimens of *C. costata*, as the sinus in '*costata*' and '*rota*' are certainly of different shapes.[100]

To others, as in these excerpts from her letters to Harmer between May 1895 and December 1897, she did not hesitate to give directives to professional taxonomists:-

> Do you not think it might be advisable with these South African specimens, to accept some that may be unnamed in the hope that either Mr. Hincks or Mr. Waters may name and publish them at some future date? As for instance, I have odd pieces of a species that seems to partake of features partly *Adeona* and partly *Adeonella*. If I am not mistaken Busk ranks those possessing true rooting processes and fenestra as *Adeona*. Now this S. African species has true fenestra but no true rooting process. Mr. Waters fancied it was *Adeona grisea*, but I have written to him and shown him that it is quite distinct from that form. It seems a pity as you intend working up a Museum collection of Bryozoa, not to receive *new* species because a specific name has not [yet] been assigned. *Localization* of all specimens is a necessity, but is a new species to be shut out

> because unnamed? Do turn this over in your mind dear Sir, and let me know your further thoughts regarding this point.[101]

or

> I am so glad that you see the necessity of keeping several slides of a species. As a general rule, I think that two slides are always required for a species with or without oecia. *Cribrilina clithridiata* W. needs several, to show the cell, the oecium, the avicularium and the central cell. In my own collection I paid special attention to the cell of origin, and I strove to make a species fully represented in all forms and varieties rather than hunt after novelties. I think the variations of a species are not sufficiently attended to.[102]

Eliza Jelly also organized many of the specimens she received into sets of slides (e.g., 50 slides of British species, 50 of South African species). She continued to disperse these collections of identified specimens to a number of institutions including the Royal Microscopical Society, where in October 1890 "attention was called to a donation of special importance, consisting of 200 slides of Marine Bryozoa, from Miss E. C. Jelly, who had recently published a catalogue of Bryozoa".[103] There are also slides from Miss Jelly at Oxford University, Cambridge University Museum of Zoology (she gave part of her personal collection to Sidney Harmer when he was at Cambridge), Sheffield, Manchester, Liverpool, and Trinity College, Dublin. These last comprise a collection of 110 bryozoans in wooden cavity slides, mainly from Port Phillip and Port Western, Australia, with other material from Western Australia, California and New Zealand (Figure 3).[104]

Abroad there are other of her specimens at the Australian Museum, Sydney, and the Zoological Museum, Copenhagen. These last were collections she sent to George Marius Reinold Levinsen (1850-1914) in Copenhagen between August 1890 and 1895 which survive there. They were referred to in Levinsen's major monograph of 1909.[105]

Eliza Jelly was honoured for her work as a bryozoologist by Levinsen by the new species *Euthyroides jellyae*.[106] Edward Wilhelm Gerard Pergens (1862-1917) had earlier, in 1894, founded a Cretaceous species, *Truncatula jellyae*, after her.[107] As, in 1896, did John Walter Gregory (1864-1932) who named a new Jurassic species *Multiclausa jellyae*.[108] After her death Sidney Frederic Harmer named *Tervia jellyae*, after her, from a Queensland specimen she had supplied to him.[109] David Alexander Brown named a further species *Exochella jellyae* in her honour.[110] More recently she has been honoured with the new genus, *Jellyella*, created by Paul Taylor and Neale Monks in 1997 for two pseudoplanktonic membraniporid species.[111]

Eliza Catherine Jelly does not fit Victorian stereotypes of the 'weaker sex'. What we know of her life through her letters and collections shows her determination, intelligence, and clarity. In many ways, she embodied those Victorian virtues, of self-confidence,

Figure 3. Five of the bryozoan cavity slides she presented to Trinity College, Dublin.

self-improvement, devotion to duty, social responsibility, and liberality of mind. As a child of a naturalist, growing up during the years of natural history's greatest popularity, she believed strongly in its benefit throughout her life:-

> The queer little fish came from Port Elizabeth, S[outh] Africa. Can you give me its name? I gave another specimen of it to a young friend of mine who is a true naturalist, and I do all I can to foster her love for the study. Her principal delight lies among shells, but I think that young eyes and minds should cultivate an all round acquaintance with Nature and Nature's objects, taking one particular branch for collecting and studying.[112]

She was interested in a wide range of natural history as her letter of 11th March 1894 to Francis Jeffrey Bell (1855-1920) proves.[113] She writes about *Gorgonia* specimens which had been sent to her from South Africa, "many months ago you very kindly offered to name some South African species of *Gorgonia* on the one condition that they were [later] given to the [Natural History] Museum".

Although always polite to her colleagues, it is also clear from her letters that in cases of taxonomic controversy she did not defer to the professionals, but defended her own knowledge and experience:-

> Mr. Hincks, as a rule, never sent me localities of a thing, but the specimen of his *T. Dawsoni* was given into my hand when I was dining there one day. He was showing me the specimens he had so named, and I protested that it was not a new species, but only *Idmonea milneana*. Later

on he examined them critically, and said that I was right. Yes, you are to consider the locality to be that given in his original description of *T. Dawsoni*.[114]

11. Her last decades

In at least the last decade of the nineteenth century, Eliza devoted her life to nursing her close friend, Edith Williams, who was suffering from breast cancer. This cannot have been a happy situation in days when pain was not so easily alleviated. Despite chronic health problems of her own, Eliza retained her sense of humour, "Ill health is a sad block to *deeds*. I seem to end in nothing but words!",[115] or:-

Please do not take the trouble and spend your valuable time in writing me thanks for each lot separately. I will tell you when I have sent you all I have to send and then you see dear Sir one thank would suffice and save you a lot of trouble![116]

We do not know what illnesses Eliza suffered from, although one letter mentions that gout had hindered her work.[117] A letter to her from the Australian bryozoologist Paul Howard MacGillivray (1834-1895) in Bendigo, Victoria[118] dated 6 June 1892, notes "I am very sorry to hear of your continued ill health, and sincerely trust that the prospect of a return to health is not as bad as you think".[119] Her death certificate merely gives the cause of her death as "old age". Miss Jelly first mentions ill-health in a letter to Harmer dated 17th July 1895, "my interest in all this work is not lessened thro' my ill health, altho' power to study, compare and mount is gone".[120] The disposals of both her bryozoan collections and her library, which also started in 1895, seem to have been instigated for this same reason. She finally decided to give up all bryozoan studies two days after her 67th birthday, due to continued illness. "Ill health has quite put an end to all work, but the interest is very keen still and always will be until the end of the chapter".[121]

When she decided to give up scientific work she decided to dispose of the remainder of her collections. The first recipient of these final disposals seems to have been Manchester Museum, although that museum's poor record-keeping makes any final determination difficult.[122] Two original wooden Jelly slide cabinets survive there, but they have in the past been both separated, and separately catalogued. This has encouraged the wrong inference that only one half of any Jelly collection has survived there. The first cabinet contains material from Genera A (*Adeona*) to L (*Lunulites*) while the second contains material from Genera M to Z. Donation registers seem not to survive at Manchester (or they may be currently unavailable). There is no evidence of the Jelly donation in those printed *Reports* which survive at that Museum either, until 1910-11. But in the *Report* for 1895-96 there is certainly the record of a gift by "Miss E.C. Jelly of *Cellaria setigera* and *M. lepida*" being made between 1st January 1895 to 30th June 1896.[123] She herself refers to some of her rarest specimens of *Retepora* and *Parmularia* already being at this, then Owens College, Museum in a letter to Harmer dated 2

September 1896,[124] so her collection must have arrived there by then. No full record was given in that Museum's *Reports* until that for 1910-11 which notes:-

> The Jelly Collection of Marine Polyzoa consisting of over 1,000 slides, has been worked over and re-arranged by Mr [Robert] Standen, [Assistant Keeper of Zoology, Manchester Museum] in two cabinets. The arrangement followed that given in Miss Jelly's book *A Synoptic Catalogue of Marine Polyzoa* and the collection as now completed forms a useful reference readily available to students.[125]

D.A. Brown worked through these Manchester Museum collections and reported that the A.W. Waters collection there also contained a considerable amount of New Zealand material from Miss Jelly's collection. He also reported that the main Jelly collection there comprised "about 1,100 mounted specimens including 203 from New Zealand localities".[125] Brown ended his introduction:-

> I cannot conclude without paying a tribute to the work of Miss E.C. Jelly, not only for her excellent *Synoptic Catalogue* (1889) but also for her vast enthusiasm displayed as a collector of Polyzoa. Outside her own large collections[s], many specimens, both Recent and fossil, mounted and labelled by her are found in the Hincks, Vine, Waters, Bracebridge Wilson, and J[ohn] W[alter] Gregory [1864-1932 - Scottish explorer and geologist][127] Collections, and Levinsen also studied a considerable amount of her material.

Yet another, whose bryozoological studies she helped, was Thomas Rogers (1827-1901).[128]

In 1897 when she finally gave up all scientific work on doctor's orders, she gave the final portion of her own collections, comprising over 300 still unnamed specimens, to Sidney Frederic Harmer at Cambridge.[129] She also made efforts to sell her collection of bryozoan literature. She gave Harmer in 1897 her MSS notebooks, which survive among the Harmer MSS at The Natural History Museum, London.[130] One of her last letters to Harmer ends by her:-

> Very heartily regretting that the severance with my favourite study of many years will bring to an end all the pleasant communications I have enjoyed with workers like yourself. I need scarcely add that the severance is one of compulsion, and not of choice; and accepted only on stringent orders of the Doctor.[131]

In spite of the ill health that had plagued her, she lived to be 85. In fact, one of the problems we have in learning more of her life stems from the fact that she outlived most of the colleagues who might have written an informed obituary. When Edith Williams died in December 1899, she left "Hatchlands" and £2,276 to Eliza, which should have kept Eliza well enough for a few years. However, by the time Eliza made her own will on 2nd December 1905 she was living at a new, smaller, house, at 34 Gloucester Road, Red Hill,

Figure 4. Her grave in Chart Lane Cemetery, Reigate, Surrey. The lower inscription reads "who died November 3rd 1914 aged 85 years. Jesus said I have the keys of death" (HST photograph).

Surrey.[132] When she died in 1914, she did so at a nursing home located at 156 Station Road, Red Hill. Her date of death was 3rd November 1914,[133] at the beginning of the first hard winter of the Great War. She was buried on 6th November in Chart Lane Cemetery, Reigate, in the same plot as Edith Williams, where their shared grave survives.

Miss Jelly's will left all her effects (valued at only £338) to Edith Frances Cooper, spinster, of the Firs, Red Hill, Surrey. It is nice to be able to report that, in October and November 1918, Miss Cooper dutifully presented a number of Eliza Jelly's remaining books to The Natural History Museum in London.[134]

12. Acknowledgements

Grateful thanks are due to Penny Barents (Sydney), Ben Bather (London), Clifford Bird (Diss), Stella Brecknell (Oxford), Roger Clark (Bristol), John Collins (London), Sammy De Grave (Oxford), Gina Douglas (London), Tabitha Driver (London), Paul Ensom (London), Patricia Francis (Bolton), Sam Hallett (Bristol), Mrs P. Hatfield (Eton College), Caroline Hensley (London), Sheila Jelley (Amersham), Colin Johnston (Bath), Mary Spencer Jones (London), Alistair Maclean (Sheffield), Henry McGhie (Manchester), Paul T. Milan (Reigate), Mary Petersen (Copenhagen), Stella Pierce (Wincanton), Derek Shorrocks (Taunton), Susan Snell (London), Ian Wallace (Liverpool), Grace Woutersz (Croydon) and Patrick Wyse Jackson (Dublin) for so helping our hunt for Eliza Jelly. We have relied too on earlier investigations by Pat Cook (London) and the late Leslie J. Pitt (London who published one of the very few notices of her work in 1971, see footnote 110). The first author's research notes have been donated to the Harmer library, NHM.

Notes

1 See: H.S. Torrens, "Mary Anning (1799-1847) of Lyme; 'the greatest fossilist the world ever knew'", *British Journal for the History of Science*, 28 (1995), 257-284.

2 H.S. Torrens, E. Benamy, E.B. Daeschler, E.E. Spamer and A.E. Bogan, 'Etheldred Benett of Wiltshire, England, the first lady geologist - Her fossil collection in the Academy of Natural Sciences of Philadelphia, and the rediscovery of "lost" specimens of Jurassic Trigoniidae (Mollusca: Bivalvia) with their soft anatomy preserved', *Proceedings of the Academy of Natural Sciences of Philadelphia*, 150 (2000), 59-123 (63-64).

3 This date comes from a later affidavit made in 1855 by Augustus George Barretté, husband of Sarah Jelly (born 1797), as to the validity of Mrs Ann Richardson (1763-1847)'s copy will, preserved at the Wiltshire Record Office. Barretté, after a career in the Navy from 1809, retired on half pay, married Sarah in 1823 and became a solicitor in Bath (W.R. O'Byrne, *A Naval Biographical Dictionary* (London, 1849), 50). Sarah Jelly was youngest daughter of Thomas Jelly (*c.* 1763?-1801) and Eliza Jelly's aunt.

4 H.S. Torrens, C.J.T. Copp and R.F. Pickford, 'The Bath Geological Collections', *Newsletter of the Geological Curators Group*, *1* (3) (1975), 88-124 (97-9).

5 D.E. Allen, *The Naturalist in Britain* (London, 1976), 22-23.

6 J.M. Edmonds, 'The First "Apprenticed" Geologist, *Wiltshire Archaeological and Natural History Magazine*, 76, 141-154.

7 J. Phillips, *Geology of Oxford and the Valley of the Thames* (Oxford, 1871), 6, 411, 414 and 415. Richardson was later one of the dedicatees of Phillips', *Memoirs of William Smith LL.D*, (London, 1844).

8 J. Evans, 'Obituary of John Phillips', *Quarterly Journal of the Geological Society of London*, 31 (1875), *Proceedings* xxxvii-xliii. For Richardson, see A.G. Davis, 'The Triumvirate: A Chapter in the Heroic Age of Geology', *Proceedings of the Croydon Natural History and Scientific Society*, 11 (1943), 123-146.

9 See R.S. Neale, *Bath 1680-1850: a Social History* (London, 1981), 73, 159-60, 182 and 210 and H.M. Colvin, *A Biographical Dictionary of British Architects 1600-1840* (London, 1978), 457.

10 See *Gentleman's Magazine*, 60 (1790), 672 and 768.

11 Registers at Somerset Record Office, Taunton.

12 At Bath Abbey on 17th July 1751, see *Publications of the Harleian Society, Registers*, 27 (1900), Bath Abbey vol. 1, 261.

13 International Genealogical Index <www.familysearch.org>

14 J.E. Jackson, *A Guide to Farleigh-Hungerford*, third edition (London, 1879), 67.

15 W.S. Mitchell, 'Notes on Early Geologists connected with the neighbourhood of Bath', *Proceedings of the Bath Natural History and Antiquarian Field Club*, 2 (1872), 324-325 and Jackson, note 14, p. 62.

16 At Bradford-on-Avon on 31st January 1785 to which Thomas Jelly junior was a witness, see Bradford Marriage Register, Wiltshire Record Office, Trowbridge and *Bath Chronicle* (3rd February 1785).

17 Prerogative Court of Canterbury, proved 1832.

18 In Prerogative Court of Canterbury, copy will at Wiltshire Record Office, Trowbridge, proved 1847 with later affidavits.

19 *Gentleman's Magazine*, 73 (1813), i 498.

20 *Historical Collections for Staffordshire*, 19 (1898), 113.

21 Copy in E. Green collection, Bristol Public Library. The Garden opened in September 1793 see *Bath Chronicle* (12th September 1793).

22 See entry in William Smith and Co's, *A list of Bankrupts... from 1786-1806* (London, 1806) and *Bath Journal* (9th February 1795).

23 S. Sydenham, 'Bath Token issues of the 18th century', *Proceedings of the Bath Natural History and Antiquarian Field Club*, 10 (2) (1903), 207-238 (209, 213-215, 221, 235 and 238).

24 T.B. Flower, *Phytologist*, 3 (1849), 581.

25 *Bath Chronicle* (20th August 1801). His PCC copy will, proved in February 1802, survives at Bath Record Office (Acc. 59/2/4).

26 Edmunds, note 6, p. 145.

27 John Skinner MSS, British Library Add Mss, 33653, 33655-6, 33663, 33665, 33668 etc.

28 Jackson, note 14, p. 67.

29 Torrens, Copp and Pickford, note 4, p. 98.

30 W. Lonsdale, *Catalogue Raisonée of Rocks, from the neighbourhood of Bath* (MSS - dated 1829 - at the Bath Royal Literary and Philosophical Institute, Bath).

31 He published the following articles (apart from a volume of *XXII Sermons* (Wigan, 1840) - copy in British Library London):- 1) 'Biographical notice of Rev. B. Richardson', *Bath and Bristol*

Magazine, 1 (1832), 303-307; 2) 'The Lansdown encrinite' [now *Ailsacrinus prattii* (Gray)], *[ditto]*, 2 (1833), 37-46; 3) 'Biographical Notice of J.S. Miller', *[ditto]*, 2 (1833), 111-122; 4) 'On the study of Nature', *[ditto]*, 2 (1833), 297-304 and 5) 'On the Fossil Shells of the genus *Modiola*, being frequently found in the Bath Oolite, enclosed in the Shells of the genus *Lithodomus*', *Magazine of Natural History*, NS 3 (1839), 551-553. He also donated fossils and minerals to the Bristol Institution in 1833 (*ex inf.* Roger Clark) and British and Foreign land shells in 1839 (C. Copp, *South West Natural Science Research Unit*, 1 (3) (1985), 12 and *ex inf.* Sam Hallett) and Jamaican geological specimens on 24th September 1839 to the Geological Society of London (*Transactions*, 6 (1) (1841), 15 - these last as Rev. Harvey Jelly).

32 According to a note in the transcript of Walcot Birth and Marriage Register by Rev. Caius Barry, Somerset Record Office, Taunton.

33 John Phillips MSS 1836/27; 1837/43; 1838/40 and 1841/09, Oxford University Museum of Natural History.

34 J. Phillips, *Figures and Descriptions of the Palaeozoic Fossils of Cornwall, Devon and West Cornwall* (London, 1841), vi-vii.

35 This was Thomas Jelly born 1793 of Savanna la Mar, West Jamaica who published three pamphlets there:- 1) *Remarks on the condition of the white and free coloured inhabitants of Jamaica* (1826) - copy in Institute of Commonwealth Studies, University of London, 2) *On the Condition of Jamaica* (1847) and 3) *On British sugar colonies* (1848) - copies of these in the British Library.

36 See death notices in *Bath Chronicle* (23rd March 1843), 3; *Felix Farley's Bristol Journal* (25th March 1843), 3 and *Gentlemans' Magazine*, 174 (June 1843), 664.

37 Phillips to H.T. De la Beche, National Museum of Wales De la Beche MSS, 29th April 1843.

38 See Death Certificate (General Register Office, London) and *Gentleman's Magazine*, NS 9 (November 1860), 561.

39 *ex inf.* Miss P. Hatfield, College archivist, 18th April 2001.

40 Marriage Certificate, General Register Office, London.

41 Death Certificate, General Register Office, London.

42 PRO RG 10/1602, folio 68.

43 PRO RG 9/1739, folio 124.

44 Proved 1866, Principal Probate Registry, London.

45 Will dated 1876, proved 1880, Principal Probate registry, London.

46 E.C. Jellie, 'A list of Land and Fresh-water Molluscs belonging to Bristol and its immediate neighbourhood', *The Naturalist*, 3 (1867), 148-149.

47 J.A. Venn, *Alumni Cantabrigienses*, part 2, vol. 3, (Cambridge, 1947), 422 and F.W. Galpin, *Account of the Flowering Plants... of Harleston*, (London, 1888), 23-4.

48 E.A. Holmes MSS Z 35 (b).

49 R. Braithwaite, *The British Moss-Flora*, vol. 1, part 4 p. 67 and part 5 p. 111 (London, 1881). Specimens she supplied from Parkgate passed into Braithwaite's own moss collection.

50 *ex inf.* Croydon Local Studies Library.

51 *ex inf.* Patricia Francis, Keeper of Botany and see E.G. Hancock, A. Howell and H.S. Torrens, 'Geological Collections and Collectors of Note - Bolton Museum', *Newsletter of the Geological Curators Group*, 1 (7) (1976), 328-330.

52 PRO RG 11/2316/150, 44.

53 See Burke's *Landed Gentry* (London, 1972), 347.

54 A. Hamilton, 'A list of Recent and Fossil Bryozoa collected in various Parts of New Zealand',

Transactions and Proceedings of the New Zealand Institute, 30 (1898), 192-199.
55 Bound in her MSS Notes on Polyzoa, Harmer library, NHM, pp. 133 and 141.
56 *Report of the 50th meeting of the BAAS* (1880), lxi.
57 See C.J. Buttler, P.N. Wyse Jackson and T. Sharpe, this volume.
58 *Report of the 52th meeting of the BAAS* (1883), 251-2 and 266.
59 See Burke's *Landed Gentry* (London, 1952), 964.
60 *Dictionary of National Biography* and J.H. Ewing (ed.) *Parables from Nature* by Margaret Gatty (London, 1903), xix. See also S.L. Sheffield, *Revealing New Worlds: Three Victorian Women Naturalists* [one of whom is Margaret Gatty], (New York and London, 2001).
61 *Report of the 53rd meeting of the BAAS* (1884), 162-3, 167-8, 172-5. Miss Jelly's Chalk fossil Bryozoa need to be traced in extant collections. One of her type specimens of a provisional new Vine species was soon noted, by J.W. Gregory, *Catalogue of the Fossil Bryozoa in the Department of Geology British Museum (Natural History) - The Cretaceous Bryozoa* (London, 1899), vol. 1, 101, as already "missing from Vine's collection". It may thus have been returned to Eliza Jelly.
62 *Report of the 54th meeting of the BAAS* (1885), 153.
63 *ex inf.* Patrick Wyse Jackson, 11th May 2001.
64 'Obituary of Thomas Hincks 1818-1899', *Year Book of the Royal Society of London* (1900-1901), 27-28.
65 See *Australian Dictionary of Biography*, 6, 1851-1890 R-Z, (Melbourne, 1976), 417-418.
66 Pre 1977 TSS in the Bryozoa section, NHM, London.
67 See D.A. Brown, *The Tertiary Cheilostomatous Polyzoa of New Zealand*, (London, 1952), 2-6.
68 A.W. Waters, 'On Fossil Chilostomatous Bryozoa from South-West Victoria, Australia', *Quarterly Journal of the Geological Society of London*, 37 (1882), 309.
69 E. Pergens, 'Nouveaux Bryozaires du Crétacé du Limbourg', *Bulletin de la Société Belge de Géologie [etc]*, 7 (1894), 177.
70 Waters, note 68.
71 See obituary notice in *Nature*, 165 (29th April 1950), 669-670 and S.J. Gould, 'Crazy Old Randolph Kirkpatrick', *Natural History*, 87 (3) (1978), 20-22.
72 R. Kirkpatrick, 'Description of a new Species of *Retepora* from Port Western, Victoria', *Annals and Magazine of Natural History*, (6) 2 (1888), 269.
73 Hamilton, note 54, p. 192.
74 *ex inf.* Gerald Turner, RMS Hon. Archivist. In case this is thought to have been some 'brave new' concession to the women this Society had admitted as Fellows from 1884, we should note that the Society took their Admission Fees and Annual Contributions but did not allow them to attend meetings (see A.T. Gage, *A History of the Linnean Society of London* (London, 1938, p. 86). Gage rightly calls this a "preposterous arrangement which endured until 1909".
75 See W. Hooper, *Reigate: its story through the ages* (Guildford, 1945), 92 and 205.
76 PRO RG12/S77, Enumeration district 7, Folio 111, 1.
77 Death Certificate, General Register Office and will (proved 23rd December 1899) in the Principal Probate Registry, London.
78 K.C. Davies and J. Hull, *The Zoological Collections of the Oxford University Museum* (Oxford, 1976), 71.
79 *ex inf.* Sammy De Grave, Oxford University Museum of Natural History.
80 Archives of the Oxford Museum of Natural History.

81 Note 62, p. 153 and 209.
82 E.C. Jelly, *A Synonymic Catalogue of the Recent Marine Bryozoa* (London, 1889), vii.
83 Jelly, note 82, p. ix.
84 Jelly, note 82.
85 But no records of this survive in the Archives of the Royal Society, *ex inf.* Martin Carr, library assistant.
86 See his obituary in *Proceedings of the Linnean Society of London*, session 111 (October 1899), 57.
87 Jelly, note 82, p. ix.
88 See P.H. Muir, 'Further Reminiscences I to IV', *Book Collector*, 5 (3), 6 (4), 7 (2), and 7 (3) (Autumn 1956 to Autumn 1958) (especially IV, 278).
89 *A Bibliography of the Foraminifera* (London, 1888).
90 *A Catalogue of British Fossil Vertebrata* (London, 1890).
91 *A Catalogue of British Jurassic Gasteropoda* (London, 1892).
92 C.D. Sherborn, 'An "Instructive" Bibliography of the Foraminifera', *Nature*, 37 (1888), 583-584.
93 J.R. Norman, *Squire: Memories of Charles Davies Sherborn* (London, 1944), 50 and 95.
94 *Journal of the Royal Microscopical Society*, part 2 (August 1890), 449.
95 *Nature*, 42 (16th October 1890), 589-590.
96 *Annals and Magzine of Natural History*, sixth series, 6 (1890), 194.
97 Hamilton, note 54, p. 192.
98 E.C. Jelly to the Curator of the Australian Museum, 5th January 1891.
99 Jelly, note 98.
100 Jelly, note 82, p. 49.
101 Jelly to Harmer, (10th June 1895), NHM Letter no. 226.
102 Jelly to Harmer, (2nd September 1896), NHM Letter no. 363.
103 Notice in Minutes of the 15th October 1890 Meeting of the Society, *Journal of the Royal Microscopical Society*, part 2 (1890), 830.
104 Registered TCD 47613-47723, 48280-48282 *ex inf.* Patrick Wyse Jackson, 16th October 2001.
105 G.M.R. Levinsen, *Morphological and Systematic Studies on the Cheilostomatous Bryozoa*, (Copenhagen, 1909).
106 Levinsen, note 105, p. 264.
107 Pergens, note 69, pp. 176-177, plate 8, fig 4.
108 J.W. Gregory, 'A Revision of the British Jurassic Bryozoa - part VI', *Annals and Magazine of Natural History*, series 6, 17 (1896), 292.
109 S.F. Harmer, 'The Polyzoa of the Siboga Expedition. Part 1. Entoprocta, Ctenostomata, Cyclostomata', *Siboga Expeditie*, 28a (1915), 143-146, pl. XI, figs 1-3.
110 Brown, note 67, pp. 67, 291-293, figs 218, 219; see also L.J. Pitt, 'An Account of the Work on Polyzoa', *Proceedings of the Geologists' Association*, 72 (1961), 171-186 (182).
111 P.D. Taylor and N. Monks, 'A New Cheilostome Genus Pseudoplanktonic on molluscs and algae', *Invertebrate Biology*, 116 (1997), 39-51. The type species was based on Queensland material she had again supplied.
112 Jelly to Harmer, (17th July 1895), NHM letter no. 228.
113 In NHM archives, DF 250/1, Dept of Zoology correspondence.
114 Jelly to Harmer, (27th October 1896), NHM letter no. 368.

115 Jelly to Harmer, (12th August 1897), NHM letter no. 500.
116 Jelly to Harmer, (10th June 1895), NHM Letter no. 226.
117 Jelly to Harmer, (22nd September 1896), NHM Letter no. 364.
118 See *Australian Dictionary of Biography*, 5, 1851-1890 K-Q, (Melbourne 1974), 155-156.
119 This letter, with two others dated 1887 and 1891, are in NHM archives.
120 Jelly to Harmer, (17th July 1895), NHM Letter no. 228.
121 Jelly to Harmer, (30th September 1896), NHM Letter no. 365.
122 The situation is not helped by the fact that these collections are in store (in Buxton, Derbyshire) as this is written and so unavailable for study.
123 This *Report* is preserved in the NHM library, London.
124 Jelly to Harmer, (2nd September 1896), NHM Letter no. 363.
125 *ex inf.* Henry McGhie, Manchester Museum.
126 Brown, note 67, pp. 2-6.
127 [P.G.H. Boswell], 'John Walter Gregory 1864-1932', *Obituary Notices of Fellows of the Royal Society of London*, 1 (1932), 53-59.
128 J.C. Melvill, 'Obituary of Thomas Rogers', *Journal of Conchology*, 10 (1902), 142-144.
129 See 'Natural Science at Cambridge', *Natural Science*, 9 (October 1896), 221 and *Cambridge University Reporter*, 1183, (26th October 1897), 141 and 1184 (2 November 1897), 159.
130 F.C. Sawyer, 'A Short History of the Libraries...', *Bulletin of the British Museum (Natural History), Historical*, 4 (2) (1971), 79-204 (145).
131 Jelly to Harmer, (11th January 1897), NHM letter, unnumbered.
132 Principal Probate Registry, London. Eliza Jelly is not listed in the 1901 British census <www.census.pro.gov.uk>
133 *Times*, (5th November 1914), 1.
134 NHM Archives DF 204/10 and 205/78.

August Emanuel Reuss – portrait of an early bryozoologist

Norbert Vávra
Institute of Palaeontology, University of Vienna, Geozentrum, Althanstraße 14, A-1090 Vienna, Austria

1. Introduction

Micropalaeontology is usually considered to have been established by at least three scientists: Christian Gottfried Ehrenberg (1795-1876) in Germany, Alcide Charles Victor d'Orbigny (1802-1857) in France, and finally by August Emanuel Reuss (1811-1873) in Bohemia (having during his lifetime been part of the Austrian Empire). Whereas Ehrenberg had designated this new field of study as 'Mikrogeologie' the term 'micropalaeontology' had been first used by H. Ford in 1883.[1] Having been one of the pioneers in this new field of research would already justify the need to study the biography and the scientific achievements of Reuss; this, however, would describe only one of many scientific aspects of his life. Reuss also did work in various fields of invertebrate palaeontology; studying corals, crustaceans, gastropods, and sponges. He contributed important studies in respect to the local geology of Bohemia, and published on different topics of mineralogy. Thus he may be described in a proper way perhaps by the old-fashioned German term 'geognost'. In the following pages, details of his studies of fossil Bryozoa will be emphasized, in addition to some general aspects of his life and times.

Figure 1. August Emanuel Reuss (1811-1873)[2]

2. Biography

For the scientific career of August Emanuel Reuss the first and - perhaps - most important influence came without any doubt from his father. Franz Ambrosius Reuss (1761-1830) was working throughout his lifetime as a physician at Bilin (now: Bílina, Czech Republic), being one of the less-known spas in Bohemia. He had, however, a very broad interest in science in general. He had taken the degree of doctor at the University of Praha not only in medical sciences but also in philosophy; in addition he had studied mineralogy at Freiberg (Saxonia, Germany) where Abraham Gottlieb Werner had been his teacher.[3] At Bílina he studied and described the water of the wells 'für deren Bekanntmachung er als Arzt und Schriftsteller in ersprießlichster Weise gewirkt hatte' [translated:'for making public them he had been acting as a physician and as a writer in a most beneficial way'].[4] This outstanding man published an extensive amount of scientific papers. Apart from various titles concerning spas, there is to be mentioned a number of publications on earth sciences;. the most remarkable being perhaps a textbook on mineralogy in eight volumes. Many of his studies were devoted to local geology and mineralogy, e.g. his *Mineralogische Geographie von Böhmen* [Mineralogical geography of Bohemia]; he had also been involved in studies about the origin of basalt - one of the crucial scientific questions of this time, the era of endless discussions and disputes between plutonism and neptunism. Realizing these various scientific achievements, it is not amazing at all that he is held among the pioneers of Bohemian geology, a standpoint expressed for example by Bartenstein in his publication on occasion of the 150th birthday of August Emanuel Reuss.[5]

This very special father with his broad knowledge of science had taught his son until he started - with the 4th form [4. 'Gramatticalklasse] ! - to attend a public school at Praha. After finishing his secondary school ['Gymnasium'] he studied philosophy (1825-1827) at the University of Praha as well as medical sciences (1827-1833).

3. Scientific career

Without having finished his medical studies completely - in fact before he had taken his doctor's degree – August Emanuel Reuss was working as a physician dealing with lots of people suffering from cholera at Praha as well as at Bílina. Having finally been appointed assistant at the clinical hospital for ophthalmology at Praha, however, he left this city for health reasons to work - like his father - at Bílina as a physician. This position offered him the opportunity to come into closer contact with the neighbourhood of this city. This was rather useful in respect to his growing interest in mineralogy, geology and palaeontology. Here he had access to the rich collections of the sovereign family Lobkowitz, which proved very useful for his studies. In respect to geological material and fossils he used collections of the Bohemian Museum, of the 'k. k. Hof-Mineralien-Cabinet' [later: Museum of Natural History, Vienna] and in the following years also the collections of the 'K. k. Geologische Reichsanstalt' [Geological Survey, Vienna]. His scientific studies concerned first geological investigations of the immediate surroundings of Bílina and Teplitz. Later he studied the Cretaceous of Bohemia and its fossils. He made also extensive field studies in Styria, Tyrol and Lombardy - very much in the style of a 'reisender Geognost' [i.e. 'travelling geognost'] as people of his time have called scientists doing field work. In the years to follow he studied 'Polyparia' [corals and Bryozoa], ostracods and Foraminifera from the Neogene of the Vienna Basin.

In 1849 Reuss became full professor of mineralogy at the University of Praha, his predecessor in office - his former teacher Zippe - having been appointed full professor for mineralogy at the University of Vienna. For the next 14 years Reuss was teaching mineralogy at Praha; in addition to his regular lecturing activity he gave extra lectures at the 'Polytechnikum' [technical university] concerning 'Geognosie' and palaeontology. During all these years he was eagerly collecting minerals, thus establishing an extensive collection at the University of Praha. Nevertheless, he found enough time to continue his own scientific studies: the fossiliferous Gosau beds (Cretaceous), fossil crabs, systematics of Foraminifera as well as problems concerning the mineralogy of the Pribram area being his main fields of scientific interest during these years.

His former teacher - Zippe - had meanwhile become full professor of mineralogy at the University of Vienna. After Zippe's death, Reuss was appointed to the same position on August 24th, 1863. At Vienna's university one of his first achievements was to enlarge the collection of minerals which had been rather small. Among his scientific activities at Vienna a monography concerning the famous salt-mines at Wieliczka (now: Poland) is to be mentioned as well as his studies on molluscs from the Neogene of Austria, investigations performed to continue the voluminous publication 'Die fossilen Mollusken des Tertiär-Beckens von Wien' [=The fossil molluscs from the Tertiary basin of Vienna], a work begun by Hoernes, which was continued and finished after this author's early and unexpected death by Reuss.[6] In this and various other fields of science he thus remained very active until his death in 1873.

4. Bryozoan research

In his publication on the bryozoan fauna from Kalksburg (Vienna) Bobies had included as some sort of introduction amongst others, a short survey of the history of bryozoan research in Austria.[7] He mentions Reuss as the very first who studied fossil Bryozoa in Austria; for these activities Bobies assumed the years 1840 - 1845, the resulting publication declared to be a 'Monographischer Versuch' [=Monographical experiment] was finally published in 1847 resp. 1848.[8] The exact date of publication remained a matter of nearly endless debates: whereas Piller proved the year 1847 to be the correct year of publication, Lagaaij, David and Pouyet, and Braga (pers. comm.) regard 1848 as the official date.[9-12] The fact is that the first page of this publication is dated with 1847 whereas the volume 2 of 'Haidingers Naturwissenschaftlichen Abhandlungen' was not printed and distributed until 1848; this fact is confirmed by a remark in volume 3 of the same series ('Vorbericht'). The exact determination of the date of publication of Reuss' studies is of great importance in connection with questions of priority and has, therefore, consequences in respect to nomenclature. For a more detailed discussion see David and Pouyet, and Piller.[13]

This extensive publication describing the 'fossilen Polyparien' - i.e. corals and Bryozoa - from the Tertiary of the Vienna Basin is characterized by a rather high scientific standard, by an enormous number of very detailed informations concerning morphology, occurrence and many other aspects of corals and bryozoans. It is worth mentioning that in this publication the very first description of *Nullipora ramosissima* is included too - at this time a rather problematical organism, later to be called *Lithothamnium* or *Lithothamnion*, to be identified as a representative of Corallinaceae, a group of red algae. Reuss attributed these fossils to the corals though he had been quite aware of the fact that many people believed them to be plants (*teste* Piller).[14] Unger (1858) was obviously the very first who clearly realized the plant nature of these specimens.[15] Reuss' publication, however, shows - at least as far as bryozoan studies are concerned - a distinct lack of basic biological understanding, a lack of knowledge of Recent material, problems with the quality of illustrations, and obvious troubles with the optical equipment available - disadvantages which could hardly have been avoided at this time realizing the equipment and information available for a pioneer in such a field of research. The 'system' he used had been mainly based on zoarial types, characteristics of single zooecia were obviously regarded to be of secondary importance only. One minor mistake in this first study had been to include material from a locality designated as 'nicht näher bekannter Fundort im Wiener Becken' [= locality in the Vienna Basin not precisely known]; as Reuss himself has confirmed in the introduction to his publication in 1874, this had been a mistake: the material erroneously included under this designation was in reality from the Paleogene of Val di Lonte (Italy)![16] This important remark in his introduction (who reads introductions with the carefulness required in such a case ?) having been overlooked by many authors, gave rise to a number of confusions, which can not be discussed in detail within the scope of this paper.

In the following years Reuss studied most of the bryozoan-bearing sediments known at this time, being partly supported by Stoliczka and others. Such studies concerned mainly Bryozoa from the Tertiary: e. g. various faunas from the Oligocene of Germany, a fauna from the Tertiary of Oberburg in former southern Styria (now Slovenia) as well as material from the Paleogene of northern Italy, and even Bryozoa from the Sarmatian of Bessarabia.[17] In addition he had studied also bryozoan faunas from the Mesozoic, e.g. from the Jurassic at Balin, near Cracow or from the Cretaceous of Saxonia.[18] In the course of years he obviously became aware of the fact that his work of 1848 badly needed a revision: new localities, new materials, and better illustrations - at least these three items were the main reasons to prepare an extensive revision, the first part of which was finally published in 1874.[19] Nearly 30 years after his first publication on this topic the first part of a revision about the 'fossilen Bryozoen des österreichisch - ungarischen Miocäns' has thus been published. Descriptions of species had been based on his latest studies, literature references were given, magnificent plates included. The illustrations still suffered, here and there, under the traditional problem of this time: that scientist and illustrator did not always co-operate really perfectly. Thus it could occur that on the basis of different specimens something like 'ideal' pictures have been created showing mainly the most important characteristics of various species. Nevertheless, Reuss' scientific achievements included in these studies and in this publication are outstanding indeed. Unfortunately Reuss did not live long enough to finish this revision: his follower Manzoni continued the publication of these faunas, using in a reverent way the material as collected, prepared, and at least partly studied already by Reuss; the plates are generally of good quality, the descriptions, however, sometimes rather short - very often too short indeed.[20, 21]

5. Collections

Reuss had been a very busy collector during his whole life; with respect to mineralogical collections this fact has been mentioned already above. Extensive collections of

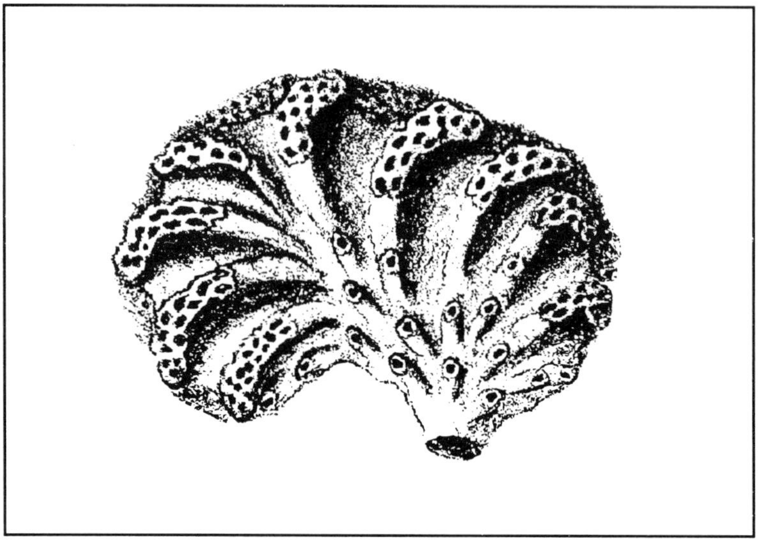

Figure 2. Original hand drawing o Tubulipora dimidiata, *Reuss (1848): pl.6, fig. 6 (Archives of the Institute of Palaeontogy, University of Vienna)*

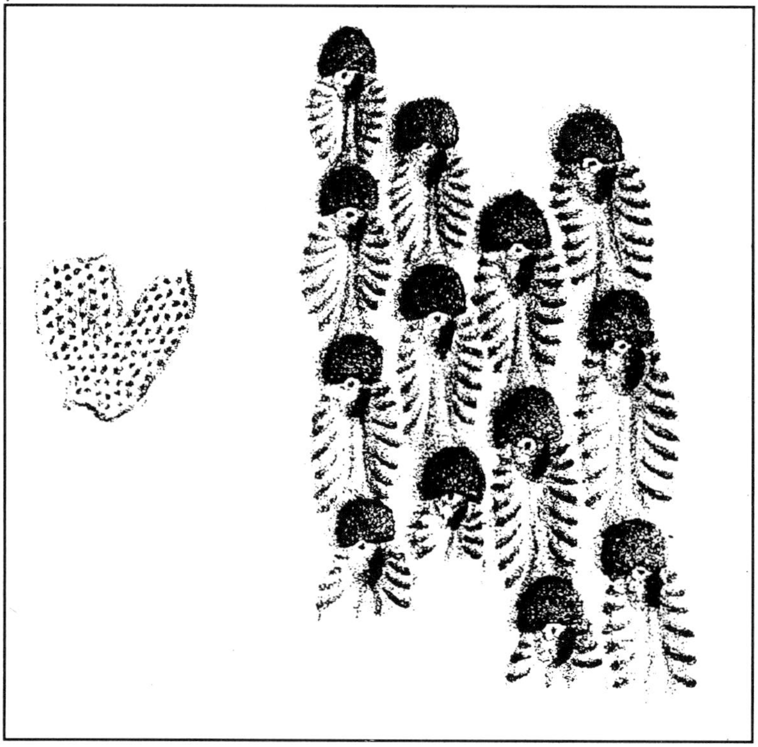

Figure 3. Original hand drawing of Umbonula macrocheila, *Reuss (1848): pl.8, fig. 14 (Archives of the Institute of Palaeontogy, University of Vienna)*

Foraminifera, Bryozoa and other invertebrate fossils are still kept at the Natural History Museum, Vienna. Realizing that Reuss has established during his life about 1000 new species, it is not amazing at all that his collection contains a true wealth of type specimens or 'Originalien' as they had been called at his time. This collection is generally in a rather good state: thus - to give a few details at least - the bryozoan material is arranged according to localities in drawers (about 20 for the Miocene), the Bryozoa being kept in small, open boxes – with most lying in small glass tubes. In each box there exists a number of labels, some of them hand-written by Reuss himself being prior to his publications. Some of them show corrections of determinations, some of them give numbers of plates and figures of his publication in 1874. The official labels show practically always inventory numbers which have been used also by David & Pouyet in their revision as well as by Vávra in his catalogue of Bryozoa from the Austrian Tertiary.[22,23] This may be important for identification of single specimens. No remarks are to be found concerning Reuss' publication of 1848 or Manzoni's of 1877 and 1878.

A considerable number of specimens have been included in various modern revisions and have also rather often been used for illustrations in the related publications: the extensive publication by David and Pouyet may be mentioned as an outstanding example in this connection.[24] For the cheilostome Bryozoa from the Italian Paleogene Braga has published a preliminary revision.[25]

Thus these rich collections of invertebrate fossils and micropalaeontological objects are still of great importance nowadays for many basic revisions concerning studies of palaeodiversity of European Tertiary faunas. In addition to this material also there exists a high number of original drawings belonging to Reuss' publications. Especially for his publication on the 'Polyparien des Tertiärbeckens von Wien' (1848) there exist 199 (small) pages with drawings of corals, bryozoans, and *Nullipora ramosissima*. Usually on each of these pages (about 10 x 16 cm, or a little bit smaller) there are drawings belonging to one species only; therefore, some pages show one figure only, others two, three or even more (Figures 2 and 3). A first preliminary sorting and identification yielded in addition: 77 pages with ostracods, 57 pages with Bryozoa from the Cretaceous ('Plänerkalk'), and 46 pages with drawings still to be identified (mostly Bryozoa). Most of these drawings show the (hand-written) name of the species, very often the name of the locality and numbers giving plates and figures. A few drawings may possibly have never been published - in these (very few) cases the drawings may be additional sources of information for taxonomists. Others may at least be useful in respect to identification of type localities etc.

All this material, not discovered until a few years ago, is kept in the archives of the Institute of Palaeontology, University of Vienna.

6. Private life

As shown in the few paragraphs above there exists a considerable amount of information concerning the scientific activities as well as the personal career of August Emanuel Reuss; as far as his private life is concerned, however, only a very few details can be given. Endeavours to obtain more details concerning his private life with the help of archives at Praha (Státní ústrední archiv v Praze - being accessible for historical studies for a few years) - unfortunately yielded no information worth mentioning. Only a single document concerning the 'Nostrifikation of the professor of mineralogy, M. D. [Dr. of medical sciences] August Emanuel Reuss....' (from 1861) has been the result of a search in these archives. From literature it is known that he had been married (since February 16th, 1841) to Anna Schubert, born at Bílina, who died in Vienna on April 26th, 1873.[26] From biographical reference books, I found out that there had been a son as well as a daughter.[27] His son (August Leopold Reuss, 1841 - 1924) started a medical career very much like his father had done during his younger years. He began his studies at the University of Praha in 1859 to continue at Vienna's university in the years 1863-1865. At the end of his studies he took two doctor's degrees in medical sciences: Dr. med. in 1865 and Dr. chir. [i.e. for surgery] in 1867. In 1870 he received his 'Habilitation' [a sort of tenure] as 'Privatdozent' for ophthalmology. In the years 1872-1922 August Leopold was then head of the department for ophthalmology at the 'Wiener Poliklinik' [Vienna's clinical hospital], and from 1909 -1918 director of the whole hospital. In sharp contrast to his father he continued to work as a physician through his whole life, however, a few publications on botany can be regarded as proof that he had broader scientific interests too. In medical science his name survives as a pioneer in various special fields of ophthalmology.

On March 14th in 1859 a daughter of A.E. Reuss was born: Jenny Reuss (Figure 4), married at the age of 18 (7th April 1877) Prof. Rudolf Hoernes (1850-1912), who was professor of palaeontology at the University of Graz (Styria).[28] They had two children: Johanna and Philipp; the son studying medical science. Jenny Reuss-Hoernes had herself been working as a writer, publishing lyrical poetry.[29] In addition she worked as co-editor of a monthly journal ('Heimgarten'), which was founded by Peter Rosegger and dedicated exclusively to dialect and popular poetry.[30] This journal existed from 1876 - 1935 and several other well-known poets, Hamerling and Anzengruber, co-operated in this project.[31]

7. Marks of honour

Realizing that Reuss had been a well-known and deeply respected scientist throughout his whole life, it is not amazing at all, the rather high number of marks of honour which he received. Some more important examples will be given below:

Doctorates *honoris causa*: University of Breslau (Dr. phil.), University of Praha, University of Vienna (Dr. phil., 1866).

Member *honoris gratia* of the 'philosophische Doktoren-Collegium'[=Faculty of

Figure 4. Jenny Reuss (born 1859), daughter of A.E. Reuss, married to Prof. R. Hoernes (1850-1912), Professor of Palaeontology at the University of Graz. (Date of picture 1907; see note 29).

philosophy] of the University of Vienna (1865).
Dean of the faculty of philosophy of the University of Praha.
Rector of the University of Praha (in 'Schiller's year', i.e. in 1859-1860)
Freeman [= 'Citizen of honour'] of the city Bílina
Member of different scientific societies:
Full member [='Wirkliches Mitglied'] of the Imperial Academy of Sciences at Vienna (since 1848 - he was therefore among the earliest members appointed)
Deutsche Akademie der Naturforscher in Halle [=German Academy of Naturalists at Halle]
Deutsche Akademie der Naturforscher, Dresden [=German Academy of Naturalists, Dresden]
Orders and decorations: 'Ritterkreuz' of the 'Franz Joseph-Orden', Orden der Eisernen Krone, 3. Classe, Ritterkreuz of the Saxonian Albrechtsorden.
Knighted in 1870: 'Ritter von Reuss'.

8. Origin of the name 'Reuss'

There are obviously two different sources from which the family name 'Reuss' can be derived: either it indicates a profession (cobbler) or it was used as an additional surname for knights, who had been in Russia.

In medieval German 'riuße' or 'altriuße' was the word for 'Flickschuster' [=cobbler]; this profession had been distinctly separated from a shoemaker: ...."daß die Reußen keinen neuen Schuh machen sollen" ... [=the cobblers are not permitted to make a new shoe] has been distinctly confirmed by medieval law at Bamberg (Germany). Names like 'Konrad der Rüße', C. der Reuße etc. are known from 13th to 16th century.

As an additional surname of knights, who had been in the country of the 'Reußen' [also: 'Reussen' = Russians], this name is very old too: e.g. a man named 'Heinrich dictus Rußo' is mentioned at about 1250, or 'Arnold der Rüße' (a knight mentioned in 1298) or Jacobus Reuß (Ruthenus) at Brno (now: Czech Republic) in 1360.[32]

9. Monument to A.E. Reuss

In Vienna there are two (small) monuments to pioneers of micropalaeontology. At the façade of the Museum of Natural History there are two portrait sculptures of Ehrenberg and d'Orbigny, but no monument of any kind for Reuss.[33] In Bilina, however, there exists something like a 'twin-monument' for Franz Ambrosius Reuss and for August Emanuel Reuss; this monument to father and son is arranged back to back.

10. Acknowledgements

For information concerning some details of Reuss' life the author wants to express his sincerest thanks to Dr Irene Zorn (Geologisches Bundesanstalt [=Geological Survey], Vienna). In connection with searches in the archives at Praha the support by Dr. K. Zagorsek (Celakovice, Czech Republic) is gratefully appreciated. Figure 1 is reproduced with the kind approval of the Mineralogické muzeum, Univerzity Karlovy v Praze, Praha.

Notes

1 U. Lehmann, *Paläontologisches Wörterbuch* (Stuttgart, 1985), 439.
2 Picture from L. Viniklár, *Vyvoj ceske prirodovedy 1869-1929* (Vydano, Prirodovedecky klub v Praze, 1931).
3 Abraham Gottlieb Werner (1749-1817), mineralogist and geologist, famous teacher of mineralogy and mining at the 'Bergakademie Freiberg' (Saxonia).
4 C.v. Wurzbach, *Biographisches Lexikon des Kaiserthums Oesterreich enthaltend die Lebensskizzen der denkwürdigen Personen, welche seit 1750 in den österreichischen*

Kronländern geboren wurden oder darin gelebt und gewirkt haben. Fünfundzwanzigster Theil. Rasner-Rhederer (Wien, 1873).
5 H. Bartenstein, 'August Emanuel Reuss. Zur Wiederkehr seines 150. Geburtstages am 8. Juli 1961', *Paläontologische Zeitschrift*, 35 (1961), 248-250.
6 M. Hörnes, 'Dis fossilen Mollusken des Tertiär-Beckens von Wien. II. Band: Bivalven ['....beendigt von Prof. Dr. August Em. Reuss'], *Abhandlungen der k. k. geologischen Reichsanstalt*, 4 (1870), 1-479.
7 C.A. Bobies, 'Bryozoenstudien I. Die Bryozoenfauna der tortonen Strandbildungen von Kalksburg bei Wien', *Jahrbuch der Geologischen Bundesanstalt*, 99 (1956), 225-258.
8 A.E. Reuss, 'Die fossilen Polyparien des Wiener Tertiärbeckens', *Naturwissenschaftliche Abhandlungen*, 2, edited by Wilhelm Haidinger (Vienna, 1848), 1-109.
9 W.E. Piller, '*Nullipora ramosissima* REUSS, 1847 - a rediscovery', *Beiträge zur Paläontologie*, 19 (1994), 181-189.
10 R. Lagaaij, 'The Pliocene Bryozoa of the Low Countries and their Bearings on the Marine Stratigraphy of the North Sea Region', Mededelingen van de geologische stichting, Serie C - V, 5 (1952), 1-233.
11 L. David and S. Pouyet, 'Revision des Bryozoaires cheilostomes miocènes du Bassin de Vienne-Autriche', Documents des Laboratoires de géologie de la faculté des sciences de Lyon, 60 (1974), 83-257. [p. 87]
12 G. Braga, Personal communication, 1993 & 1995 (letters).
13 See David and Pouyet, note 10, and Piller, note 8.
14 See Piller, note 8.
15 F. Unger, 'Beiträge zur näheren Kenntnis des Leithakalkes, namentlich der vegetabilischen Einschlüsse und der Bildungsgeschichte desselben', *Denkschriften der Kaiserlichen Akademie der Wissenschaften, Mathematisch-naturwissenschaftliche Klasse*, 14 (1858), 13-35.
16 A.E. Reuss, 'Die fossilen Bryozoen des österreichisch-ungarischen Miocäns', *Denkschriften der Kaiserlichen Akademie der Wissenschaften, Mathematisch-naturwissenschaftliche Klasse*, 33 (1874), 141-190; A.E. Reuss, 'Die fossilen Bryozoen des österreichisch-ungarischen Miocäns', *Sitzungsberichte der Kaiserlichen Akademie der Wissenschaften, Mathematisch-naturwissenschaftliche Klasse*, 68 (1874), 219-222.
17 A.E. Reuss, 'Ueber Anthozoen und Bryozoen des Mainzer Tertiärbeckens', *Sitzungsberichte der Kaiserlichen Akademie der Wissenschaften, Mathematisch-naturwissenschaftliche Klasse*, 50 (1865), 197-210; A.E. Reuss, 'Die Foraminiferen, Anthozoen und Bryozoen des deutschen Septarienthons. *Ein Beitrag zur Fauna der mitteloligozänen Tertiärschichten, Denkschriften der Kaiserlichen Akademie der Wissenschaften, Mathematisch-naturwissenschaftliche Klasse*, 25 (1866), 117-214; A.E. Reuss, 'Über einige Bryozoen aus dem deutschen Unteroligozän', *Sitzungsberichte der Kaiserlichen Akademie der Wissenschaften, Mathematisch-naturwissenschaftliche Klasse*, 55 (1867), 216-234; A.E. Reuss, 'Die fossilen Foraminiferen, Anthozoen und Bryozoen von Oberburg in Steiermark', *Denkschriften der Kaiserlichen Akademie der Wissenschaften, Mathematisch-naturwissenschaftliche Klasse*, 23 (1864), 1-38; A.E. Reuss, ' Paläontologische Studien über die älteren Tertiärschichten der Alpen. II. Abtheilung. Die fossilen Anthozoen und Bryozoen der Schichtengruppe von Crosara', *Denkschriften der Kaiserlichen Akademie der Wissenschaften, Mathematisch-naturwissenschaftliche Klasse*, 29 (1869), 215-298; A.E. Reuss, 'Über tertiäre Bryozoen von Kischenew in Bessarabien', *Sitzungsberichte der Kaiserlichen Akademie der Wissenschaften, Mathematisch-naturwissenschaftliche Klasse*, 60 (1870), 505-513.

18 A.E. Reuss, ' Die Bryozoen, Anthozoen, Spongiarien des braunen Jura von Balin bei Krakau', *Denkschriften der Kaiserlichen Akademie der Wissenschaften*, 27 (1867), 1-26; A.E. Reuss, ' Die Bryozoen und Foraminiferen', in H.B. Geinitz 'Das Elbthalgebirge in Sachsen, Erster Theil, Der untere Quader', *Palaeontographica*, 20/1 (1871-1875), 95-144; A.E. Reuss, 'Die Foraminiferen, Bryozoen und Ostracoden des Pläners', in H.B. Geinitz 'Das Elbthalgebirge in Sachsen. Zweiter Theil. Der mittlere und obere Quader', *Palaeontographica*, 20/2 (1872-1875), 73-157.

19 Reuss, note 16.

20 Angelo Manzoni (d. 1895), palaeontologist who studied mainly fossil Bryozoa, echinoderms and sponges.

21 A. Manzoni, 'I Briozoi fossili del Miocene d'Austria ed Ungheria, II', *Denkschriften der Kaiserlichen Akademie der Wissenschaften, Mathematisch-naturwissenschaftliche Klasse*, 37 (1877), 49-78; A. Manzoni, 'I Briozoi fossili del Miocene d'Austria ed Ungheria, III', *Denkschriften der Kaiserlichen Akademie der Wissenschaften, Mathematisch-naturwissenschaftliche Klasse*, 38 (1878), 1-24.

22 David and Pouyet, note 11.

23 N. Vávra, 'Bryozoa tertiaria', in *Catalogus Fossilium Austriae*, Vb/3, edited by H. Zapfe (Vienna, 1977), 210.

24 David and Pouyet, note 11.

25 G. Braga, 'Reuss' collection of cheilostome bryozoans from Venetia stored in Natur-Historisches Museum Wien: a proposed revision', in *Bryozoaires actuels et fossils: Bryozoa living and fossil*, edited by F.P. Bigey & J.-L. d'Hondt (Nantes, 1991), 599 (49-59).

26 H.B. Geinitz, 'Dr. August Emanuel v. Reuss', *Leopoldina*, 9 (1873-74), 67-72.

27 E. Obermayer-Marnach, *Österreichisches Biographisches Lexikon, 1815-1950, 42. Lieferung*, (Vienna, 1985); F. Czeike, *Historisches Lexikon Wien*, 5 volumes, (Vienna, 1992-1997), 3.447.

28 J. G. Haditsch, 'Ein Besuch auf dem Evangelischen Friedhof Graz-St. Peter: Totengedenken an einige Bergleute und Erdwissenschaftler', *Berichte der Geologischen Bundesanstalt*, 51 (2000), 88-96.

29 J. Hoernes geb. v. Reuss, *Tempi passati. Dichtungen von Jenny Reuß,* (Graz, 1898), 180; J. v. Reusz-Hoernes, 'Vom Baume der Erkenntnis' (Breslau, 1907), 117.

30 Peter Rosegger (1876-1918; original spelling: P. Roßegger; pseudonyms: P.K., Hans Malser), Austrian writer, son of a farmer, writing lyrics in local dialect, folkloristic descriptions and novels (partly autobiographical), influenced by B. Auerbach (Germany), A. Stifter and L. Anzengruber (both: Austria). Novels about differences between rural and urban life-style. Using his influence in younger years in respect to social questions, also politically engaged, in his older years tending to cultural pessimism.

31 Robert Hamerling (1830-1889) [real name: Rupert Johann Hammerling], teacher, writer of epic poetry, poems; satirist and translator; his epic poem 'Ahasverus in Rom' (1865) had great success at one time.

Ludwig Anzengruber (1839-1889): bookseller, actor at a travelling theatre, clerk at Vienna's police, writer and editor; writer of dramas and comedies (in an early stage of naturalistic style) as well as of rural novels.

32 H. Bahlow, *Deutsches Namenlexikon. Familien-und Vornamen nach Ursprung und Sinn erklärt*, (Hamburg, 1988), 576.

33 F. Hauer, *Allgemeiner Führer durch das k. k. Naturhistorische Hofmuseum.* 2nd edition, edited by F. Steindachner (Vienna, 1912), 375.

Ernst Gustav Gotthelf Marcus (1893–1968) and Eveline Agnes du Bois-Reymond Marcus (1901–1990)

Judith E. Winston
Virginia Museum of Natural History, 1001 Douglas Avenue, Martinsville, VA 24112, USA

1. Introduction

The names of Ernst and Eveline Marcus are so well-known to bryozoan workers that it surprised me when preparing this paper to find little published biographical material available beyond a few brief obituaries. Fortunately, some additional sources have made it possible to create a more complete picture of their lives for the IBA history session. Meeting several current and former members of the Zoology Department at the University of São Paulo during the 16th International Congress of Zoology in Athens in September 2000, made me aware of Ernst's still vivid reputation as a great teacher and the founder of biological science at that university. One of those present, Cláudio Gonçalves Tiago, Centro de Biologia Marinha, University of São Paulo, generously offered to interview some older and retired colleagues concerning their remembrances of the Marcus's years in Brazil, to translate and e-mail their comments to me. He also found and sent copies of additional biographical material published in Brazil.

Another rich source was the correspondence of the American bryozoan worker, Mary Dora Rogick, passed on to me by Timothy Wood in the hope that I could find a permanent archive for it. Mary Rogick was a prodigious correspondent, who not only maintained letter writing relationships with bryozoan workers around the world, but kept files with carbon copies of her letters, as well as her correspondents' replies. The first few items in the Marcus file are reprint requests: two 1935 postcards, one in Ernst's handwriting and one in Eveline Marcus's, from the Zoological Institute in Berlin, requesting copies of Mary's papers, and a 1937 request from Ernst written from the University of São Paulo, asking to be placed on her permanent mailing list:

> May I ask you to be so kind to send me a copy of your studies on fresh water Bryozoa V and also of all your future publications, because I am still working on Polyzoa. In a few months I hope to be able to revenge myself, sending you a copy of my paper on some marine brazilian Polyzoa, that is in print now. Very cordially yours, Ernst Marcus.[1]

From that time on letters were exchanged between the Marcuses and Mary several times yearly until Mary's death in1964.

Bryozoan worker Mary Dora Rogick (Figure 3), a professor at the College of New Rochelle, New York, was known by her colleagues for the humorous drawings and cartoons with which she illustrated her letters and papers. Eveline Marcus proved up to returning the challenge of replying in kind (Figure 1). The Marcus's letters, written by Eveline in her idiosyncratic style and spelling, were also illustrated by her with charming colour pencil drawings. The drawings, which illuminate the margins of fragile airmail letters or are interwoven with the text, illustrate animals, plants, the Marcus's research subjects, or other aspects of their lives in Brazil.

Figure 1. Characteristic illustrated signatures from the correspondence of Mary Dora Rogick and Ernst and Eveline Marcus: Mary Rogick (from letter to E. and Ev. Marcus, 4th September 1944), Ernesto and Eveline (from letter to M. Rogick, 27th February 1941), the Marcuses and their cat (from letter to M. Rogick, 29th July 1940), dancing Seal and Bear (from letter to M. Rogick, 3rd March 1938), Bear and Seal in frontal view (from letter to M. Rogick, 27th December 1937).

2. The years in Germany

Ernst Marcus was born in Berlin, Germany, on 8th June 1893, the son of Georg Marcus, a district court judge,[2] and Regina Schwartz.[3] As a child he lived near the famous Berlin Zoo, where he learned to observe and appreciate all kinds of animals. He was educated at the Kaiser Friedrich Gymnasium in Charlotenburg and entered the Friedrich-Wilhelms-Universität (University of Berlin) in 1912 to study zoology. As a boy he had collected beetles, and he originally intended to become a coleopterist. He had begun his doctoral research in the Entomology Department at the Berlin Museum when his studies were interrupted by World War I. Ernst served in a cavalry regiment during the war[4] and was awarded the Iron Cross, I & II Class.[5] Once the war was over he returned to the University. He received his Doctorate in 1919, with a thesis on Coleoptera.

After graduation he continued to work at the Museum where, according to Eveline, the museum's director assigned him the Bryozoa collection to study.[6] In doing so, he was on his own, with no resident specialist to learn from, and only an old and inadequate collection of literature to use. Despite these drawbacks Ernst plunged wholeheartedly into the taxonomic study of bryozoans, working up collections from a number of German, Swedish and Danish expeditions to the Pacific, Australia, the Aru Islands (off New Guinea), and South Africa, in addition to various collections from Europe. In doing so he gained a valuable knowledge of the global bryozoan fauna, as well as facility in taxonomic writing. His work on collections made in the Pacific Ocean by Dr. Theodor Mortensen of the Zoological Museum, University of Copenhagen, also made him a Danish friend with whom he corresponded and eventually visited.

In 1923 he obtained the 'Privat-Dozent'. This credential permitted him to teach courses at the university level, and he was appointed assistant to Professor Karl Heider, the invertebrate embryologist at the Zoological Institute of the University of Berlin.[7]

Ernst was 31 when he married Eveline Agnes du Bois-Reymond in 1924. Eveline, 23 at the time of their marriage, was also from Berlin, the daughter of René du Bois Reymond, a physiology professor at the University of Berlin, and the granddaughter of Emile du Bois-Reymond, one of the founders of electrophysiology. Like Ernst, Eveline was devoted to zoology. By training she was a laboratory technician and zoologist, skilled in the new art of microphotography. According to those who knew her well, she had delighted in small invertebrates since childhood, when she had studied them under her father's microscope.[8] Figure 2 taken in Berlin in 1926 shows the young couple with student bryozoologist Ehrhardt Voigt and the American Ray S. Bassler.

The marriage of Ernst and Eveline also became a scientific partnership that lasted until Ernst's death. She drew the illustrations for his papers, Ernst did the references and literature survey, both participated in the studies themselves.[9] Visit(s) to the marine

Figure 2. Left to right: Ernst and Eveline Marcus with Ehrhard Voigt and Ray Bassler, Berlin, 1926 (Photograph from the Bassler Photo Collection, Department of Paleobiology, National Museum of Natural History, Smithsonian Institution)

station at Büsum on the North Sea coast resulted in the Marcus's first studies of living marine bryozoans, and they also began to study freshwater species.

Ernst's scientific career progressed satisfactorily for some time. He was appointed Associate Professor at the Zoological Institute in 1929 at the age of 34. However, the institution of Nazi laws in 1933 soon resulted in the dismissal of all German Jewish scientists from their jobs. Most Jews were dismissed immediately, but Ernst and others who had received military honors during the war were protected until Hindenberg's death in 1935.[10] Correspondence between Ernst and Theodor Mortensen of the Zoological Museum, University of Copenhagen shows that he wrote to Mortensen in 1933 to say that he was afraid, but would not leave Germany. However, it is clear that the Marcuses soon started making plans to emigrate. He visited the Copenhagen museum for two or three days in January-February 1934, and stayed with the Mortensen family. Ernst and Eveline began studying Danish, and Ernst began writing to Mortensen in Danish in December 1934.[11] That he was completing a monograph on the Danish bryozoan fauna for Mortensen's *Danmarks Fauna* series was a compelling reason for learning Danish, of course, but a letter from Eveline to Mary Rogick[12] mentions that they also thought they might find refuge there.

In 1935 Ernst was dismissed from the University. However, he did not remain unemployed long. He wrote to Mortensen in March 1936 to say that he had finished the

Figure 3. Mary Dora Rogick (photograph taken from copy made by J.D. Soule in 1969 from original in the album of G. Bobin, Paris)

linguistic revision of the *Danmarks Fauna* bryozoans thanks to the help of a visiting Danish student. Later that month he sent Mortensen another letter saying that he had received a telegram from São Paulo with an offer of a professorship.[13] He had been recommended for the Chair of Zoology at the new University of São Paulo in Brazil. According to Eveline,[14] the offer came about through the efforts of the Society for the Protection of Science and Learning, Ltd, a group which tried to find jobs for displaced Jewish scientists.

The new University of São Paulo had been established in 1934. The university was well funded, able to import library resources and lab equipment and even provide thc students' textbooks. It was patterned after European institutions, and its organizers hired faculty primarily from Europe, even to the extent of recruiting for each discipline from a particular country, e.g., France for philosophy, and sociology, Germany for botany, chemistry and zoology, etc.[15] Ernst Bresslau, an embryologist and turbellarian specialist from the University of Cologne was originally invited to found the Zoology Department. His sudden death in 1935, shortly after he had arrived in Brazil, provided an opening for Ernst Marcus.

3. The years in Brazil

The Marcuses arrived in Brazil on 1st April 1936.[16] They plunged immediately into the study of Brazilian bryozoans, utilizing both preserved material collected by others and living animals and larvae they collected on the coast near Santos (50 km from São Paulo), and they began publishing a series of papers in the Zoology Department's journal, *Boletim Zoologia*. They were concerned about whether bryozoan workers would be able to read Portuguese, writing to Mary:

> If you are familiar with Latin, you can perhaps find out of the "Bryozoarios marinhos brasileiros" which I hope to send to you in the course of September; the English summary is surely quite insufficient for systematic matter. I beg you to write frankly, whether it is possible with help of my wife's nice drawings to understand the diagnoses and discussions, though they are written in Portuguese. The embryological studies, with which we now are occupied, will be summarized much more amply.[17]

Mary reassured them:

> I certainly do admire your ability to write in and to use so many different languages so well. The Portuguese articles do not appear to be so very hard to translate–at least, not nearly so difficult as German. . . . The drawings in the Brazilian Marine Bryozoa paper are exceedingly clear and the diagnostic characters are well brought out. You are most lucky to have such an able and willing assistant as Mrs. Marcus! With the aid of the excellent drawings and also some slight knowledge of Bryozoa in general, it should not be difficult for anyone to understand the taxonomic discussion in the Brazilian paper. I wish more writers would follow your plan of including a list of definitions of terms in large papers,–it is a most helpful feature.[18]

To which Eveline and Ernst replied

> We are very glad that you don't find it impossible to work with our Portuguese paper; the old Sir Sidney Harmer unfortunately has much trouble with reading it. But most scientists, especially the older generation, were brought up with so much Latin that they ought to be able to understand Portuguese. I myself always blamed my teachers for the exaggerated time we had to spend on Latin but now I see the use of it. I expect you won't enjoy my Danish paper from the linguistic point of view. Unfortunately, science as well as we will have to wait til 1940 until it gets printed. They lack money for the print in Denmark though the Bryozoa for "Danmarks fauna" lie ready since two years. (MM to MDR, Dec. 27, 1937).

Their work on the marine Bryozoa seems to have progressed rapidly, for in March, 1938 they wrote

> We are now beginning to study the anatomy of the marine Bryozoa from Santos, because we have found and classified nearly all species that occur there in shallow water. For preserving material we had a week's trip to the seashore and have been bathing in Bryozoa. We also could

and parrots are quite strange enough. But we miss the singing of black-birds, thrushes and finches, although we have not as much sense for music as you seem to have, in spite of the records, that seals and bears are so very fond of music!

Very sincerely yours

Eveline and Ernst Marcus

Figure 4. Eveline and Ernst as Seal and Bear, complete with microscopes and phylactolaemates (from letter to M. Rogick, 27th July 1938).

observe some colonies alive, unfortunately they all died very soon because of the warm water, but we have seen some of the marvelous larva alive with their brilliant colors that lose themselves completely in alcohol. We hope to get some colored pictures printed in our next paper.[19]

As in many countries, the students who took up science at São Paulo were generally not from the top of the social hierarchy, but included women, children of immigrants, and country people. They may have been lacking in formal biological training, but they were enthusiastic about the new world of science and the opportunities it opened to them.[20] The Marcuses returned their students' enthusiasm. They had no children of their own and seem to have treated their students and students' children as their extended family:

> We have so much to do with the education of our 3 little girls, now they are beginning to grow up, that I often think of the old woman that lived in a shoe. 1.) Marta is modest and does not claim much of our time--if she does not just have a ms to be looked at. (She has written a kind of report on her year's work at Yale.) 2.) Gabriella is doing her first morphological steps with anatomy of Saccocirrus gabrielleae with sectioning, staining, and reconstruction. The latter needs near accompanying and often help. Number 3, Diva is building her thesis on the embryology of Bugula flabellata. We had preserved one beautiful colony richly beset with embryos in 1938 and had patiently waited for a student to study it. When Diva accepted the place of our third assistant, she had to begin to elaborate her thesis and chose this subject. She made nice models of Plasticine of the segmentation and nice drawings of the totality of the fertile colony. Now she is finishing text and the definitive drawings and also needs much of our time. In the July holidays we will all together go to the seashore to have some rest and some new material.[21]

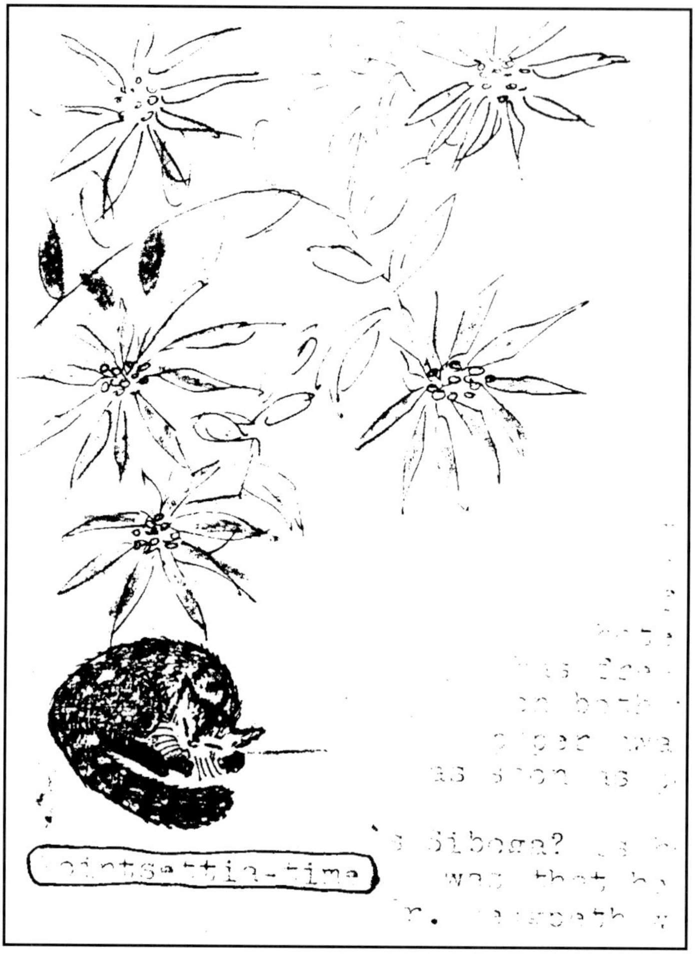

Figure 5. 'Poinsetta-time', the Marcus cat with Brazilian flora (from letter to M. Rogick, 11th June 1946)

By all accounts Ernst and Eveline were devoted to each other. The signatures of their letters to Mary show their fond characterizations of each other as favourite zoo animals (Figure 4). Eveline is a Saima Seal, Ernst, a Red Panda (often a pipe-smoking Red Panda). They were also devoted to their pets (Figure 5):

> . . . we dedicate all our free time to research and publication with exception of the hours our cat requires. He is a yellow beauty and very tame, only every night he strays about with his black lady-love.[22]

However, even pet animals were subjected to their scientific scrutiny:

> His yellow fur seems to grow in the rhythm of the northern hemisphere, where his ancestors

> came from: he had a beautiful mane all along his back in December, and now his coat is much thinner in spite of our winter. It is true that until now the weather is exceptionally warm but nevertheless he ought to put on his winter coat, it might turn cold over night. After only one year of observation this fact is NOT ready for publication but if his mane grows again at the end of this year we at least can, but need not state the cycle of the fur for ONE South American Tom-cat.[23]

Their love of zoos continued strong, and they seem to have interested Mary Rogick in zoos, also, for in 1937 and 1938 she visited the New York Zoological Park, the Brooklyn Zoo, the Central Park Zoo, and the Aquarium and sent photographs she had taken of some of the animals to the Marcuses who replied:

> We get a lot of information about your North America by means of the *National Geographic* magazine and have read the article about procuring animals for the Washington Zoo with special interest. Ernst grew up very near and therefor most of his time IN the Berlin Zoo and was acquainted with most of its inhabitants. Here in SP they only have a small private zoo but a very fine garden with domestic animals and also a few native wild ones among which the huge Tapir is so tame that we could stroke her fur. She is quite black with white tipped ears. Of the brasilian colored birds we do not see many but for us Europeans hummingbirds and parrots are quite strange enough.[24]

The Marcus's scientific productivity burgeoned in their new hunting grounds around São Paulo. Mary Rogick, who had a very heavy teaching load, as well as the care of her ailing elderly mother, wrote wistfully:

> The amount of work that you have done in the last two or three years is truly amazing. About all I can do is to publish one little paper each year. This year I have 238 girls in my classes (General Zoology, Embryology and Teaching of Science). I spend most of my time grading their papers and drawings. My research I have had to leave for the vacations and summers.[25]

She asked the Marcuses how they managed to produce so many papers. Eveline described their São Paulo life and work schedule:

> Our working technique has several secrets that we can easily reveal, as they are not so easy to copy: lessons claim three mornings weekly from 8-12, with the three previous afternoons for preparation (for one half hour of lecture and 2-1/2 of dissection or work with the microscope). I, woman, have nothing whatever to do with the teaching, and little with our household of two persons that are more concerned in much than in complicated food (of course it shall also be tasty). We have a Yugo-slavian housekeeper that comes from the frontier of Rumania and speaks German. She has been with us for 8 years now, and does everything necessary, including washing and ironing in 4 half days weekly. So all my free time and three full days (except of course lots of conversation at the Department and some instruction for the younger members of the staff, viz. Miss Zuccari) of his go to our own research work. Sundays serve for collection. Sao Paulo, 750 m over sea level, city of one to two millions of inhabitants, spreads over ca. 15 x 15 km of rolling country with deep cut valleys. Our hired house lies half way between the

Figure 6. The coast at Ihlabela, São Sebastio Island, Brazil, showing some marine animals collected there; note the ciliate capital R on Rogick. During this visit the Marcuses stayed at a weekend house belonging to the sister of Marta Vanucci, one of their graduate students (from letter to M. Rogick, 19th June 1947)

"city" and the boundaries, so that we are at the outskirts of the town with a 15-minute tram-ride. We love to walk, and in our reach there are some rivers and many mostly very dirty, rivulets, a bit of wood, about 1000 m. square, and in the further range hills up to 1100 m. Unfortunately they are beginning to make use of the rivers for electric power and are damming them and turning them so that the water is often quite putrefied. . . . The illustration of our papers is exclusively my dominion, and with a double set of genes of the painter and engraver Chodowiecki from two great grandmothers, and 30 years of experience, I do not spend too much time with drawing. F. ex. the figures in Danmark's fauna, that are reduced to 1/2 of their original size, took 3-4 hours each. I admire the enormous, ant-like assiduity of your statoblast-dotting. The effect is marvelous and much superior to many of my drawings. We have very good binocular microscopes as well as Greenoughs both in the Department and at home, so the microscopical work does not strain the eyes, even with many hours of work every day. (Waldo Schmitt considered it a South American hobby of luxurious apparatus, when he saw our microscopes, but it's not, it is only common sense and self-preservation). The scientific libraries of Sao Paulo are very rich; beside the department that has a number of complete sets of reviews (Quarterly Journal of Microscopical Science, Biol. Bull, Zeitschrift Wissenschaftl. Zoologie, etc.) the Instituto Butantan, the Museu Paulista, and the Instituto Biologico have got collections of books so that most of the papers we want are available. A great number of separate copies of Bryozoan papers is in our private library, bought before we left Europe, and an equally rich one on Turbellaria the Dept. bought from the widow of our predecessor Bresslau. We even have several acquainted Russians that love to translate Russian articles for us. Can anyone combine so many favorable factors for scientific research?[26]

In 1940 Ernst and Eveline became Brazilian citizens, an event which they celebrated by buying a gold medal with a symbol of the Brazilian Republic.[27]

Ernst's scientific interests had always been broad. A 1928 directory listed them as 'Tardigrada, Tiergeographie, Coleoptera, Bryozoa (Systematik, Verbreitung, Embryol., Physiol.)'.[28] With a vast unstudied neotropical fauna before them, the couple's scope of work continued to increase (Figure 6 shows some examples of their interests). When they first arrived in Brazil the Marcuses had inquired about the presence of freshwater bryozoans, but without much success:

> We did not yet find Lophopodella here but we confess we have not looked for freshwater Bryozoa and of course the people that "naturally have met them so often" can't find them again if we ask them to give us some.[29]

However, when World War II began marine work became impossible for them because of their German origins,[30] so the Marcuses turned to the study of freshwater bryozoans, as well as to research on other freshwater and terrestrial groups, particularly the Oligochaeta and Turbellaria. They also utilized collections they had previously made of animals associated with marine bryozoans, such as pycnogonids:

> With the same you receive our *Boletim Zoologia* 4, this time not dealing with Bryozoa, but with all the Pantopoda, that we in the course of our studies could pick from our Bryozoa.[31]
>
> The Pantopoda were also an old "faible" of ours. Among the tufts of Bryozoa we had gathered so great a number during the first years here, that they were worth the while. But it seems that their biology is nearly inaccessible. They want living Hydroids for dinner, and we can only offer them bananas and coffee.[32]

Both Mary Rogick and the Marcuses wrote of the changes brought by World War II:

> The war is just beginning to make itself felt in our school system. Some of its effects are delayed deliveries of scientific equipment, shortening of courses so that students can graduate sooner in some of the schools. Some schools plan to arrange a number of courses during the summer so that students can finish in 3 or 3-1/2 years instead of 4 the work for their B. A. degrees. About 4 months ago I received a reprint from Dr. Anna B. Hastings so evidently she is still alive and well. I have written her since but have not yet had a reply and probably won't for some time. I have not communicated with Borg or some others in Europe since the war.[33]
>
> The war has cut nearly all our correspondence with family and scientific relations, with exception of England and North America. There is a rather great colony of old and new acquaintances here, and we have enough occupation with zoological research, not to feel the gap too strongly. . . . with very good wishes for you and your country, we remain very truly yours, Eveline and Ernesto Marcus.[35]

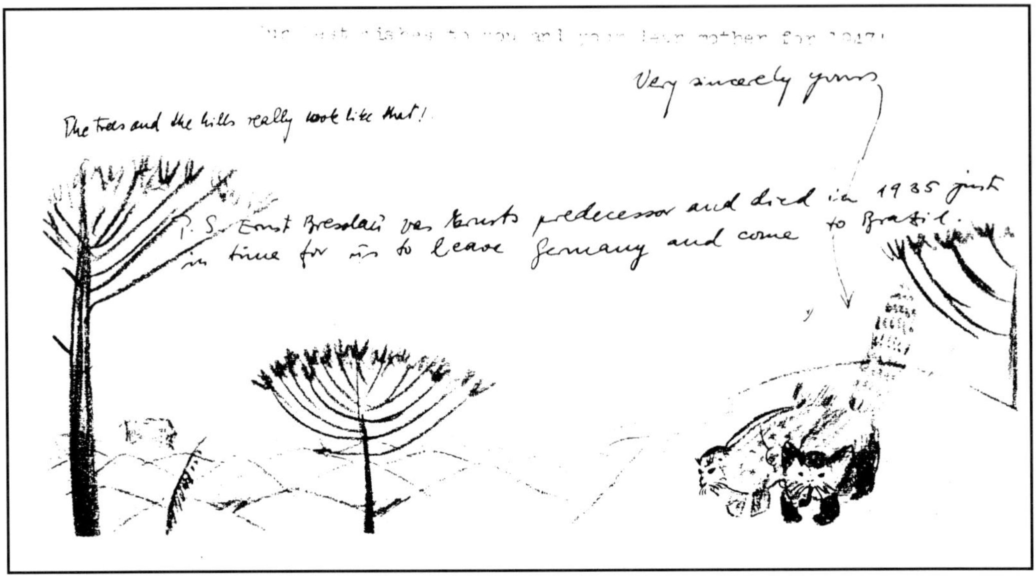

Figure 7. Seal and pipe-smoking Bear, the Marcuses on vacation in hill country (from letter to M. Rogick, 13th January 1947), according to Eveline this was their first vacation since arriving in Brazil more than 10 years earlier, even so, they couldn't resist collecting some Fredericella *in the hotel's pond).*

> Under separate cover I have sent you and your faithful coworkers some fairly recent reprints. I had them packed as early as last July but hesitated to send them on account of the war because shipping space is so important. . . . The cancellation of the annual meeting of the American Association for the Advancement of Science was a disappointment since the meeting was to have been held at New York City. . . . This is the first cancellation of the annual meetings since our Civil War days (1865) and was due to the war.

In 1945, Ernst became a Full Professor of Zoology, presenting a large thesis about microturbellarians to the University in support of his promotion.[36] They wrote to Mary:

> We fear you will have to buy a Danish-English dictionary next, because now we hope you will soon receive your copy of the Bryozoa in "Danmark's Fauna" (No. 46) written in Danish 400 pages and ca. 200 figs in 1935 and published in 1940. Our copies were on the train in Dec. 1941, via Siberia and never found their way. But we promise to stop publishing in further languages; Ernst has just signed his definitive nomination in his position here, so we hope we won't need to learn more languages.[37]

And, in fact, when Ernst was offered a position back in Germany after the war he declined:

> In the course of the last year we had the surprise to receive the invitation to return to Germany to the chair of Eugen Korschell at Marburg but we are not willing to begin with reconstruction

a third time, once after 1918, once here, and now once more. So we decided to stay here, where we have plenty to eat and plenty to investigate. The facilities for studying marine fauna are a great attractive for us.[38]

The end of World War II brought a return to normal times. Ernst and Eveline took a real holiday at a hotel in the hills, their first since moving to Brazil (Figure 7).

The post-war economic boom also brought rapid changes and growth to both New Rochelle and São Paulo.

It seems so unusual to go into stores and groceries in town and to find their shelves filled again with all sorts of canned foods and goods. . . .The black-outs have been lifted long ago and very few things (sugar) are being rationed any more. . . .The housing shortage is terrific. There have been no vacancies in many of the apartment buildings about town for a long time. No one dares move.[39]

The Marcuses replied:

If you find it unusual to buy your food without points now, we find it much more astonishing that we are here and now beginning to have our bread rationed. Until now only sugar was really scarce, but now wheat is going to Europe. and Brazil is not at all autarc. The housing problem is much like yours, also for building we need foreign materials, so that the number of new houses built during the war did not suffice for the many workmen that came to town.[40]

As the post-war boom progressed, both North and South American biologists noted with dismay the effect of increasing human population density on the animals they studied:

We are always on the lookout for *Lophopodella iheringi*, though it is every year more difficult to collect fresh water-fauna near S. Paulo. The city is growing too rapidly, about one house every ten minutes or so, and the sewage problem has not been solved better than through the natural ways. Downstream from town the rivers are a mess of decaying substances, and animal life is restricted to Chironomids and Tubificids. How now and then a fish still manages to breathe in the stinky soup is incomprehensible.[41]

New Rochelle is growing fast. Many new (but very expensive and luxurious) apartment houses are springing up around our waterfront. Soon it will be difficult to take classes on field trips along the shore because of so much new building. Automobile traffic is becoming frightful. The chief criticism seems to be that there is not enough reasonably priced housing which poor or middle class professional people can afford. There are plenty of expensive places in which to live and the new housing is practically all of that type.[42]

Your description of the growing New Rochelle might be applied to the city of S. Paulo too. Enormous scy-scrapers (sic) of most elegant apartments and scarce housing for the poor. LIFE certainly did not picture the living quarters around the town, though they are picturesque in their

Figure 8. Seal and Bear on the Brazilian coast, with some of their favourite marine animals, including tardigrade, opisthobranch molluscs and bryozoans (from letter to M. Rogick, 6th February 1956).

variety. The hills are covered with most primitive shacks built of old boards and the like, it is true, sometimes crowned by a television antenna. The nearby swamp, where we had collected numbers of *microturbellaria*, new for science, has also been filled up and is going to be sold for nice family homes, and the scientific *turbellaria* have to look for new lodgings, poor things.[43]

New Rochelle is experiencing quite a building boom and also urbanization. Many new apartments and also new industries have arisen in this area. Our waterfront is far more occupied and bordered by large apartment buildings. Collecting of marine specimens in this area has been rendered almost impossible. In previous years we could collect about 50 spp. of marine animals without even getting our feet wet. Now, about all we can collect is Mytilus, Modiolus, Littorina, Fucus, Enteromorpha, Nassa, Mya, and a few other forms.[44]

When Ernst Marcus joined the University of São Paulo Zoology Department it was housed in a former private home called the Street Mansion, located in Alameda Glette near Princess Isabelle Square in São Paulo.[45] Although luxuriously built, with marble staircases, the building soon became too small for the growing department. In 1954 the department was moved to a new building on a new university campus 5 km out of town. The extra space in the new building was much appreciated, but the developing campus still had a few drawbacks:

In the meantime we have every day the exciting expedition to the Department that has moved to the incipient University Campus 3 years ago. The building is so to say ready, but the roads have only been begun 2 years ago, and the bus varies its circles on different dirt tracks. If it only is dry dust, it is not so bad, but now it has been raining, and we slide through the red soap, fortunately in a plain, where there are precipices, there is a little sand. They have peeled off all

Figure 9. Libbie Hyman arriving to visit the Marcuses in São Paulo (from letter to M. Rogick, 6th February 1956)

the humic layer and left the red laterit, and erosion works on it, we have grand Canyons in all the colors of the rainbow that would make your Grand Canyon blush, through they are in minor scale, and when it rains the book gurgles and everything drips, it is so natural!![46]

In the post-war years Mary and thc Marcuses continued to correspond a couple of times a year, but they wrote less frequently about bryozoans. Although the Marcuses continued to produce papers on bryozoans (1946, 1949, 1953, 1955*a*, 1955*b*, 1957, 1962, 1968), they had also become increasingly involved with research on other invertebrate groups, first with Turbellaria, then with Eveline's favourite animals, opisthobranch molluscs (Figure 8):

Many thanks for *Emballotheca* with the beautiful ample illustration. We are very far away from the Bryozoa with our studies now in fact our sea-slugs eat Bryozoa! One had its gut full of avicularia bitten off from the Bugula on which it is found. Another digests by intracellular vacuoles that contain teeth of the gizzards of Amathia.[47]

Mary missed their contributions to bryozoan work and wrote to them:

When are you going to work again on the Bryozoa? I miss your absence from the group. How about some physiological, embryological, or ecological experiments with the group? Or studies with the electron microscope or polarized light? Such work is badly needed and you are the most logical and best qualified team for the work.[48]

Shortly afterwards, a new opportunity did draw them briefly back to the Bryozoa:

Furthermore we can inform that your claim for our return to Bryozoa is granted, at least temporarily, because we have received such a charming sample of easily attainable Cupularia-like colonies. So we took advantage of the July-holidays and went dredging for more. Our furnisher took us, and at the point he indicated, he lowered a small bottom sampler, about 6 inches square, and from the bottom of the bay 5 m below us he brought up a scoop with 30 colonies of dark red discs in pure sand. What do they eat there? We brought 'em home alive, but they did not say much. They evaginated shyly, and the big vibracularia made some tentative very slow movements, but that was all; they did not show anything of ravishing dance Bassler had imagined, it was disappointing. However, we have an ample material, and it will justify a morphological study to be undertaken after the finishing of the Californian opisthobranches.[49]

Although the Marcuses do not seem to have traveled far themselves during those years, they corresponded with invertebrate biologists from around the world, including Theodor Mortensen, O. Nordgaard, Sidney F. Harmer, Anna B. Hastings, Libbie Hyman, R. S. Bassler, Raymond C. Osburn, Fritz Wiebach, and Patricia L. Cook, as well as Mary Rogick. One of their students, Marta Vanucci Mendes, spent a year in the U.S. on a fellowship to Yale University. During this time she met Mary Rogick and took the invertebrate course at the Marine Biological Laboratory in Woods Hole, in which Mary was one of the instructors. A number of scientists also visited the Marcuses in São Paulo, including Helena Fyfe from New Zealand and Waldo Schmitt and Libbie Hyman from the U.S.[50] Libbie Hyman's visit (Figure 9) was in support of her her work on volume V of *The Invertebrates*. The Marcuses wrote to Mary of Libbie's visit:

Your kind letter of Jan. 9 was received together with one from Lima Peru informing us, that Libbie Henrietta would be here the 23. We expected her at the airport and attending her wishes went to the beach with her to show her living Bryozoa. We found living Watersipora quite common, but the most beautiful species with red tentacles; and enormous tufts of Zoobotrion that impress more by size than by beauty, and bulks of Schizoporella that kindly evaginated under the microscope. Libbie is not in the best state of health, and the January warmth is a strain for her. Nevertheless we hope she enjoys herself a little. Just now we are expecting her back from a 4-days trip to the Iguassu Falls that are larger than the Niagara and that she wanted to see, only 900 km. from S. Paulo.[51]

Figure 10. Ernst and Eveline in old age (Photograph courtesy of Professor Dr Ehrhard Voigt, Hamburg).

Ernst retired from his position as Director of the Zoological Institute of the University in 1963, when he was 70. Eveline described their life after retirement in one of the last letters written before Mary Rogick's death:

> Research can go on as usual, and we will have more time on the beach. . . . Our pleasures outside zoology are f. ex., long walks, generally in the town, where the different ethnic groups are more or less united in different centres and suburbs. Italians and Japonese are the most picturesque ones. Further we enjoyed reading the French plays of Marcel Pagnol, that we found in one of the Oceanographic Institutes' libraries, Marius, Fanny, and the others. Friends of ours let us hear records of most variable types: Russian church songs, German silly songs, Bach, Elvis Presley, Negro spirituals, so that in our old age we learn to appreciate music.[52]

It is clear that the Marcus's zest for life and knowledge of all kinds had not diminished. Although music may had to wait for retirement, their interest in literature was long standing. Claudio Tiago wrote[53] of having received from a former Marcus student a 1945 copy of John Steinbeck's *Cannery Row* that had originally belonged to the Marcuses. In the book was pasted a portion of a letter to them from Florence and Joel Hedgpeth, announcing the death of Ed Ricketts, real-life model for the book's protagonist.

Ernst Marcus died at the age of 75, on 30th June 1968. Eveline Marcus continued to publish scientific papers, mostly on opisthobranchs, on her own, and traveled extensively to visit mollusc workers and other colleagues around the world[54] until her death in 1990. Ernst and Eveline are buried together in the Cemitírio da Paz, São Paulo.[55]

In an obituary published in Brazil Ernst's colleagues at São Paulo remembered him with affection as a good companion, a friend to his collaborators and students, and a person who lived for science.[56] Throughout Ernst's career it is clear that the Marcuses continued to make students (including at least 11 graduate students[57]) and colleagues their family, closely following their lives. For example, the death of Gabriella Zuccari in a motor vehicle accident before she could even finish her Ph.D. greatly saddened them. Ernst kept her picture hanging in his department office[58] and named a bryozoan in her honour.[59] The couple entertained friends and students at their small two-storey house on Soracaba Street (now renamed Ernst Marcus Street) where visitors remember that there was always a table loaded with cakes and other goodies.[60]

A former student remembers that as a teacher Ernst 'frequently imitated animals to show how they acted – like taking flight, munching (how they used their teeth), swimming, etc. He had difficulties in speaking because during World War I he was hit by a bomb in the trenches and when he recovered he stuttered a bit, but in spite of it, his teaching was of the highest quality'.[61] Another student remembers that Marcus, with characteristic thoroughness, had started taking Portuguese lessons before leaving Europe, so that he began to teach in Portuguese as soon as he arrived at the University (although students sometimes had difficulty understanding his strong German accent).[62] Courses he taught over the years included Invertebrate and Vertebrate Zoology, as well as advanced courses on Embryology and Developmental Mechanisms.[63]

The Marcus's joint scientific productivity was impressive, totaling about 220 titles, including many long papers, containing large numbers of illustrations, keys, and new taxa. According to a 1977 estimate they described 222 new species, and 22 new genera of opisthobranchs alone -- almost half of the opisthobranch species then known from the tropical Western Atlantic fauna had been described by the Marcuses.[64] The bryozoan portion of the Marcus opus (see Bibliography below) consists of 53 papers. Their taxonomic papers described more than 50 new species and 5 new genera. Other papers made major contributions to bryozoan biology and development.

4. Acknowlegments

This paper could not have been completed without the help of those listed below. Thanks to all who helped with research or provided their recollections of Ernst and Eveline: Karen Bille Hansen, Claus Nielsen, Andrei Ostrovsky, Mary Petersen (Denmark), Sabina Hackenthal, Gero Hillmer, Rolf Kohring, Priska Schaefer, Joachim Scholz, Ehrhardt

Voigt (Germany), Claudio Gonçalves Tiago, Tagea Kristina Simon Björnberg, Erika Schlenz, Claudio Froehlich (Brazil), Jo Ann Sanner, Timothy Wood (USA). Thanks also to Janet Roetken, VMNH, for her painstaking transcription of the Marcus-Rogick correspondence.

Bibliography of bryozoan papers by Ernst and Eveline Marcus

1 E. Marcus, 'Einiges Material marine Bryozoen des Berliner Zoologischen Museums', *SitzBer. Ges. Naturf. Freunde, Berlin*, 1919 (1919), 255-284.

2 E. Marcus, 'Notizen über einiges Material mariner Bryozoen des Berliner Zoologischen Museums', *SitzBer. Ges. Naturf. Freunde, Berlin*, (1917) (1919), 255-284.

3 E. Marcus, 'Mittelmeer-Bryozoen aus der Sammlung des zoologischen Museums zu Berlin', *SitzBer. Ges. Naturf. Freunde, Berlin*, no.2/3 (1920), 65-108.

4 E. Marcus, 'Bryozoen von den Auckland- und Campbell-Inseln. Papers from Dr. Th. Mortensens' Pacific Expedition, 1914-16, VI', *Videnskabelige Meddelelser Dansk naturhistorisk ForeningVidenskabelige Meddelelser Dansk naturhistorisk Forening*, 73 (1921*a*), 85-121.

5 E. Marcus, 'Bryozoa von den Juan Fernandez-Inseln', in *The natural history of Juan Fernandez and Easter Island*, edited by C. Skottsberg, (1921*b*), chapter (1).

6 E. Marcus, 'Results of Dr. E. Mjöbergs Swedish Scientific Expeditions to Australia, 1910-13, XXIV, Bryozoen', *Kungl. Svenska Vetenskaps-Akademiens Handlinger Stockholm*, 61 (1921*c*), 1-34.

7 E. Marcus, 'Über die Verbreitung der Meeresbryozoen', *Zool. Anz. Leipzig*, 53 (1921*d*), 205-221.

8 E. Marcus, 'Einiges über Bau und Entwicklung der Meeresbryozoen', *Schrift d.Zool. Stat. Büsum f. Meeresku.* no. 3-4 (1921*e*), 22-7.

9 E. Marcus, 'Bryozoen von den Aru-Inseln', *Abh. Senckenbergischen Naturforschenden Gesellschaft Frankfurt am Main*, 35 (1922*a*), 421-446.

10 E. Marcus, 'Indopacifische Bryozoen aus dem Riksmuseum in Stockholm', *Arkiv för Zoologi*, 14 (1922*b*), 1-23.

11 E. Marcus, 'Referat über die historische und moderne Auffassung des Baues und ser systematischen Stellung der Bryozoengattung *Adeona*', *Verh. zool. bot. Ges. Wien*, 72 (1922*c*), 42-61.

12 E. Marcus, 'Südafrikansiche Bryozoen aus der Sammlung des Gothenburger Museums, nebst 1 westafr. Species', *Göteborgs Musei Zoologiska Avdelning*, 15 (1922*d*), 1-14.

13 E. Marcus, 'Hydrostatik bei Meeresbryoozen', *Verhandlungen Deutschen zoologischen Gesellschaft Leipzig*, 28 (1923), 39-40.

14 E. Marcus, 'Zur vergleichenden Embryologie der Bryozoen', *Mittelungen aus den zoologische Museum in Berlin*, 11 (1924), 157-166.

15 E. Marcus, 'Bryozoa', in P. Schulze, *Biol. Tiere Deutschlands, pt. 47 (lief. 14)* (Berlin, 1925*a*).

16 E. Marcus, 'Über *Stirpariella mortenseni* und das Genus *Stirpariella*. Papers from Dr. Th. Mortensens' Pacific Expedition, 1914-1916, XXXII', *Videnskabelige Meddelelser Dansk naturhistorisk Forening*, 81 (1925*b*), 37-55.

17 E. Marcus, 'Über *Victorella symbiotica* Rouss. (Ectopr. Gemnol. Ctenost.)', *Zool. Anz. Leipzig*, 62 (1925*c*), 129-133.

18 E. Marcus, 'Zum Polymorphismus der Bryozoen', *Verh. D. zool. Ges. Leipzig*, 30 (1925*d*), 152-

159.
19 E. Marcus, 'Beobachtungen und Versuche an lebenden Meeresbryozoen', *Zool. Jahrb. Jena, Syst.*, 52 (1926*a*), 1-102.
20 E. Marcus, 'Beobachtungen und Versuche an lebenden Süsswasser-Bryozoen', *Zool. Jahrb. Jena, Syst.*, 52 (1926*b*), 279-350.
21 E. Marcus, 'Bryozoa', in G. Grimpe and E. Wagler, *Die Tierwelt der Nord- und Ostsee.* (Leipzig, 1926*c*), chapter VIIC.
22 E. Marcus, 'Bryozoa', in W. Junk, *Tabulae Biologicae* 1926d), chapter (=suppl 2).
23 E. Marcus, 'Sinnesphysiologie und Nervensystem der Larve von *Plumatella fungosa* (Pall.)', *Verh. D. zool. Ges. Leipzig*, 31 (1926*d*), 86-90.
24 E. Marcus, 'Über *Lophopus crystallinus* (Pall.)', *Zool. Jahrb. Jena Anat.*, 58 (1934), 501-606.
25 E. Marcus, 'Über marine Bodentiere an der Küste St. Helenas', *Forsch. Fort-schr. Jahrg.*, 11 (1935), 228.
26 E. Marcus, 'Sôbre alguns phenómenos de vida dos Bryozoarios Marinhos', *Arch. Inst. Biol. São Paulo*, 7 (1936*a*), 203-208.
27 E. Marcus, 'Sôbre o systema natural dos Bryozoarios', *Bol. Biol. São Paulo*, 2 (1936*b*), 129-135.
28 E. Marcus, 'Bryozoarios marinhos Brasileiros I', *Boletins da Faculdade de Filosofia, Ciências e Letras,* 1 (1937), 3-224.
29 E. Marcus, 'Bryozoarios marinhos Brasileiros II', *Boletins da Faculdade de Filosofia, Ciências e Letras,* 4 (1938*a*), 1-196.
30 E. Marcus, 'Bryozoarios perfuradores de conchas', *Arquivos Instituto Biologico São Paulo*, 9 (1938*b*), 273-296.
31 E. Marcus, 'Bryozoen von St. Helena', *Videnskabelige Meddelelser Dansk naturhistorisk Forening*, 101 (1938*c*), 183-252.
32 E. Marcus, 'Bryozoarios marinhos Brasileiros III', *Boletins da Faculdade de Filosofia, Ciências e Letras.* 13 (1939*a*), 11-353.
33 E. Marcus, '*Sawayella polzoorum*, gen. nov., spec. nov. dos Schizogregarinaria', *Arq. Inst. Biol. São Paulo*, 10 (1939*b*), 259-278.
34 E. Marcus, 'Mosdyr (Bryozóa eller Polyzóa)', in *Danmarks Fauna no. 46, Copenhagen* (Copenhagen, 1940*a*).
35 E. Marcus, 'Sôbre a interferencia da propagaçao vegetiva na gamogonia', *Ann. Acad. Bras. Sci. Rio*, 12 (1940*b*), 1-16.
36 E. Marcus, 'Bryozoarios Marinhos do Litoral Paranaense', *Arquivos do Museu Paranaense*, 1 (1941*a*), 7-36.
37 E. Marcus, 'Sôbre Bryozoa do Brasil', *Boletins da Faculdade de Filosofia, Ciências e Letras.* 22 (1941*b*), 3-208.
38 E. Marcus, 'Sôbre desenvolvmiento do Bryozoario *Synnotum aegyptiacum*', *Arq. Cir. Clin. Exp. São Paulo*, 5 (1941*c*), 227-234.
39 E. Marcus, '*Cryptopolyzoon evelinae*, a new ctenostomatous bryozoan from the Brazilian coast', *Proc. 8th Amer. Sci. Congr. Washington* , 3 (1942*a*), 477-479.
40 E. Marcus, 'Sôbre Bryozoa do Brasil. II', *Boletins da Faculdade de Filosofia, Ciências e Letras.* 25 (1942*b*), 57-106.
41 E. Marcus, '*Beania cupulariensis* Osb. (Bryozoa Chilost.) nova para Brasil', *Communicaciones del Zoologicas Museo de Historia natural de Montevideo*, 1 (1944), 1-3.
42 E. Marcus, 'On a new Brazilian form of *Fredericella sultana*, etc', *Communicaciones del*

Zoologicas Museo de Historia natural de Montevideo, 2 (1946), 1-10.
43 E. Marcus, 'Some Bryozoa from the Brazilian coast', *Communicaciones del Zoologicas Museo de Historia natural de Montevideo*, 3 (1949), 1-33.
44 E. Marcus, 'Systematical remarks on the bryozoan fauna of Denmark', *Videnskabelige Meddelelser Dansk naturhistorisk Forening*, 112 (1950*a*), 1-34.
45 E. Marcus, 'A new Loxosomatid from Brazil', *Boletins da Faculdade de Filosofia, Ciências e Letras.* (1950*b*), 193-206.
46 E. Marcus, 'Notas sôbre Briozoos marinhos Brasileiros', *Arq. Mus. Nac.*, 42 (1953), 273-342.
47 E. Marcus, 'Polyzoa. Percy Sladen Trust Exp. Lake Titicaca', *Trans. Linn Soc. London ser. 3*, 1 (1955), 355-357.
48 E. Marcus, 'New Entoprocten aus der Gegend von Santos', *Zool. Anz.*, 159 (1957), 68-75.
49 E. Marcus, 'On the evolution of the animal phyla', *Quarterly Review of Biology*, 33 (1958), 24-58.
50 E. Marcus, 'Neue brasilianische Loxosomen', *Zoolog. Beitr. N. F.*, 14 (1968), 203-212.
51 E. Marcus and Ev. Marcus, 'On some lunulitiform Bryozoa', *Boletins da Faculdade de Filosofia, Ciências e Letras.* 261 (1962), 281-324.
52 Ev. Marcus, 'Bryozoa', *Manual de identificação dos invertebrados límnicos do Brasil* , 66 (1984), 1-26.
53 Ev. Marcus, 'Notes on some Brazilian Bryozoa Ectoprocta', *Boletim Zoologia, Univers. S. Paulo*, 8 (1985), 137-143.

Notes

1 Ernst Marcus to Mary Rogick, 10th June 1937.
2 K. Herter, 'Ernst Marcus', *Sitzungberichte der Gesellschaft Naturforschender Freunde zu Berlin (N. F.)*, 8 (1968), 81-92.
3 ex inf. Sabine Hackenthal, obtained from archives of Humboldt University, Berlin
4 His regiment was the 2. Garde Dragoner Regiment Kaisern Alexander, according to Ehrhard Voigt in interview with Priska Schaefer, 11 July 2001.
5 N.B. Eales, 'Obituary – Ernst Marcus', *Proceedings of the Malacological Society of London*, 38 (1969), 371-373.
6 Ev. Marcus, 'Ernst Marcus', *Atti della Societa Italiana di Scienze Naturali e del Museo Civico di Storia Naturale di Milano*, 108 (1968), 7-12.
7 Ev. Marcus, note 6, p. 7.
8 D.D. Corrêa, 'Dr. Eveline du Bois Reymond-Marcus', *Hydrobiologia*, 227 (1991), xxiii-xxvi.
9 Corrêa, note 8, p. xxiii, and Ev. Marcus, note 6, p. 8.
10 'Ernst (Ernesto) Marcus', in Biographical Etymology of Marine Organism Names (BEMON), http://www.tmbl.gu.se/libdbtaxonpersonetymol/index.htm
11 e-mail communication from C. Nielsen to J. Winston, July 2001, summarizing correspondence between Marcus and Mortensen found by K. Bille Hansen and her daughter in the Danish Royal Archives.
12 Ev. and E. Marcus to M. Rogick, 20th January 1960.
13 Nielsen, not 11.
14 Ev. Marcus, note 6, p. 8.
15 S. Schwartzman, *A Space for Science: the Development of the Scientific Community in Brazil*, (University Park, PA, 1991), 1-286.

16 E. Mendes, 'Ernst Marcus', http: www.usp.br/iea/revista/revista22/marcus.html (1994).
17 Ev. and E. Marcus to M. Rogick, 15th August 1937.
18 M. Rogick to Ev. And E. Marcus, 18th November 1937.
19 Ev. and E. Marcus to M. Rogick, 3rd March 1938.
20 S. Schwartzman, note 15.
21 Ev. and E. Marcus to M. Rogick, 19th June 1947.
22 Ev. and E. Marcus to M. Rogick, 21st April 1940.
23 Ev. and E. Marcus to M. Rogick, 29th July 1940.
24 Ev. and E. Marcus to M. Rogick, 27th July 1938.
25 M. Rogick to Ev. and E. Marcus, 21st March 1940.
26 Ev. and E. Marcus to M. Rogick, 11th June 1946.
27 D. Corrêa, op. cit., xxiii.
28 G.C. Hirsch, ed. *Index Biologorum: Investigatores, Laboratoria, Periodica* (Berlin,1928), 189.
29 Ev. and E. Marcus to M. Rogick, 3rd March 1938.
30 D. Corrêa, op. cit., xxiii.
31 Ev. and E. Marcus to M. Rogick, 15th February 1941.
32 Ev. and E. Marcus to M. Rogick, 17th April 1941.
33 M. Rogick to Ev. and E. Marcus, 31st December 1941.
34 Ev. and E. Marcus to M. Rogick, 5th April 1942.
35 M. Rogick to Ev. and E. Marcus, 12th February 1943.
36 D. Corrêa, note 8, p. xxiii,
37 Ev. and E. Marcus to M. Rogick, 18th August 1945.
38 Ev. and E. Marcus to M. Rogick, 21st February 1949.
39 M. Rogick to Ev. and E. Marcus, 26th April 1946.
40 Ev. and E. Marcus to M. Rogick, 11th June 1946.
41 Ev. and E. Marcus to M. Rogick, 14th February 1950.
42 M. Rogick to Ev. and E. Marcus, 29th October 1957.
43 Ev. and E. Marcus to M. Rogick, 24th February1958.
44 M. Rogick to Ev. and E. Marcus, 14th June 1963.
45 *Inf. ex.* Claudio G. Froehlich, in e-mail from C.G. Tiago to J. Winston, 23rd August 2001.
46 Ev. and E. Marcus to M. Rogick, 18th July 1957.
47 Ev. and E. Marcus to M. Rogick, 26th August 1955.
48 M. Rogick to Ev. and E. Marcus, 5th June 1959.
49 Ev. and E. Marcus to M. Rogick,19th July 1959.
50 Compiled from names mentioned in the Rogick-Marcus correspondence or mentioned in Ev. Marcus, note 6, p. 10
51 Ev. and E. Marcus to M. Rogick, 6th February 1956.
52 Ev. and E. Marcus to M. Rogick, 11th July 1963.
53 In e-mail to J. Winston, 23rd August 2001.
54 M. Edmonds, 'Obituary – Eveline du Bois Reymond Marcus, 1901-1990', *Journal of Molluscan Studies*, 57 (1991), 141-142, and personal communication to P. Schaefer by E. Voigt (11th July 2001) noting that Eveline had visited him in Germany three times, in 1969, 1981 and 1984.
55 E. Mendes, note 16.
56 P. Sawaya, 'Professor Doctor Ernst Gotthelf Marcus', *Boletim, Zoologia e Biologia Marinha,*

Univerisidade de São Paulo, Instituto de Biociências, no. 1, n.s. no. 27 (1970), 5-18.

57 Including Diva Diniz Côrrea, Michel Pedro Sawaya, Marta Vanucci Mendes, Gabriella Zuccari, Caludo Gilberto Froehlich, Eudoxia Maria Froehlich, Water Narchi, Gilberto Righi, Edmundo Ferraz Notato, Tagea Kristina Simon Björnberg, Liliana Forneris according to Erika Schlenz in e-mail from C.G. Tiago to J. Winston, 23rd August 2001.

58 Schlenz, note 57.

59 *Lacerna gabriellae* 1955.

60 Tagea K.S. Björnberg quoted in e-mail from C.G. Tiago to J. Winston, 23rd August 2001.

61 Björnberg, note 60.

62 E. Schlenz in e-mail (Schlenz, part 2) from C.G. Tiago to J. Winston, 24th August 2001.

63 Schlenz (part 2), note 62.

64 Edmunds, note 54, p. 141.

From Rondelet to Stockholm: four centuries of bryozoological research

Patrick N. Wyse Jackson* **and Mary E. Spencer Jones†**
**Department of Geology, Trinity College, Dublin 2, Ireland*
†Department of Zoology, The Natural History Museum, Cromwell Road, London SW7 5BD, UK.

1. Introduction
2. Chronology of bryozoological research 1555-1965

1. Introduction

This listing, which is largely a personal choice, is presented in an attempt to demonstrate the range of bryozoological research carried out in the 411 years from 1855 to 1965. It commences with the work of Rondelet who first observed bryozoans, and we have taken 1965 as a convenient year to conclude our survey as it marks the foundation of the International Bryozoology Association. This organisation is now the prime agency for the promotion of scientific and social interaction between bryozoologists who meet every three years during its international conference. 1965 is also a useful year in that it allows us to discuss the work of deceased bryozoologists and others, without venturing too far into the dangerous realm of the more recent - living scientists who might be upset or bewildered at the inclusion or exclusion of themselves or others from this listing. In a small number of cases we have strayed from this necrological course to include some living scientists, as it would be ridiculous to exclude giants of bryozoology such as Ehrhard Voigt simply because they been blessed with longevity - of life and more importantly in this context - of research on bryozoans.

We have compiled our list from an examination of primary literature where possible. We have also drawn extensively on published accounts of the development of bryozoan studies (Bassler 1953, Ryland 1970, Boardman *et al.* 1983),[1] and from Nickles and Bassler's invaluable bibliography published in 1900,[2] as that on the Order Trepostomata by McKinney (1973).[3] Additional information has been gleaned from the contributions to the International Bryozoology Association - Society for the History of Natural History meeting in Dublin in July 2001 published herein.[4] W.A.S. Sarjeant's monumental 10 volume bibliography of geologists has proved invaluable,[5] as has Hans G. Hansson's

Biographical Etymology of Marine Organism Names (BEMON).[6] We gratefully acknowledge the contributions of the compilers of these works, which have eased the compilation of this listing of highlights of bryozoological research.

References to further papers or books of interest pertaining to individual entries are given in the notes immediately following the list. We have not attempted to follow up on every entry in this way, but point out some otherwise obscure references.

2. Chronology of bryozoological research 1555-1965

1555 **Guillaume Rondelet** 1507-1566, a French physician and naturalist who described the first known bryozoan.[7]

1599 **Ferrante Imperato** 1550-1625, a naturalist and apothecary of Naples (Figure 1). In his *Historia Naturale* he published the first use of familiar bryozoan generic names *Retepora* and *Frondipora*. He also discussed the animal nature of these zoophytes.

Figure 1. Ferrante Imperato's private museum in Naples.

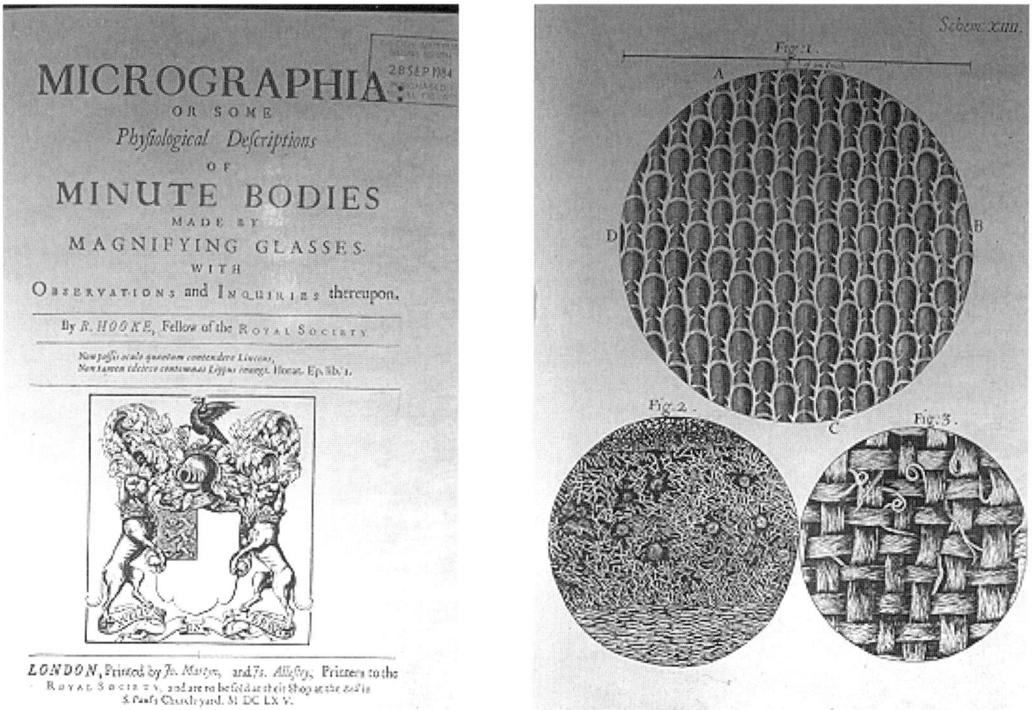

Figure 2. Title page of Hooke's Micrographia *published in 1665 (left) and plate showing structure of* Flustra *(right).*

1665 **Robert Hooke** 1635-1703, was a noted inventor and architect, as well as being interested in natural history. He invented the iris diaphragm in cameras. He illustrated the structure of *Flustra* in his *Micrographia: or, Some physical descriptions of minute bodies made by magnifying glasses, with observations and enquiries thereupon*, published by Jo Martyn and Jo Allestry in London (Figure 2).

1699 **Edward Lhwyd** 1670-1709 was born in Wales, and later was keeper of the Ashmolean Museum in Oxford. He is largely remembered for his work on the natural history of Britain and his books on Irish placenames. In the former volume he published the first illustration of a British fossil bryozoan - "*Alcyonium*" from the Greensand (Cretaceous) of Farington.

1727 **Jean-André Peyssonnel** 1694-1759, a French naturalist, who discussed the animal affinities of zoophytes in a paper delivered to the Academy of Sciences in Paris. However, his paper was communicated by Réaumur (*q.v.*) who concealed the name of the author (he may have been attempting to save Peyssonnel from ridicule). In fact Réaumur at the same meeting delivered a paper which opposed the views of his younger colleague. Peyssonnel's paper was never published but its contents can be deduced from a reading of Réaumur's.[8]

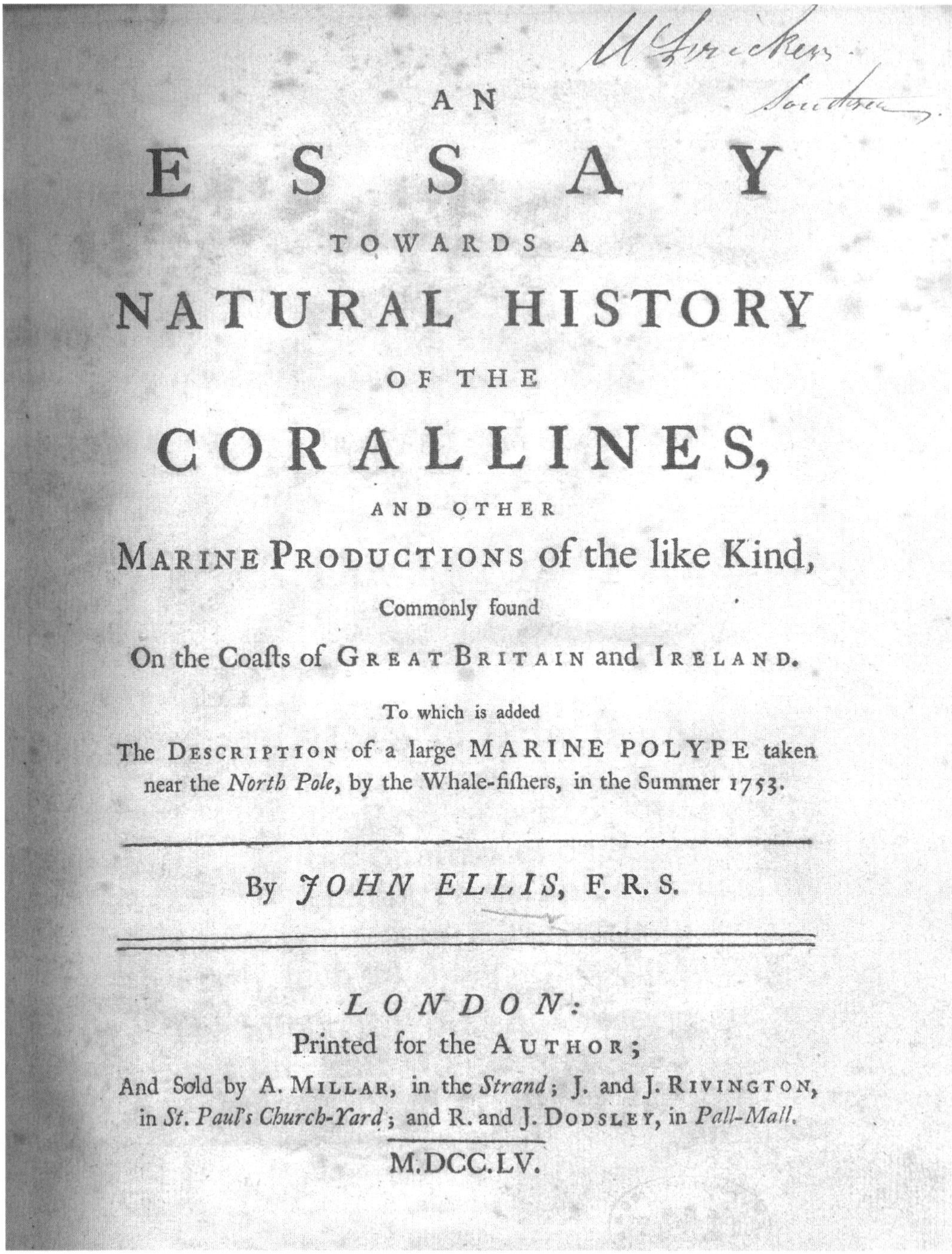

AN

ESSAY

TOWARDS A

NATURAL HISTORY

OF THE

CORALLINES,

AND OTHER

MARINE PRODUCTIONS of the like Kind,

Commonly found

On the Coafts of GREAT BRITAIN and IRELAND.

To which is added

The DESCRIPTION of a large MARINE POLYPE taken near the *North Pole*, by the Whale-fifhers, in the Summer 1753.

By *JOHN ELLIS*, F. R. S.

LONDON:

Printed for the AUTHOR;

And Sold by A. MILLAR, in the *Strand*; J. and J. RIVINGTON, in *St. Paul's Church-Yard*; and R. and J. DODSLEY, in *Pall-Mall.*

M.DCC.LV.

Figure 3. Title page of John Ellis' Essay towards a natural history of the Corallines.

1742 René Antoine Ferchault de Réaumur 1683-1757, a French naturalist and physicist. He was the first to observe the lophophore in marine bryozoans, and also described statoblasts. On his death his collection of natural history was donated to the Académie des Sciences in Paris.[9]

1744 Abraham Trembley 1710-1784, a Swiss-born naturalist and tutor. In his *Mémoires, pour servir à l'histoire d'un genre de polypes d'eau douce à bras en forme de cornes* published in Leiden in 1744, Trembley gave a description of freshwater bryozoans discovered by him three years earlier.[10]

1754-1755 John Ellis 1710-1776, an Irish merchant who worked out of London. His book *Essay towards a natural history of the Corallines* published in London (Figure 3) was the first English text to note the animal affinities of zoophytes. A joint work *The natural history of many curious and uncommon Zoophytes* with Daniel Carlsson Solander (1733-1782) was published posthumously in London in 1786. The genus *Ellisina* Norman, 1903 was named for him.[11]

1758 Carl Linnaeus (von Linné) 1707-1778, erected the Order Zoophyta into which he placed corals and bryozoans. In the 10th edition of his *Systema Naturae* he also applied binomial nomenclature to John Ellis' species.

1759 The British Museum opened in Montagu House, Bloomsbury, London. The familiar South Kensington building was opened to the public in 1881 and was known as the British Museum (Natural History) until the 1980s when it was restyled the less cumbersome The Natural History Museum. Major bryozoans collections are housed in the Departments of Palaeontology and Zoology; these include material from the *Challenger* (*q.v.*) and *Discovery* expeditions of the 1800s. The museum also holds an important bryozoan research library, which included the papers of George Busk (*q.v.*), S.F. Harmer (*q.v.*) and others.

1810 Jean Vincent Félix Lamouroux 1776–1825 (Figure 4), a French marine biologist, who published a paper on the classification of marine polypide corals (bryozoans). His principal contribution to natural history however, was his recognition and division of marine algae into green, red and brown.[12]

Figure 4. Jean Vincent Félix Lamouroux 1776–1825.

1824 **Jean René Constant Quoy** 1790-1869 and **Joseph Paul Gaimard** 1796–1858, undertook several expeditions, including that of the *Astrolabe* (1826 -1829) that collected bryozoans from the Pacific Ocean.

1827 **Robert Edmond Grant** 1793-1874, a Scottish comparative anatomist, published one of the first papers to deal with the microstructure and physiology of Bryozoa. Grant described the structure of *Flustra*, and observed movement of tentacular cilia. This paper was first published in English in two parts in the *Edinburgh New Philosophical Journal*, and subsequently that year in German and in French a year later.

1828 **Samuel George Morton** 1799-1851, was a Philadelphia-born medical doctor who produced the first description of fossil bryozoans (*Eschara*, *Flustra* and *Retepora*) from North America. These were from Big Timber Creek, New Jersey. This report was published first in 1828 in a volume authored by Lardner Vanuxem (1792-1848) and Morton on the Mesozoic, Tertiary and Recent geology and fossils of the Atlantic coast of the USA, and then again the following year in the *Journal of the Academy of Natural Sciences of Philadelphia*. Morton was also interested in the origins of Native Americans, and suggested that they were of a separate origin to Europeans. It was later demonstrated that he falsified his measurements on skulls to suggest that Europeans had a large cranial capacity which he thought suggested they were intellectually superior.

1829 **Karl Eduard von Eichwald** 1794–1876, published the first of his many contributions on Palaeozoic bryozoans from Russia.

1830 **John Vaughan Thompson** 1779–1847, was an army surgeon, who when based in Cork in Ireland, published his celebrated monograph in which he coined the name "Polyzoa" (Figure 5). In 1834 he moved to Sydney, Australia where he served as a director of a prison hospital. He also became involved in an unsavoury libel action.[13]

1831 **Christian Gottfried Ehrenberg** 1795-1876 (Figure 6), published the term "Bryozoa". For many year debate raged between zoologists and palaeontologists as to which of the two similar names "Polyzoa" and "Bryozoa" was correct to use. The latter was globally accepted. Ehrenberg is now principally remembered for his work on diatoms and other microscopic organisms. He was the subject of a number of papers brought together in a recent special issue of *The Linnean*.[14]

1839 **William Lonsdale** 1794-1871, described bryozoans in Sir Roderick Impey Murchison's *Silurian System*. A specimen of *Dianulites* in thin section, now in The Natural History Museum, London, probably represents the earliest example of a

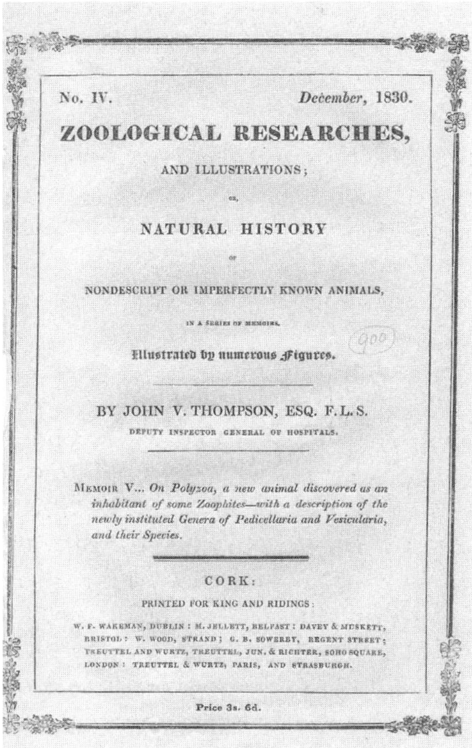
No. IV. December, 1830.

ZOOLOGICAL RESEARCHES,

AND ILLUSTRATIONS;

OR,

NATURAL HISTORY

OF

NONDESCRIPT OR IMPERFECTLY KNOWN ANIMALS,

IN A SERIES OF MEMOIRS,

Illustrated by numerous Figures.

BY JOHN V. THOMPSON, ESQ. F.L.S.

DEPUTY INSPECTOR GENERAL OF HOSPITALS.

MEMOIR V... *On Polyzoa, a new animal discovered as an inhabitant of some Zoophites—with a description of the newly instituted Genera of Pedicellaria and Vesicularia, and their Species.*

CORK:

PRINTED FOR KING AND RIDINGS:

W. F. WAKEMAN, DUBLIN: M. JELLETT, BELFAST: DAVEY & MUSKETT, BRISTOL: W. WOOD, STRAND; G. B. SOWERBY, REGENT STREET; TREUTTEL AND WURTZ, TREUTTEL, JUN. & RICHTER, SOHO SQUARE, LONDON: TREUTTEL & WURTZ, PARIS, AND STRASBURGH.

Price 3s. 6d.

Figure 5. Cover of J.V. Thompson's monograph on Polyzoa.

Figure 6. Christian Gottfried Ehrenberg, author of the term "Bryozoa".

bryozoan thin section. Lonsdale later reported on material from Russia, New Jersey, and Sussex, England.

1846 **Smithsonian Institution** established. It contains the National Museum of Natural History (1857) which is a constituent part of the Smithsonian Institution in Washington, D.C. The museum has become one of the major repositories for fossil and Recent bryozoan collections, and contains most of the major collections made in North America. Staff who specialised in the study of bryozoans included E.O. Ulrich, R.S. Bassler, and more recently, R.S. Boardman and A.H. Cheetham.[15]

1847 **James Hall** 1811-1898, was perhaps the most influential palaeontologist of his generation in North America. In this year he published his celebrated *Palaeontology of New York* - it contained descriptions of many Lower Palaeozoic bryozoans. It was the first of many works and papers to contain information on bryozoans.[16]

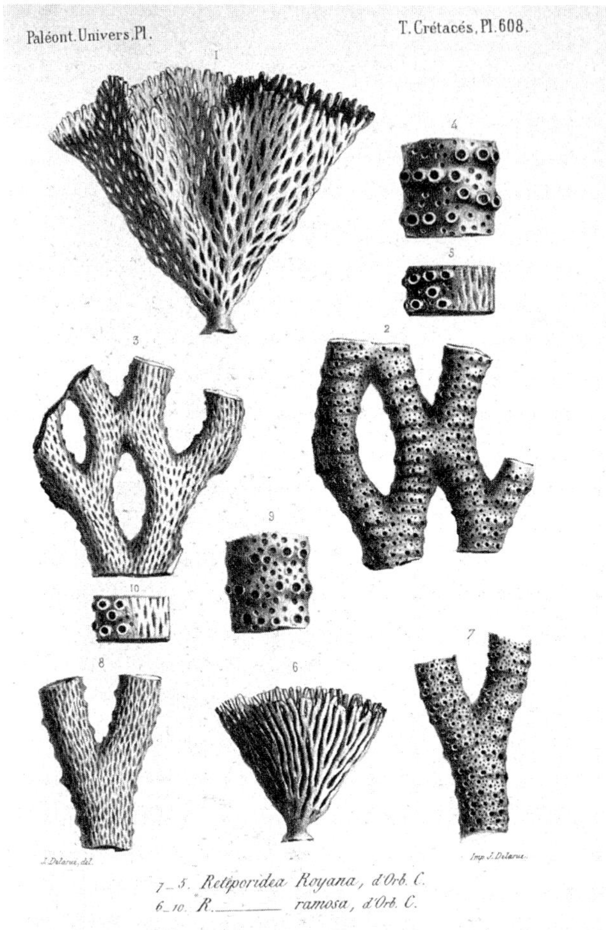

Figure 7. Plate from d'Orbigny's Paléontologie française *showing two bryozoan species from the Cretaceous.*

1849 George James Allman 1812-1898, was born in Cork and became Professor of Botany at Trinity College, Dublin, and later Professor of Natural History at Edinburgh. He was President of the Linnean Society of London 1874-1880, and received the Linnean Medal in 1896. Allman carried out extensive studies on freshwater bryozoans between 1843 and 1880, including many on the physiology, breeding, and larval state of the group. His celebrated paper on their nervous system was published by the British Association for the Advancement of Science.

1850-1852 Alcide Dessalines d'Orbigny 1802-1857, is widely regarded as one of the greatest micropalaeontologists of the nineteenth century. He travelled extensively in the two Americas in the 1830s, and described bryozoans from South American waters. Later he produced a useful account of systematic palaeontology. But he is

Figure 8. George Busk (1807-1886) author of the Orders Ctenostomata, Cyclostomata and Cheilostomata.

largely remembered for his *Paléontologie française* of 1850-1852 which ran to 1192 pages and included descriptions and fine illustrations (Figure 7) of many Cretaceous bryozoans.

1852 **George Busk** 1807-1886 (Figure 8), wrote on bryozoans between 1849 and 1886, and became one of the greatest authorities on the group. The Orders Ctenostomata, Cyclostomata, Cheilostomata were erected by Busk in a paper that described the bryozoans collected during the expedition of H.M.S. *Rattlesnake* to Australia. He also made contributions to the debate on hominid evolution, and was Secretary and Vice-President of the Linnean Society of London. He received the Royal Society of London's Royal Medal in 1871.

1856 **George James Allman** (see also 1849) erected the Classes Gymnolaemata and Phylactolaemata. These were published in a monograph of the Ray Society of London which documented all freshwater bryozoans known at that time.

Figure 9. H.M.S. Challenger.

1859 **George Busk** (see also 1852 and 1884-1886), his *Fossil Polyzoa of the Crag* was published by the Palaeontographical Society. It remains the only monograph on bryozoans published by this organisation established to report on British fossil species.

1859 **Paul Howard MacGillivray** 1834-1895, described new Recent bryozans from Australia. McGillivray's publication record spanned five decades from 1859 until 1895. He was responsible for the reports on bryozoans published in Frederick M'Coy's *Prodromus of the zoology of Victoria*, published in Melbourne between 1879 and 1890. He was the brother of John MacGillivray (1821-1867) who wrote on Recent bryozoans from Aberdeen in 1842.

1861 **Thomas Hincks** 1818-1899, was born in Exeter, England. He was a Unitarian minister who served in Cork, Dublin and latterly in England. In a paper to the British Association for the Advancement of Science he reported on ovicells in bryozoans.

1866 **Fredrik Adam Smitt** 1839-1904, worked from 1871 in the Natural History Museum in Stockholm. In a valuable paper of 1866 he noted polymorphism in bryozoans.

1873-1876 H.M.S. *Challenger* (Figure 9), the ship in which an important series of scientific voyages in the Atlantic Ocean and elsewhere were undertaken under the direction of Charles Wyville Thomson. These resulted in a huge volume of new biological material becoming available for scientific study.

1872 Ferdinand Stoliczka 1838-1874, was employed as a palaeontologist with the Geological Survey of India. He produced a number of publications on Cretaceous bryozoans from the sub-Continent.

1873 Frederick Wollaston Hutton 1836-1905, emigrated to New Zealand in 1866 where he became provincial geologist of Otago before taking up an appointment as Professor of Natural Science and Curator of the University Museum at the University of Otago in 1876. He resigned four years later. He also served as Director of the Canterbury Museum from 1887 until 1905. He was President of the New Zealand Institute in 1904 and 1905. Hutton wrote a large number of papers on New Zealand Recent molluscs, fish and bryozoans. Some of his research papers remain at Canterbury. He is commemorated by the Hutton Medal awarded biennially by the Royal Society of New Zealand.[17]

1874 Henry Alleyne Nicholson 1844-1899, a Scottish palaeontologist who worked on North American, British and Australian bryozoans. He pioneered the use of thin sections. He authored two important book on monticuliporids and corals, and his material is now in the Royal Scottish Museum, Edinburgh and in the University of Aberdeen, where he was Professor of Natural History from 1882 to his death.[18]

1877 Jules Barrois born 1852, wrote several papers on bryozoan embryology between 1875 and 1886, of which his monumental memoir published by the l'Institut de Zoologie in Lille is perhaps the most important (Figure 10).

1878 Fredrik Adam Smitt (see also 1866) described Arctic bryozoans.

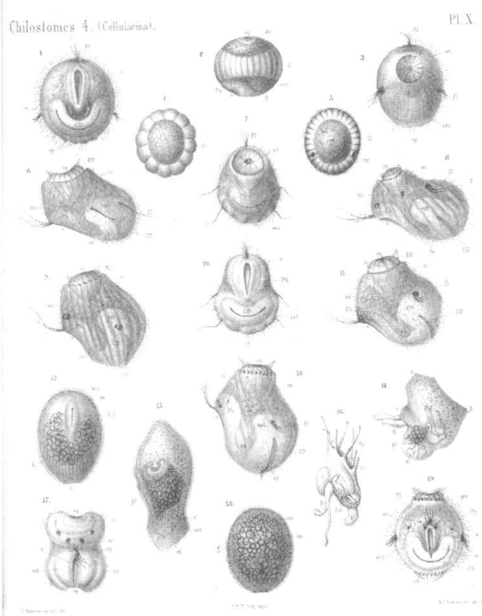

Figure 10. Plate from Jules Barrois' memoir on bryozoan embryology published by the l'Institut de Zoologie in 1877.

1880 Thomas Hincks (see also 1861) published his authorative book *A history of British marine Polyzoa* in two volumes.

1882 Edward Oscar Ulrich 1857-1944, was raised near Cincinnati, and from a young age became involved in the Cincinnati Society of Natural History. He became a dealer in fossils from the area, and was a mentor to R.S. Bassler (*q.v.*). Ulrich was later employed by the U.S. Geological Survey from 1897 until 1932, and saw out his days at the American Museum of Natural History. In a paper to the *Journal of the Cincinnati Society of Natural History* he erected the sub-Order Trepostomata (it was later elevated to full ordinal status).[19]

1883 James Hall (see also 1847) produced a paper on the growth of *Fenestella*.

1884-1886 George Busk (see also 1852 and 1859) authored the reports on the bryozoans collected during the voyages of H.M.S. *Challenger* (*q.v.*).

1884 George Robert Vine 1825-1893, erected the Order Cryptostomata. Vine was an amateur, who made his livelihood as a corset-maker. He was responsible for a series of papers on the fossil bryozoans of Great Britain, published by the British Association for the Advancement of Science.[20]

1890 Edward Oscar Ulrich (see also 1882), published his monumental monograph on the Palaeozoic bryozoans of Illinois, published by the Geological Survey of Illinois. The fifty lithographs, which are particularly beautiful as bryozoan illustrations go, were drawn on stone by Ulrich himself and by Charles Schuchert.

1893 Alphonse Peron 1834-1908, was a French soldier who described Cretaceous bryozoans from Tunisia in north Africa in a paper published in 1893. His material is in the Sorbonne and the Museum of Natural History, Paris.

1893 Grenville Arthur James Cole 1859-1924, was an English-born petrologist who worked in Dublin, and applied microscopical techniques to demonstrate that the unusual superstructure associated with the Carboniferous taxon *Hemitrypa hibernica* was a integral part of the colony zoarium, and not an overgrowth.[21]

1900 John Milton Nickles 1859-1945 and **Ray Smith Bassler** 1878-1961, published their indispensible *Synopsis of North American Fossil Bryozoa*. Nickles is particularly remembered for his valuable contribution to geological research as the compiler of the *Bibliography of North American Geology* series which were published as *Bulletins of the U.S. Geological Survey*.[22]

1904-1912 Edgar Roscoe Cumings 1874-1967 , was Professor of Geology at Indiana

Figure 11. S.F. Harmer.

University from 1905 until his retirement in 1944 as Emeritus Professor. He produced a series of valuable papers on stenolaemate astogeny, wall-structure of Ordovician bryozoans and their affinities. As has been pointed out elsewhere in this volume his 1912 paper was: "especially significant because it settled the long-standing controversy in favour of the "stony" or "monticuliporoid" trepostomes being bryozoans rather than corals."[23]

1913 Edgar Roscoe Cumings (see 1904-1912) and **Jesse James Galloway** 1882-1962, authored a fine paper on the morphology and histology of trepostomes and laid the foundations for present-day ulturastructure studies. Galloway was a student of Cumings who went on to have a distinguished academic carccr at Indiana University (1913-1916; 1931-1955) which was divided by a period spent at Columbia University (1916-1931)

1915-1957 Sidney Frederic Harmer 1862-1950 (Figure 11), who was born at Heigham near Norwich, England, produced a large body of work on Recent bryozoans between 1893 and 1957, including reporting on those collected during the Dutch *Siboga* Expedition. This monumental piece of work appeared in four volumes between 1915 and 1957 - the final volume was published posthumously and completed by A.B. Hastings (*q.v.* 1943). Harmer was educated at University College, London (1878-1881) and Cambridge (1881-1884). He was Superintendent of Cambridge University Museum of Zoology from 1892, and was appointed Keeper of Zoology at the British Museum (Natural History) in London in 1907. He rose to become Director in 1919, and was knighted the following year. He was President of the Linnean Society of London 1927-1930 (he received the Linnean Medal in 1934). He established the journal *Natural History Magazine* in 1927, and retired soon afterwards to Cambridge.

1917-1933 Ferdinand Canu 1863-1932 and **Ray Smith Bassler** (see also 1900, 1935 and 1953) collaborated on a number of important monographs, that reported on a diverse range of bryozoans: North American Early Tertiary bryozoans (1917 and 1920), North American later Tertiary and Quaternary bryozoans (1923), fossil and Recent bryozoa of Mexico (1928), the bryozoans of the Galapagos Islands (1929) and finally the bryozoans of the Vincentown limesand of New Jersey (1933). All were published by the United States National Museum, of which Bassler was Head Curator in the Department of Geology.[24]

1923-date Ehrhard Voigt born 1905, of the University of Hamburg, has produced nearly 150 papers on Cretaceous and other bryozoans published over nine decades. He is currently working on several projects including an examination of growth form and ovicells of actinoporid genera (Cyclostomata) from the Upper Cretaceous, and studies on European Upper Cretaceous - Danian species of Hoplitaechmella (Cheilostomata).

1926 Folke Arvidsson Borg 1892-1950. A Swedish zoologist, he gained his Ph.D. from Uppsala in 1926, and in that year erected the Class Stenolaemata. He wrote extensively on Recent cyclostome bryozoans

1926-date Vasiliy Petrovich Nekhoroshev 1893-1977, **Aleksandra Ivanovna Nikoforova** 1894-1939 and **Lyudmila V. Nekhorosheva** - a family of father, mother and daughter who have collectively contributed many papers and monographs that advanced the understanding of the systematics and distribution of Palaeozoic bryozoans of Russia and the former Soviet Union.

1926-1966 Madeleine Alberta Fritz 1896-1990 - authored many papers on Palaeozoic bryozoans of Canada.

1935 Ray Smith Bassler (see also 1900, 1917-1933 and 1953), authored the *Fossilium Catalogus* (Part 67, Bryozoa).

1936 Leo W. Stach in a now classic paper published in the *Journal of Geology*, discussed the correlation of colony form with habitat, and laid the foundations of bryozoan palaeoecological studies.[25] This was Stach's only paper in bryozoology; he worked all his life in the mineral and petroleum industry, often in remote parts of the world, and frequently advised the United Nations and government agencies. He endowed a studentship in geology at the University of Melbourne in 1995. His life story was told in the book *In Search of Recognition* by Kerry Collison, published in 1999.[26]

1938 Lars Silén 1910-1999, was born Luleå, Sweden. In a 1938 paper he described in detail the structure and functions of avicularia. He became Professor of Zoology at the University of Lund in 1956, but that year moved as Professor of Zoology at Stockholm University where he remained until his retirement in 1975.

1941-1957 Joan M. Crockford, produced a series of monographs on Australian Carboniferous and Permian bryozoans, in which she erected the genera *Minilya* (1944) and *Stenodiscus* (1945).

1943 Anna Birchall Hastings, 1902-1977, was educated at Cambridge and was appointed an Assistant in the Department of Zoology of the British Museum (Natural History) in 1927. She resigned in 1942 on her marriage to Henry Dighton Thomas of the Department of Palaeontology in the same institution. They had one son. A person of stong opinions, Hastings remained at South Kensington in an unpaid research capacity for many years. She reported on the bryozoans from the Antarctic collected during the voyage of H.M.S. *Discovery*.

1950-1952 Raymond Carroll Osburn 1872-1955, after a lengthy career with the Ohio State University as Chairman of Zoology and Entomology, Osburn embarked on a late second career as a Research Associate in the Allan Hancock Foundation of the University of Southern California. There he wrote three major and important monographs on the bryozoa of the Pacific coast of America. The genus *Osburnostylus* Bassler, 1952 was named for him.[27]

1950-1965 King-chin Yang, published a series of papers detailing the Palaeozoic bryozoans of China

1952 David Alexander Brown, his monograph on New Zealand Tertiary cheilostome bryozoans was published by the British Museum (Natural History).

1953 Ray Smith Bassler (see also 1900, 1917-1933 and 1935) - authored the volume on Bryozoa of the *Treatise on Invertebrate Paleontology* (Figure 12). This series, instigated by Raymond C. Moore in the 1950s, and published by the Geological Society of America and the University of Kansas, now comprises over fifty

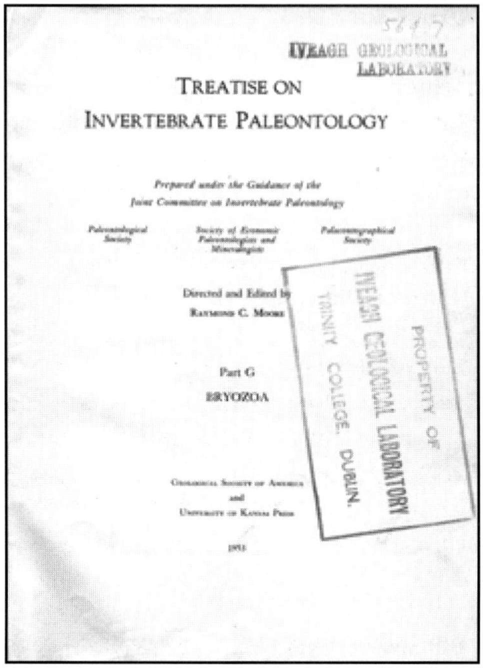
TREATISE ON
INVERTEBRATE PALEONTOLOGY

Part G
BRYOZOA

Figure 12. Bryozoan volume in the series Treatise on Invertebrate paleontology *published by the Geological Society of America and the University of Kansas in 1953.*

volumes and further volumes are planned. Bassler's volume was the first to be published, and although now seen to be limited in terms of quality of the generic descriptions and illustrations given, was an ambitious attempt to document all fossil genera.

1954-date Richard Stanton Boardman, spent his scientific career at the Smithsonian Institution. He has written many papers on fossil and recent bryozoans, especially on the trepostomes. He was the coordinating editor and an author (with others) of the revised volume on bryozoans of the *Treatise on Invertebrate Paleontology* series published in 1983.

1955-date Iraida Pavlovna Morozova has authored numerous papers and monographs on Palaeozoic bryozoans of Russia and the former Soviet Union. Her most recent monograph was a systematic account of fenestrate genera.[28]

1957 George Evert Condra 1869-1958 and **Maxim Kondra Elias** 1889-1982,[29] established the Order Fenestrata in their paper: '*Fenestella* from the Permian of West Texas', published as *Memoir* 70 of the Geological Society of America. Condra was Director of the Nebraska Geological Survey and a faculty member of the University of Nebraska.

1957 Helen Margaret Duncan 1910-1971, of the United States Geological Survey authored a valuable synthesis on bryozoan palaeoecology, published by the Geological Society of America. She also wrote on Devonian trepostomes.[30] The genus *Helenopora* Ross, 1987 was named for her.[31]

1960-date Sumio Sakagami, has produced a large body of work on Upper Palaeozoic bryozoans of Japan and south-east Asia.

1962 Gilbert Powell Larwood 1939-1997, his monograph on Cretaceous cribrimorph bryozoans was published in 1962 in the *Bulletin of the British Museum (Natural History) geology* 6, 1-285. For many years he lectured in palaeontology at the University of Durham. Gilbert was a founding member of the International Bryozoology Association (*q.v.* 1965), and was responsible for hosting two of the International Bryozoology Association conferences in 1971 and 1980, and he edited, or co-edited, four of the conference proceedings.[32]

1964 Galina Grigorevna Astrova 1906-1973 - erected the Order Cystoporata. She was also the author of an indispensible monograph on trepostome bryozoans, published in Moscow in 1978.

1965 The International Bryozoology Association (IBA) was established at a meeting in Stockholm, Sweden in 1965.[33] The Association convened for its first international conference in Milan in 1968 and has since met regularly every three years. It produces an annual newsletter. The published proceedings of the 12 international conferences constitute an invaluable source of literature for bryozoologists, and document the major advances made in the knowledge of fossil and Recent bryozoans over the last 34 years.[34]

Notes

1 R.S. Bassler, 'Part G. Bryozoa', in *Treatise on Invertebrate Paleontology*, edited by R.C. Moore. Geological Society of America and University of Kansas Press (1953), 253 pp; J.S. Ryland, *Bryozoans*. Hutchinson University Library, London, (1970); R.S. Boardman, A.H. Cheetham, D.B. Blake, J.E. Utgaard, O.L. Karlkins, P.L. Cook, P.A. Sandberg, G. Lutaud and T.S. Wood, 'Bryozoa, Part G, revised, volume 1: Introduction, Order Cystoporata, Order Cryptostomata', in *Treatise on Invertebrate Paleontology*, edited by R.A. Robison. Geological Society of America and University of Kansas Press (1983), 625 pp.

2 J.M. Nickles and R.S. Bassler 'A synopsis of American fossil Bryozoa including bibliography and synonomy', *Bulletin of the United States Geological Survey*, 173 (1900), 1-663.

3 F.K. McKinney 'Bibliography and catalogue (1900-1969) of the Trepostomata (Phylum Ectoprocta). *Southeastern Geology Special Publication*, 4 (1974 [dated 1973]), 1-147.

4 P.N. Wyse Jackson and M.E. Spencer Jones, *Abstracts International Bryozoology Association and Society for the History of Natural History meeting - From John Ellis to bryostatins: the history of study of fossil and living bryozoans.* Trinity College, Dublin, Ireland, 21st July 2001.

5 W.A.S. Sarjeant, *Geologists and the History of Geology*. Macmillan Reference Books, London and Basingstoke (Volumes 1-5 (1980)); Robert E. Krieger Publishing Company, Malabar (Supplement 1, Volumes 1-3 (1987)); (Supplement 2, Volumes 1-2 (1996)).

6 This is available on the web: www.tmbl.gu.se/libdb/taxon/personetymol/index.htm

7 d'Hondt, this volume.

8 An account of Peyssonnel and Reaumur's entanglements is given by George Johnston, A history of the British Zoophytes. 2nd edition. Volume 1. John Van Voorst, London, (1847), pp. 409-411.

9 d'Hondt, note 7.

10 Cadée, this volume.

11 For more information on Ellis and his work see P.F.S. Cornelius and J.W. Wells, 'Ellis & Solander's 'Zoophytes', 1786, six unpublished plates and other aspects', *Bulletin of the British Museum (Natural History) historical series*, 16 (1) (1988), 17-87, and J. Groner and P.F.S. Cornelius, *John Ellis*. Pacific Grove, The Boxwood Press (1996), 315 pp.

12 d'Hondt, note 7.

13 P.N. Wyse Jackson, 'Distribution of Irish Marine Bryozoa, together with biographical notes relating to the chief researchers in the group', *Bulletin Irish Biogeographical Society*, 14(2) (1991), 129-184.

14 *The Linnean* published by the Linnean Society of London and edited by D.M. Williams and R. Huxley in 1998.

15 E.L. Yochelson, 'The role and development of the Smithsonian Institution in the American geological community', in *Geologists and ideas; a history of North American geology* edited by E.T. Drake and W.M. Jordan, Geological Society of America Centennial special volume, 1 (1985), 337-354.

16 See *Earth Sciences History* Volume 6(1) 1987 for a special issue devoted to the life and work of James Hall.

17 Futher information on Hutton's work in New Zealand is contained in the following two recent publications: B.A. Marshall, 'Molluscan and brachiopod taxa introduced by F.W. Hutton in The New Zealand journal of science', *Journal of the Royal Society of New Zealand*, 25 (4) (1995), 495-500; B.C. Russell, 'Type specimens of New Zealand fishes described by Captain F.W. Hutton, F.R.S. (1836-1905)', *Journal of the Royal Society of New Zealand*, 26 (2) (1996), 215-236.

18 M.J. Benton, 'H.A. Nicholson, invertebrate palaeontologist: bibliography and catalogue of his type and figured material', *Royal Scottish Museum Information Series, Geology*, No. 7 (1979), 94 p.

19 Cuffey, Davis and Utgaard, this volume.

20 Buttler, Wyse Jackson and Sharpe, this volume.

21 P.N. Wyse Jackson, 'On rocks and bicycles: a biobibliography of Grenville Arthur James Cole', *Bulletin of the Geological Survey of Ireland*, 3 (1990), 151-163.

22 Cuffey et al., note 18.

23 This paper was E.R. Cumings, 'Development and systematic position of the monticuliporoids', *Geological Society of America Bulletin*, 23 (1912), 357-370. For further information on Cumings see Cuffey *et al.*, note 19.

24 Sanner, this volume.
25 L.W. Stach, 'Correlation of zoarial form with habitat', *Journal of Geology*, 44 (1936), 60-65.
26 *In Search of Recognition* by Kerry Collison, published in 1999 by the Kambing Publishing Company.
27 Soule, this volume.
28 I.P. Morozova, *Bryozoans of the Order Fenestellida (morphology, system, historical development).* Transactions of PIN, Russian Academy of Sciences, 277 (2001).
29 A. Lesnikowska, 'Maxim Konrad Elias (1889-1982)' in Historical perspective of early twentieth century Carboniferous paleobotany in North America, edited by P.C. Lyons, E.D. Morey and R.H. Wagner. *Memoir Geological Society of America*, 185 (1995), pp. 169-173.
30 See R.H. Flower, 'Memorial to Helen M. Duncan, 1910-1971', *Geological Society of America Memorials*, 5 (1977).
31 *Duncanopora* is decribed in J.R.P. Ross, 'New chaetetiform trepostome Bryozoa from the Upper Mississippian of the western United States', *Palaeontology*, 31 (1987), 46-51.
32 P.D. Taylor and A.H. Cheetham, 'Gilbert Powell Larwood(1939-1997)', in *Proceedings of the 11th International Bryozoology Conference*, edited by A. Herrera Cubilla and J.B.C. Jackson, Smithsonian Tropical Research Institute, Republic of Panama, (2000), unpaginated.
33 Cheetham, this volume.
34 Reguant, this volume.

828992